Methods in Molecular Biology

Series Editor
John M. Walker
School of Life and Medical Sciences
University of Hertfordshire
Hatfield, Hertfordshire, AL10 9AB, UK

For further volumes:
http://www.springer.com/series/7651

Histone Deacetylases

Methods and Protocols

Edited by

Sibaji Sarkar

Boston University School of Medicine, Boston, MA, USA

Editor
Sibaji Sarkar
Boston University School of Medicine
Boston, MA, USA

ISSN 1064-3745 ISSN 1940-6029 (electronic)
Methods in Molecular Biology
ISBN 978-1-4939-3665-6 ISBN 978-1-4939-3667-0 (eBook)
DOI 10.1007/978-1-4939-3667-0

Library of Congress Control Number: 2016933776

Cover illustration: Sarah Heerboth, Vanderbilt University School of Medicine and Karolina Lapinska, Boston University School of Medicine.

Printed on acid-free paper

This Humana Press imprint is published by Springer Nature
The registered company is Springer Science+Business Media LLC New York

Preface

The concept of epigenetics has shifted since it was first described in the 1970s, when gene function and expression could not always be correlated with changes in DNA sequences. After completion of the genome project, it became clear that changes above the genome play a significant role in gene expression. Further, the alteration of gene expression was found to be involved in the progression of many diseases. Epigenetic changes include modifications of histones and DNA bases (e.g., methylation of adenosine in lower eukaryotes and cytosine in mammals). The last two decades saw the development of technologies that enabled detailed examination of these molecular modifications. Histone acetylation confers open chromatin, which favors transcription. In contrast, histone deacetylation results in a more closed chromatin structure, inhibiting transcription. There are four classes of histone deacetylases (HDACs), which are involved in the deacetylation of histones in the nucleus and in the deacetylation of cytoplasmic signaling proteins. Class I HDACs include HDAC 1, 2, 3, and 8; class II HDACs include HDAC 4, 5, 6, 7, 9, and 10; class III HDACs are called sirtuins (SIRTs), of which there are seven; class IV consists only of HDAC 11. Inhibitors of histone deacetylases are currently at the center stage of drug development against neurological disorders, cancer, metabolic disorders, and cardiovascular diseases.

This volume of Methods in Molecular Biology specifically provides different methodologies for all classes of histone deacetylases. In addition, this volume includes procedures which discuss class I and II histone deacetylase inhibitors, SIRT inhibitors, and bromodomain inhibitors. This volume is divided into three parts. Part I describes methodologies used to detect the activity, function, or chromatin location of HDACs 1 through 11. Part II focuses on the methodologies for cloning and characterizing the acetylation of SIRTs 1–7. Finally, Part III describes methods related to histone deacetylase inhibitors.

Compiling such a diverse field of histone deacetylase methodologies was an enormous and challenging task. I would like to acknowledge the participation of my current and previous students in this endeavor, who have made this work possible. Ms. Karolina Lapinska (Boston University School of Medicine) managed the overall production of this volume. Other students who helped tremendously in this process included Ms. Genevieve Housman (currently at Arizona State University), Ms. Sarah Heerboth (currently at Vanderbilt University School of Medicine), Ms. Meghan Leary (Boston University School of Medicine), and Ms. Amber Willbanks (Boston University). Their help was instrumental in the production of this volume.

I think that this volume of Methods in Molecular Biology will be extremely useful for the investigators working on epigenetics, molecular biology, and genetics.

Boston, MA *Sibaji Sarkar*

Contents

Contributors

Ebru Aydin • *Department of Surgery and Cancer, Imperial College London, Hammersmith Hospital Campus, London, UK*

Bo Bai • *Department of Pharmacology and Pharmacy, Li Ka Shing Faculty of Medicine, The University of Hong Kong, Hong Kong, China*

Mina E. Bekheet • *Laboratory of Cancer Biology, Department of Oncology, University of Oxford, Oxford, UK*

Elena Bernasconi • *Lymphoma and Genomics Research Program, IOR Institute of Oncology Research, Bellinzona, Switzerland*

Francesco Bertoni • *Lymphoma and Genomics Research Program, IOR Institute of Oncology Research, Bellinzona, Switzerland; IOSI Oncology Institute of Southern Switzerland, Bellinzona, Switzerland*

Danica Chen • *Program in Metabolic Biology, Nutritional Science and Toxicology, University of California, Berkeley, CA, USA*

Jie Chen • *Department of Immunology, H. Lee Moffitt Cancer Center and research Institute, Tampa, FL, USA*

Susanna Chiocca • *Department of Experimental Oncology, European Institute of Oncology, IFOM-IEO Campus, Milan, Italy*

Eleonora Ciarlo • *Infectious Diseases Service, Department of Medicine, Centre Hospitalier Universitaire Vaudois and University of Lausanne, Lausanne, Switzerland*

Simona Citro • *Department of Experimental Oncology, European Institute of Oncology, IFOM-IEO Campus, Milan, Italy*

Ileana M. Cristea • *Department of Molecular Biology, Princeton University, Princeton, NJ, USA*

Susan Deng • *Department of Immunology, H. Lee Moffitt Cancer Center and Research Institute, Tampa, FL, USA*

J. Ronald Doucette • *Department of Anatomy and Cell Biology, College of Medicine, University of Saskatchewan, Saskatoon, Canada*

Sylvie Duclaud • *Département de Biologie Structurale Intégrative, Institut de Génétique et Biologie Moléculaire et Cellulaire (IGBMC), Université de Strasbourg (UDS), CNRS, INSERM, Illkirch Cedex, France*

Eugenio Gaudio • *Lymphoma and Genomics Research Program, IOR Institute of Oncology Research, Bellinzona, Switzerland*

Ana R. Gomes • *Department of Surgery and Cancer, Imperial College London, Hammersmith Hospital Campus, London, UK*

Todd M. Greco • *Department of Molecular Biology, Princeton University, Princeton, NJ, USA*

Amanda J. Guise • *Department of Molecular Biology, Princeton University, Princeton, NJ, USA*

Shaoping Ji • *Laboratory of Molecular Cell Biology, College of Pharmacy and Nutrition and Neuroscience Research Cluster, University of Saskatchewan, Saskatoon, Canada; Department of Molecular Biology and Biochemistry, Medical School, Henan University, Henan, China*

MATTAKA KHONGKOW • *Department of Surgery and Cancer, Imperial College London, Hammersmith Hospital Campus, London, UK*
KHAI CHENG KIEW • *Department of Surgery and Cancer, Imperial College London, Hammersmith Hospital Campus, London, UK*
TESSA KNOX • *Department of Immunology, H. Lee Moffitt Cancer Center and Research Institute, Tampa, FL, USA*
SURINDER KUMAR • *Department of Pathology, University of Michigan, Ann Arbor, MI, USA*
IVO KWEE • *Lymphoma and Genomics Research Program, IOR Institute of Oncology Research, Bellinzona, Switzerland; Dalle Molle Institute for Artificial Intelligence (IDSIA), Manno, Switzerland; Swiss Institute of Bioinformatics (SIB), Lausanne, Switzerland*
MAIJA LAHTELA-KAKKONEN • *School of Pharmacy, University of Eastern Finland, Kuopio, Finland*
ERIC W.-F. LAM • *Department of Surgery and Cancer, Imperial College London, Hammersmith Hospital Campus, London, UK*
SASIWAN LAOHASINNARONG • *Department of Surgery and Cancer, Imperial College London, Hammersmith Hospital Campus, London, UK*
LIN LI • *Department of Medicine, The Rosalind and Morris Goodman Cancer Research Center, McGill University, Montreal, QC, Canada*
MIAO LI • *Department of Pathogen Biology, College of Basic Medical Sciences, China Medical University, Shenyang, Liaoning, People's Republic of China*
MARITZA LIENLAF • *Department of Immunology, H. Lee Moffitt Cancer Center and Research Institute, Tampa, FL, USA*
DAVID B. LOMBARD • *Department of Pathology, Institute of Gerontology, University of Michigan, Ann Arbor, MI, USA*
ANDY W.C. MAN • *Department of Pharmacology and Pharmacy, Li Ka Shing Faculty of Medicine, The University of Hong Kong, Hong Kong, China*
MARTIN MAREK • *Département de Biologie Structurale Intégrative, Institut de Génétique et Biologie Moléculaire et Cellulaire (IGBMC), Université de Strasbourg (UDS), CNRS, INSERM, Illkirch Cedex, France*
ANDRIANA MARGARITI • *Centre for Experimental Medicine, Queen's University Belfast, Belfast, UK*
ROMMEL A. MATHIAS • *Department of Biochemistry and Genetics, La Trobe Institute for Molecular Science, La Trobe University, Melbourne, Australia*
RUIN MOADDEL • *Bioanalytical and Drug Development Unit, National Institute on Aging, National Institutes of Health, Baltimore, MD, USA*
ADIL J. NAZARALI • *Laboratory of Molecular Cell Biology, College of Pharmacy and Nutrition and Neuroscience Research Cluster, University of Saskatchewan, Saskatoon, Canada*
HEIDI OLZSCHA • *Laboratory of Cancer Biology, Department of Oncology, University of Oxford, Oxford, UK*
PATRICIO PEREZ-VILLARROEL • *Department of Immunology, H. Lee Moffitt Cancer Center and Research Institute, Tampa, FL, USA*
RAYMOND J. PIERCE • *Center for Infection and Immunity of Lille (CIIL), INSERM U1019 – CNRS UMR 8204, Université de Lille, Institut Pasteur de Lille, Lille Cedex, France*

JOHN POWERS • *Department of Immunology, H. Lee Moffitt Cancer Center and Research Institute, Tampa, FL, USA; Department of Malignant Hematology, H. Lee Moffitt Cancer Center and Research Institute, Tampa, FL, USA*
MINNA RAHNASTO-RILLA • *Bioanalytical and Drug Development Unit, National Institute on Aging, National Institutes of Health, Baltimore, MD, USA; School of Pharmacy, University of Eastern Finland, Kuopio, Finland*
THIERRY ROGER • *Infectious Diseases Service, Department of Medicine, Centre Hospitalier Universitaire Vaudois and University of Lausanne, Lausanne, Switzerland*
CHRISTOPHE ROMIER • *Département de Biologie Structurale Intégrative, Institut de Génétique et Biologie Moléculaire et Cellulaire (IGBMC), Université de Strasbourg (UDS), CNRS, INSERM, Illkirch Cedex, France*
EVA SAHAKIAN • *Department of Immunology, H. Lee Moffitt Cancer Center and Research Institute, Tampa, FL, USA; Department of Malignant Hematology, H. Lee Moffitt Cancer Center and Research Institute, Tampa, FL, USA*
TAJITH B. SHAIK • *Département de Biologie Structurale Intégrative, Institut de Génétique et Biologie Moléculaire et Cellulaire (IGBMC), Université de Strasbourg (UDS), CNRS, INSERM, Illkirch Cedex, France*
BIN SHAN • *College of Medical Sciences, Washington State University at Spokane, Spokane, WA, USA*
SEMIRA SHEIKH • *Laboratory of Cancer Biology, Department of Oncology, University of Oxford, Oxford, UK*
JIYUNG SHIN • *Program in Metabolic Biology, Nutritional Sciences and Toxicology, University of California, Berkeley, CA, USA*
NORIYUKI SUGO • *Neuroscience Laboratories, Graduate School of Frontier Biosciences, Osaka University, Osaka, Japan*
NICHOLAS B. LA THANGUE • *Laboratory of Cancer Biology, Department of Oncology, University of Oxford, Oxford, UK*
ALEJANDRO VILLAGRA • *Department of Immunology, H. Lee Moffitt Cancer Center and Research Institute, Tampa, FL, USA*
JIAN-QIU WANG • *Institute of Aging Research, Department of Basic Medical Science, School of Medicine, Hangzhou Normal University, Hangzhou, China*
YU WANG • *Department of Pharmacology and Pharmacy, Li Ka Shing Faculty of Medicine, The University of Hong Kong, Hong Kong, China*
KOU-JUEY WU • *Research Center for Tumor Medical Science, Graduate Institute of Cancer Biology, China Medical University, Taichung, Taiwan*
MEI-YI WU • *Department of Biochemistry and Molecular Medicine, The George Washington University, Washington, DC, USA; Department of Molecular and Cellular Biology, Baylor College of Medicine, Houston, TX, USA*
MIN-ZU WU • *Gene Expression Laboratory, Salk Institute for Biological Studies, La Jolla, CA, USA*
RAY-CHANG WU • *Department of Biochemistry and Molecular Medicine, The George Washington University, Washington, DC, USA; Department of Molecular and Cellular Biology, Baylor College of Medicine, Houston, TX, USA*
NOBUHIKO YAMAMOTO • *Neuroscience Laboratories, Graduate school of Frontier Biosciences, Osaka University, Osaka, Japan*
JUNYAO YANG • *Cardiovascular Division, King's College London, London, UK*

XIANG-JIAO YANG • *Department of Medicine, The Rosalind and Morris Goodman Cancer Research Center, McGill University and McGill University Health Center, Montreal, QC, Canada*
JAY SZE YONG • *Department of Surgery and Cancer, Imperial College London, Hammersmith Hospital Campus, London, UK*
LINGFANG ZENG • *Cardiovascular Division, King's College London, London, UK*
YAN ZHUANG • *Department of Medicine, Tulane University School of Medicine, New Orleans, LA, USA*

Part I

Class I, II, and, IV Histone Deacetylases

Chapter 1

A Sensitive and Flexible Assay for Determining Histone Deacetylase 1 (HDAC1) Activity

Mei-Yi Wu and Ray-Chang Wu

Abstract

Histones acetylation and deacetylation constitute part of the so-called "histone code" and work in concert with other posttranslational modifications to determine the activity of genes. Deacetylation of histone is carried out by a class of enzymes, known as histone deacetylases (HDACs). The action of HDAC is countered by histone acetyltransferases. Although histone is the best characterized substrate of HDACs, increasing evidence also indicates that non-histone proteins are equally important subtract of HDACs. Since HDACs play an important role in normal physiological and pathophysiological conditions, a sensitive and flexible deacetylation assay that can reliably detect HDAC activity and identify potential novel targets of HDACs is critical.

Key words HDAC1, Deacetylation, Immunoaffinity purification, Core histone, Posttranslational modification, Colorimetric assay

1 Introduction

Acetylation is one of the best characterized histone posttranslational modifications that form the histone code [1]. Acetylation of histone involves the transfer of an acetyl group from acetyl coenzyme A to the lysine residue of the histone and is carried out by the histone acetyltransferases (HATs). Acetylation of histone removes the negative charge from the histone and correlates with an open chromatin state that permits the access of DNA by transcriptional factors. Since histone acetylation creates a permissive state for transcriptional activation, it is expected that removal of acetyl group by another class of enzymes known as histone deacetylases (HDACs) plays an equally important role in gene expression. In fact, deacetylation of histone allows the histones to interact DNA more tightly, thus representing a repressive state of gene transcription.

HDACs are evolutionarily conserved from yeast to plants and animals [2, 3]. In humans, a total of 18 HDACs has been identified so far. The classification of human HDACs is based on their

Sibaji Sarkar (ed.), *Histone Deacetylases: Methods and Protocols*, Methods in Molecular Biology, vol. 1436, DOI 10.1007/978-1-4939-3667-0_1, © Springer Science+Business Media New York 2016

sequence homology to the three histone deacetylases of *Saccharomyces cerevisiae*, Rpd3, HDA1, and Sir2. Based on the sequence homology, human HDACs are divided into the following four classes (Table 1).

Class I HDACs are highly homologous to the yeast transcriptional regulator RPD3 protein and include HDAC1, -2, -3, and -8. HDAC1 is the first HDAC to be purified and cloned using trapoxin, an inhibitor and a subtract mimic of HDAC activity [4]. HDAC1 and HDAC2 are the most phylogenetically related members of the Class I HDACs, a result of a duplication of an ancient gene [5]. Class I HDACs are ubiquitously expressed in all tissues and are found almost exclusively in the nucleus. The only exception is HDAC3 which can also be found in the cytoplasm [6, 7]. Class II HDACs are closely related to the yeast deacetylase HDA1 and include HDAC4, -5, -6, -7, -9, and -10. Class II HDACs can

Table 1
Classification of Human HDACs

Classification	Member	Subcellular Localization
Class I HDAC (yeast Rpd3 homologue, Zn-dependent)	HDAC1	Nucleus
	HDAC2	Nucleus
	HDAC3	Nucleus/cytoplasm
	HDAC8	Nucleus
Class II HDAC (yeast HDA1 homologue, Zn-dependent)	HDAC4	Nucleus/cytoplasm
	HDAC5	Nucleus/cytoplasm
	HDAC6	Mostly cytoplasm
	HDAC7	Nucleus/cytoplasm
	HDAC9	Nucleus/cytoplasm
	HDAC10	Nucleus/cytoplasm
Class III HDAC (yeast Sir2 homologue, NAD+-dependent)	SIRT1	Nucleus/cytoplasm
	SIRT2	Cytoplasm
	SIRT3	Nucleus/mitochondria
	SIRT4	Mitochondria
	SIRT5	Mitochondria
	SIRT6	Nucleus
	SIRT7	Nucleolus
Class IV HDAC (homologous to Class I and II, Zn-dependent)	HDAC11	Nucleus/cytoplasm

be further divided into two subclasses: Class IIa (HDAC4, -5, -7, and -9) and Class IIb (HDAC6 and -10). Class II HDACs are shown to express in a more tissue-restricted manner, and can shuttle between the cytoplasm and nucleus in response to cellular signaling [8]. Class III HDACs include seven members, Sirt1–7 [9]. In contrast to Class I and II HDACs whose activity can be inhibited by HDAC inhibitors, such as trichostatin A, are referred to as the "classical" HDACs, Class III HDACs are resistant to these HDAC inhibitors but are absolutely dependent on NAD+ for their activity. Class IV contains only one member, HDAC11, and shares some homology to both Class I and II HDACs [10].

Alternatively, the HDACs can be divided into two families based on their mechanism of action: zinc and NAD+ dependent. In this case, the classical RPD3/HDA1 family (HDAC1–11) belongs to the Zinc-dependent HDACs, whereas the Sirt1–7 are the NAD+-dependent HDACs.

The diversity of HDACs indicates that they are involved in the function of different tissues and regulate the activity of various biological processes. While Class III HDACs are mainly involved in metabolism and aging [9], Class I HDACs play a crucial role in cellular proliferation, differentiation and cancer [11, 12]. For example, HDAC1 not only plays an important role in the normal development and physiological function of the heart and nervous systems [13], but also is involved in cell cycle progression and cell death [14]. Ablation of HDAC1 which resulted in embryonic lethality due to significant defect in cellular proliferation underlines the essential function of HDAC1 [14]. Although data from gene ablation or silencing experiments clearly support the biological importance of HDAC1, increasing evidence strongly suggest that the activity of HDAC1 in the cells is further regulated by post-translational modifications, such as phosphorylation, in response to different stimuli and signaling pathways [15–19]. Therefore, it is important to preserve or recapitulate the posttranslational modification state of HDAC1 when evaluating its deacetylase activity. Herein, we describe a sensitive and flexible assay for determining HDAC1 activity by combining HDAC1 expression in mammalian cells and immunoaffinity purification of enzymatically active HDAC1.

2 Materials

2.1 Expression of Flag-HDAC1 in Mammalian Cells

1. Mammalian expression plasmid pCMV-Flag-HDAC1 was generated by inserting the HDAC1 cDNA into the pCMV-Tag vector. The construction of pCMV-Tag-Flag-HDAC1 has been described previously [20].

2. 293T cells (ATCC CRL-3216), which express the SV40-T antigen are a highly transfectable derivative of HEK293 cells (*see* **Note 1**).
3. 293T growth medium, Dulbecco's Modified Eagle's Medium (DMEM) supplemented with 10 % fetal bovine serum, 2 mM L-glutamine, and 1 % Penicillin/Streptomycin.
4. Trypsin (0.05 %, w/v)–EDTA (0.53 mM) solution.
5. Hemocytometer.
6. 15 ml conical tubes.
7. Cationic liposome transfection reagents, Opti-MEM I reduced serum medium.
8. One time phosphate buffered saline (PBS, 140 mM NaCl, 2.7 mM KCl, 10 mM Na_2HPO_4, and 1.8 mM KH_2PO_4, pH 7.4), diluted from 10× PBS concentrate.

2.2 Cell Extract Preparation

1. Whole cell lysis buffer (50 mM Tris–HCl (pH 8.0), 150 mM NaCl, 0.5 % NP-40).
2. Phosphatase Inhibitor Cocktail (without addition of EDTA).
3. Cell lifter.
4. 1.5 ml Eppendorf tubes.
5. Vortex.

2.3 Immunoblotting and Immunoaffinity Purification

1. Anti-Flag M2 affinity gel, anti-Flag M2-peroxidase (HRP) antibody.
2. Three times Flag peptide.
3. Washing buffer (20 mM HEPES, pH 7.6, 70 mM KCl, 1 mM DTT, 0.1 % NP-40, 8 % Glycerol).
4. One time HDAC1 assay buffer (25 mM Tris–HCl, pH 8.0, 150 mM NaCl, 2.7 mM KCl, 1 mM $MgCl_2$, 10 % glycerol).
5. Bradford protein assay.
6. Trans-Blot transfer membrane.
7. TBST buffer (50 mM Tris–HCl, pH 7.5, 150 mM NaCl, 0.1 % Tween 20).
8. ECL Prime Western blotting detection reagent.
9. Protein concentrators, 10K MWCO (88513, optional).
10. Chromatography column (optional).

2.4 Deacetylation Assay

1. Histone purification kit.
2. Five times HDAC1 assay buffer.
3. Five times SDS sample buffer (60 mM Tris–HCl, pH 6.8, 25 % glycerol, 2 % SDS, 14.4 mM 2-mercaptoethanol, 0.1 % bromophenol blue).

4. Anti-acetyl-histone H3, anti-acetyl-histone H4, anti-histone H3, anti-histone H4, acetylated-lysine antibodies.
5. Colorimetric HDAC assay kit.
6. 96-well half area microplate.
7. Monochromator-based multi-mode microplate reader.

3 Methods

3.1 Expression of Flag-tagged HDAC1 in 293T Cells

In order to obtain sufficient quantity of Flag-HDAC1 for in vitro deacetylation assay, it is necessary to achieve high levels of protein expression. Therefore, a highly transfectable mammalian cells, 293T cells, and a mammalian expression vector, pCMV-Tag, will be used. Since the biological activity of HDACs, including HDAC1, is subject to regulation by posttranslational modifications, expression of HDAC1 in mammalian 293T cells not only maintains the important posttranslational modification state, but also offers the feasibility for comparing HDAC1 activity in response to extracellular stimuli.

1. Seed the 293T cells one day before transfection. To disperse 293T cells, wash the cells carefully with 1× PBS once. Add 1 ml of trypsin–EDTA solution to the plate. Incubate the plate in 37 °C incubator for 5 min. Add 5 ml of fresh growth medium to plate and collect the cells into a 15 ml conical tube. Pellet the cells by centrifugation at 800 ×*g* for 3 min. Aspirate off the supernatant and resuspend the cells in 10 ml of fresh growth medium by pipetting. Count the cell manually using a hemocytometer. Seed 6×10^6 cells per 10-cm plate. Grow the cells in 37 °C incubator with 5 % CO_2 overnight.
2. The next day, prepare the transfection reaction by combining the DNA and liposome-based transfection reagent diluted in Opti-MEM I. Follow the procedure as instructed by the manufacturers for use of these reagents. The 293T cells will be transfected with pCMV-Tag-Flag-HDAC1 (10 μg) or the same amount of parental vector as a control. Transfection can be carried out in medium with or without serum. If serum-free condition is used for the transfection, replace the medium containing the DNA transfection mixture with complete growth medium after 6 h. If the transfection is carried out in complete growth medium, replace with fresh growth medium after incubation overnight.
3. For optimal expression of Flag-HDAC1, wait for additional 24 h before starting the treatment. Treat the transfected cells with potential inhibitors or various stimuli as desired. Incubate the cells at 37 °C incubator with 5 % CO_2 for the desired duration or allow the cells to grow for additional 24 h if no treatment is performed.

3.2 Preparation of Cell Extract for Immunoblotting and Immunoaffinity Purification

Harvest the transfected cells for immunoblotting to confirm expression of Flag-HDAC1 before proceeding to immunoaffinity purification of Flag-HDAC1 protein by M2 antibody agarose.

1. Wash the transfected cells with 10 ml of 1× PBS once. Be careful not to dislodge cells from the plate during the wash. After aspirating off the PBS completely, add 1 ml of fresh PBS to the plate. Scrape the cells off the plate using a cell lifter and collect the cells into a 1.5 ml microcentrifuge tube. Pellet the cells by centrifugation at 1500 × *g* for 5 min using an Eppendorf 5415R microcentrifuge. Aspirate off the PBS completely, and lyse the cells by adding 1 ml of whole cell lysis buffer (supplemented with protease and phosphatase inhibitors). Place the tube on ice for 30 min, and vortex the tubes briefly to mix every 5–10 min. Pellet the cell debris by centrifugation at 15,700 × *g*, 4 °C for 15 min in an Eppendorf 5415R microcentrifuge. Transfer the clear supernatant into a new 1.5 ml microcentrifuge tube.
2. Determine the protein concentration by Bradford protein assay.

3.3 Detect the Ectopic Expression of Flag-HDAC1 by Immunoblotting

It is recommended that an immunoblotting be performed to confirm expression of Flag-HDAC1 before proceeding to immunoaffinity purification. After determining the protein concentrations, resolve the samples by 8 % SDS-PAGE and transfer to nitrocellulose membranes. After blocking in TBST with 5 % milk, dilute the primary antibodies in TBST buffer with 5 % milk and add to the membranes. Incubate the membrane with the antibodies for 1 h at room temperature or overnight at 4 °C. Add horse peroxidase-conjugated second antibodies and incubate at room temperature for 1 h. After extensive wash with TBST, develop all blots using the ECL prime Western blotting detection reagent and visualize by chemiluminescence.

3.4 Immunoaffinity Purification of Flag-HDAC1

After expression of Flag-HDAC1 is confirmed by immunoblotting as described above, the cell lysate which contains Flag-HDAC1 can be used for immunoaffinity purification using anti-Flag M2 agarose (*see* **Note 2**).

1. For binding of antigen and antibody, use 30 μl (bead volume) of anti-Flag M2 agarose for every 500 μg of cell lysate. After combining the M2 agarose with cell lysate in 1.5 ml microcentrifuge tubes, place the tubes on a nutator mixer for 1.5 h at 4 °C.
2. Pellet the M2 agarose beads by centrifugation at 15,700 × *g* for 1 min at 4 °C. Remove the supernatant carefully to avoid disturbing the agarose beads. Add 1 ml of washing buffer to the agarose beads, and place the tubes on a nutator mixer for 10 min at 4 °C. Pellet the M2 agarose beads by centrifugation. Repeat this wash four more times.

3. Wash the M2 agarose beads once using the 1× HDAC1 assay buffer. Pellet the M2 beads by centrifugation and carefully remove the supernatant without disturbing the beads.
4. Although the Flag-HDAC1 that remains bound on the M2 antibody beads can be used in the deacetylation assay, it is recommended that the Flag-HDAC1 be eluted with Flag peptide prior to deacetylation assay. For competitive elution of Flag-HDAC1, add two bead volumes (60 μl) of 1× HDAC1 buffer containing 3x Flag peptide (100 μg/ml) to the M2 agarose beads. Incubate the M2 beads at 4 °C for 30 min to elute the Flag-HDAC1. Pellet the agarose beads by centrifugation, and carefully transfer the supernatant to a new 1.5 ml microcentrifuge tube. Repeat the elution step once for complete recovery of Flag-HDAC1. The eluate which contains the Flag-HDAC1 from multiple tubes can be pooled. Measure the concentration of purified Flag-HDAC1. If necessary, the purified Flag-HDAC1 can be concentrated using Protein Concentrators (10K MWCO). The purified Flag-HDAC1 is ready for deacetylation assay and should be stored at −80 °C in small aliquot. Frequent freeze and thaw should be avoided to prevent loss of enzymatic activity.

3.5 Deacetylation Assay Using Purified Flag-HDAC1

To determine the activity of purified Flag-HDAC1, the following two nonradioactive deacetylation assays can be used.

3.5.1 Deacetylation Assay Using Core Histone as Substrate and Immunoblotting as Detection Method

Histones are the best characterized substrate of HDACs. Although using histone radiolabeled with [^{3}H] as substrate offers sensitivity, preparation of correctly labeled histones is a time consuming process. Here, a sensitive nonradioactive deacetylation assay which uses core histone as substrate and takes advantage of acetylation-specific antibodies against histones is adopted and described below.

1. Core histone preparation kits are commercially available from several vendors, including Abcam, Epigentek, Enzo, and Active Motif. We used kit that offers the option to further separate H2A/H2B fraction from the H3/H4 fraction if necessary while preserving the posttranslational modifications of histones, including acetylation. Follow the instruction from the manufacturers to isolate the core histone. The core histone can be used for deacetylation assay after dialysis against 1× HDAC1 assay buffer for 2 h at 4 °C. Concentrate the core histone by precipitation if necessary. Measure the protein concentration, aliquot the core histone and store at −80 °C (*see* **Note 3**).
2. Reconstitute the deacetylation reaction in a 1.5 ml microcentrifuge tube by adding purified Flag-HDAC1 (up to 500 ng), core histone (1 μg) and adjust the final volume to 40 μl using 1× or 5× HDAC1 assay buffer. Heat-inactivated Flag-HDAC1

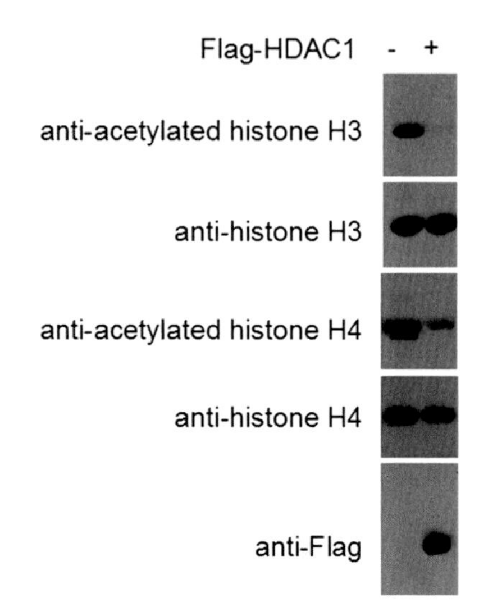

Fig. 1 In vitro deacetylation assay using purified core histone and acetyl histone-specific antibodies. After the deacetylation reaction, the samples (10 μl from each reaction) were separated by 12 % SDS-PAGE and used for immunoblotting by the indicated antibodies. As evidenced by immunoblotting using the acetyl histone-specific antibodies, addition of purified Flag-HDAC1 (indicated by "+") reduced the levels of acetylation in H3 and H4 compared to that of control (indicated by "−"). Histone H3, H4, and Flag-HDAC1 were detected by the indicated specific antibodies

(95 °C, 10 min), or the eluate from cells transfected with parental pCMV-Tag vector can be used as a control.

3. Incubate the reaction tube at 37 °C for 30 min. Stop the reaction by adding 10 μl of 5× SDS sample buffer and inactivation at 95 °C for 5 min.
4. Separate the sample by 8 % SDS-PAGE. Perform immunoblotting using acetylation-specific antibodies against H3 or H4 (*see* **Note 4**). By comparing to the acetylation levels in the control, the activity of Flag-HDAC1 can be determined (Fig. 1). The advantage of this assay is its flexibility as it can be used to compare the Flag-HDAC1 activity from cells treated with inhibitors or extracellular stimuli.

3.5.2 Colorimetric Deacetylation Assay

The colorimetric deacetylation assay utilizes a short peptide that contains an acetylated lysine residue as a substrate for the HDACs. Once deacetylated by the HDACs, the lysine residue will then react with substrate present in the developing solution, resulting in release of a chromophore which generates a yellow colored solution with maximal absorbance at 405 nm. Colorimetric deacetylation assay kits are available from several sources. Refer to the manufacturer's manual for more information on preparation of buffer, developing solution and detailed protocols.

1. It is recommended that different concentrations of the Flag-HDAC1 from each treatment be tested. Dilute the purified Flag-HDAC in 1× HDAC1 assay buffer if necessary. Calculate the volume of Flag-HDAC1 needed and maintain the final volume of the reaction constant at 50 μl. Reconstitute the deacetylation reaction in a 96-well half area microplate by adding desired volume of HDAC assay buffer, 10 μl of HDAC substrate (diluted to 5 mM in HDAC assay buffer), and desired concentration of Flag-HDAC1. In addition, use the HDAC assay standard provided in the kit to generate a standard curve to calculate the activity of Flag-HDAC1 from each sample (*see* **Note 5**).
2. Incubate the microplate at 37 °C for 30 min. In the meantime, follow the instruction in the manual to prepare a standard cure using the HDAC assay standard on a separate microplate. Incubate the microplate containing the HDAC assay standard at 37 °C for 30 min.
3. During the incubation period, prepare the HDAC assay developing solution according to the instructions.
4. After the end of incubation period, stop the reaction by adding 50 μl of assay developing solution to each well. Incubate the microplate at room template for 15 min.
5. Read the absorbance at 405 nm using a monochromator-based multi-mode microplate reader. Calculate the activity using the standard curve.

4 Notes

1. 293T cells is used for expression of Flag-HDAC1 because of high transfection efficiency. In The pCMV-Flag-HDAC1 mammalian expression vector is used because the pCMV-Tag series of vectors contain the SV40 origin of replication that allows the propagation of plasmids in SV40 Large T antigen expressing cells (such as 293T cells).
2. Immunoaffinity purification of Flag-HDAC1 can be performed in batch using 1.5 ml microcentrifuge tube as described above in Subheading 3.4, or it can be performed using the chromatography column. To pack the column, transfer the immunocomplex to the chromatography column after incubation at 4 °C for 2 h. Allow the agarose beads to drain completely. Wash the agarose beads in the column thoroughly with 10 ml of washing buffer each time. Repeat the wash four more times, followed by one wash with 10 ml of HDAC1 assay buffer. Elute the Flag-HDAC1 with two bed volumes of 1× HDAc1 assay buffer containing 3x Flag peptide (100 μg/ml). Repeat the elution once. Spin the column at 170 ×*g* for 30 s to collect all the solution at the end of the second elution.

3. To increase the levels of acetylation in the core histone, the cells can be treated with HDAC inhibitors, such as sodium butyrate (3–5 mM), for 24 h before purification of core histone.
4. To determine whether a protein of interest is a potential target of HDAC1, substitute the core histone with the protein of interest in the deacetylation assay described in Subheading 3.5.1. In this case, acetylated lysine antibodies can be used to determine the change of acetylation levels on the protein of interest.
5. For colorimetric HDAC assay, HeLa nuclear extracts included in the kit can be used as a positive control for the presence of HDAC enzymatic activity. However, it should be noted that the HDAC detected in the HeLa nuclear extract is not from HDAC1 alone.

Acknowledgements

Work in the authors' laboratory is supported by grants CA187857 and CA188471 from National Cancer Institute, and McCormick Genomic and Proteomic Center at the George Washington University.

References

1. Strahl BD, Allis CD (2000) The language of covalent histone modifications. Nature 403:41–45
2. Leipe DD, Landsman D (1997) Histone deacetylases, acetoin utilization proteins and acetylpolyamine amidohydrolases are members of an ancient protein superfamily. Nucleic Acids Res 25:3693–3697
3. Gregoretti IV, Lee YM, Goodson HV (2004) Molecular evolution of the histone deacetylase family: functional implications of phylogenetic analysis. J Mol Biol 338:17–31
4. Taunton J, Hassig CA, Schreiber SL (1996) A mammalian histone deacetylase related to the yeast transcriptional regulator Rpd3p. Science 272:408–411
5. de Ruijter AJ, van Gennip AH, Caron HN, Kemp S, van Kuilenburg AB (2003) Histone deacetylases (HDACs): characterization of the classical HDAC family. Biochem J 370: 737–749
6. Takami Y, Nakayama T (2000) N-terminal region, C-terminal region, nuclear export signal, and deacetylation activity of histone deacetylase-3 are essential for the viability of the DT40 chicken B cell line. J Biol Chem 275:16191–16201
7. Yang WM, Tsai SC, Wen YD, Fejer G, Seto E (2002) Functional domains of histone deacetylase-3. J Biol Chem 277:9447–9454
8. Witt O, Deubzer HE, Milde T, Oehme I (2009) HDAC family: what are the cancer relevant targets? Cancer Lett 277:8–21
9. Haigis MC, Guarente LP (2006) Mammalian sirtuins—emerging roles in physiology, aging, and calorie restriction. Genes Dev 20:2913–2921
10. Gao L, Cueto MA, Asselbergs F, Atadja P (2002) Cloning and functional characterization of HDAC11, a novel member of the human histone deacetylase family. J Biol Chem 277:25748–25755
11. Senese S, Zaragoza K, Minardi S, Muradore I, Ronzoni S, Passafaro A, Bernard L, Draetta GF, Alcalay M, Seiser C, Chiocca S (2007) Role for histone deacetylase 1 in human tumor cell proliferation. Mol Cell Biol 27: 4784–4795

12. Zupkovitz G, Grausenburger R, Brunmeir R, Senese S, Tischler J, Jurkin J, Rembold M, Meunier D, Egger G, Lagger S, Chiocca S, Propst F, Weitzer G, Seiser C (2010) The cyclin-dependent kinase inhibitor p21 is a crucial target for histone deacetylase 1 as a regulator of cellular proliferation. Mol Cell Biol 30:1171–1181
13. Haberland M, Montgomery RL, Olson EN (2009) The many roles of histone deacetylases in development and physiology: implications for disease and therapy. Nat Rev Genet 10:32–42
14. Lagger G, O'Carroll D, Rembold M, Khier H, Tischler J, Weitzer G, Schuettengruber B, Hauser C, Brunmeir R, Jenuwein T, Seiser C (2002) Essential function of histone deacetylase 1 in proliferation control and CDK inhibitor repression. EMBO J 21:2672–2681
15. Karwowska-Desaulniers P, Ketko A, Kamath N, Pflum MK (2007) Histone deacetylase 1 phosphorylation at S421 and S423 is constitutive in vivo, but dispensable in vitro. Biochem Biophys Res Commun 361:349–355
16. Pflum MK, Tong JK, Lane WS, Schreiber SL (2001) Histone deacetylase 1 phosphorylation promotes enzymatic activity and complex formation. J Biol Chem 276:47733–47741
17. Rush J, Moritz A, Lee KA, Guo A, Goss VL, Spek EJ, Zhang H, Zha XM, Polakiewicz RD, Comb MJ (2005) Immunoaffinity profiling of tyrosine phosphorylation in cancer cells. Nat Biotechnol 23:94–101
18. Sun JM, Chen HY, Davie JR (2007) Differential distribution of unmodified and phosphorylated histone deacetylase 2 in chromatin. J Biol Chem 282:33227–33236
19. Tsai SC, Seto E (2002) Regulation of histone deacetylase 2 by protein kinase CK2. J Biol Chem 277:31826–31833
20. Wu MY, Fu J, Xiao X, Wu J, Wu RC (2014) MiR-34a regulates therapy resistance by targeting HDAC1 and HDAC7 in breast cancer. Cancer Lett 354:311–319

Chapter 2

Detection of Sumo Modification of Endogenous Histone Deacetylase 2 (HDAC2) in Mammalian Cells

Simona Citro and Susanna Chiocca

Abstract

Small ubiquitin-related modifier (SUMO) is an ubiquitin-like protein that is covalently attached to a variety of target proteins and has a significant role in their regulation. HDAC2 is an important epigenetic regulator, promoting the deacetylation of histones and non-histone proteins. HDAC2 has been shown to be modified by SUMO1 at lysine 462. Here we describe how to detect SUMO modification of endogenous HDAC2 in mammalian cells by immunoblotting. Although in this chapter we use this method to detect HDAC2 modification in mammalian cells, this protocol can be used for any cell type or for any protein of interest.

Key words HDAC2, SUMO, UBC9, Immunoblotting, Posttranslational modification, PTM

1 Introduction

Histone deacetylase 2 (HDAC2) belongs the human class I HDACs that regulates gene expression by deacetylation of histones and non-histone proteins. Among class I HDACs (HDAC1 and HDAC2, HDAC3 and HDAC8), HDAC1 and HDAC2 are believed to regulate most of the changes observed in histone acetylation. HDAC2, with high homology to yeast RPD3, was identified in a yeast two-hybrid screen as a corepressor that binds the YY1 transcription factor in Edward Seto's laboratory [1]. Transcriptional repression by YY1 was shown to be mediated by tethering HDAC2 to DNA as a corepressor. Mammalian HDAC2 was then cloned and characterized by the same group [2].

HDACs activity is regulated by their binding to corepressor complex partners and depends on their posttranslational modifications (PTMs) (reviewed in Ref. [3]). In particular, HDAC2 has been shown to be phosphorylated by CK2 [4], acetylated by CBP in response to cigarette smoke treatment [5], ubiquitinated for proteasomal degradation [6, 7], nitrosylated [8, 9] and carbonylated [10]. Sumoylation is a PTM that involves the covalent attachment of

Sibaji Sarkar (ed.), *Histone Deacetylases: Methods and Protocols*, Methods in Molecular Biology, vol. 1436,
DOI 10.1007/978-1-4939-3667-0_2, © Springer Science+Business Media New York 2016

SUMO (small-ubiquitin-like modifier) to lysine residues of proteins. Four SUMO proteins (SUMO1–4) have been identified in humans (reviewed in Ref. [11]). Covalent conjugation of SUMO proteins to their substrates requires an enzymatic cascade, comprising the sequential action of three enzymes: a modifier activating enzyme (E1), the E2 conjugating enzyme (UBC9), and a member of the ligases (E3s). Many proteins modified by SUMO contain an acceptor lysine residue within the consensus motif ψKxE (where ψ is an aliphatic branched amino acid) [12]. The covalent attachment of SUMO to target proteins regulates many functional properties of target proteins, such as protein localization, binding, and transactivation functions of transcription factors [13, 14]. Recently Brandl et al. [15] showed that HDAC2 is covalently modified by SUMO1 at lysine 462 and this modification promotes deacetylation of p53 at lysine 320. p53 acetylation blocks its recruitment into promoter-associated complexes and its ability to regulate the expression of genes involved in cell cycle control and apoptosis. Thus, HDAC2 sumoylation inhibits p53 functions and attenuates DNA damage-induced apoptosis. Moreover, HDAC2 has been shown to have a sumoylation-promoting activity in a deacetylase-independent manner. Indeed, HDAC2 seems to promote sumoylation of the eukaryotic translation initiation factor 4E (elF4E), stimulating the formation of the eukaryotic initiation factor 4F (elF4F) complex and the induction of protein synthesis of a subset of elF4E-responding genes [16].

This chapter describes the detection of HDAC2 sumoylation by immunoblotting. SUMO is typically conjugated only to a small fraction of the target protein, yet having a global and significant effect, but, as a result, the detection of the modification can be difficult. Here we show how to increase the amount of endogenous modified HDAC2 by overexpressing SUMO and the E2 conjugating enzyme UBC9 in mammalian cells. We describe how to prepare lysates from mammalian cells using denaturing lysis buffer designed to block the action of desumoylating enzymes and all the best settings to improve the detection of this modification, as previously shown for HDAC1 [17]. This method is the quicker way to detect HDAC2 sumoylation, and can be used to check changes in this modification upon different stimuli (e.g., upon treatment with the proteasomal inhibitor MG132 or deacetylase inhibitors). Although only conjugation of SUMO1 to HDAC2 has been previously described [15], this protocol shows that also conjugation of SUMO2 can be detected. This chapter focuses on the detection of sumoylated HDAC2 in mammalian cells; however, this protocol can be adapted to any cell type to detect SUMO modification for any intracellular protein.

2 Materials

2.1 Cell Culture

1. HeLa, MCF7, and U2OS cell lines (American Type Culture Collection).

2. 100 mm treated tissue culture dishes.
3. Dulbecco's modified Eagle's medium (DMEM) supplemented with antibiotics, 2 mM L-glutamine, and 10% fetal bovine serum (FBS).
4. Trypsin–EDTA.
5. Sterile phosphate buffered saline (PBS).

2.2 Transfection

1. DMEM with L-glutamine, without FBS and antibiotics.
2. Fugene® 6 transfection reagent (Promega).
3. Plasmid DNA: pcDNA3/HA-N vector containing human SUMO1 (described in Ref. [18]), pcDNA3/HA-N vector containing human SUMO2 (described in Ref. [19]), pcDNA3 vector containing human UBC9 (described in Ref. [20]), and pcDNA3 empty vector.

2.3 Cell Lysis

1. SDS lysis buffer:
   (a) Solution I (5% SDS, 0.15 M Tris–HCl pH 6.8, 30% glycerol), stored at room temperature.
   (b) Solution II (25 mM Tris–HCl pH 8.3, 50 mM NaCl, 0.5% NP40, 0.5% deoxycholate, 0.1% SDS), stored at 4 °C.
2. Protease inhibitors: 100 μg/mL phenyl-methyl-sulfonyl fluoride (PMSF), 1 μg/mL leupeptin, 1 μg/mL aprotinin (*see* **Note 1**).
3. 5 mM NEM (*see* **Note 2**).
4. Five times loading buffer (312 mM Tris–HCl pH 6.8, 10% SDS, 40% glycerol, 20% β-mercaptoethanol, bromophenol blue).
5. Protein quantitation assay: Bradford or Lowry.

2.4 Immunoblotting

1. Resolving gel buffer: 1.5 M Tris–HCl pH 8.8.
2. Stacking gel buffer: 0.5 M Tris–HCl pH 6.8.
3. 30% acrylamide–bis solution.
4. Ammonium persulfate (APS): 10% solution in water (*see* **Note 3**).
5. *N*,*N*,*N*,*N*′-tetramethyl-ethylenediamine (TEMED).
6. Resolving portion of polyacrylamide gel: acrylamide–bis solution, resolving solution, 0.1% APS, and TEMED.
7. Stacking portion of polyacrylamide gel: acrylamide–bis solution, stacking solution, 0.1% APS, and TEMED.
8. SDS-PAGE running buffer: 2 M glycine, 0.25 M Tris–HCl pH 8.3, 0.02 M SDS.
9. Benchmark prestained molecular weight standards.
10. Immobilon® Polyvinylidene difluoride (PVDF) membrane (Millipore).
11. Western blot transfer buffer: 0.025 M Tris, 0.192 M glycine, and 20% methanol.

12. Tris buffered saline (TBS; 1×): 1.5 M NaCl, 0.1 M Tris–HCl, pH 7.4, containing 0.1 % Tween 20 (TBST).
13. Blocking solution: 5 % low-fat milk in TBST.
14. Enhanced chemiluminescence (ECL) solutions.
15. Antibodies: anti-HDAC2 rabbit polyclonal (Abcam, ab7029), anti-HA mouse monoclonal (Covance), anti-UBC9 rabbit polyclonal (Santa Cruz, sc-10759).
16. Blot Stripping solution: 62.5 mM Tris–HCl, pH 6.7, 100 mM β-mercaptoethanol, and 2 % SDS.

3 Methods

3.1 Transfection of Human Cancer Cell Lines HeLa, MCF7, and U2OS (See Note 4)

1. Plate cells the day before transfection in 100 mm tissue culture dishes in order to obtain a 60 % of confluence the day of transfection.
2. For each transfection sample prepare one tube of transfection reagent: add 24 μl of Fugene® 6 in 600 μl of serum and antibiotic-free DMEM, and incubate at room temperature for 5 min. Add 5–10 μg of each DNA to the solution, and incubate for 20 min at room temperature to promote the formation of membrane delimited vesicles containing DNA.
3. Once waiting for the complex to form, remove old media from cells, wash once with PBS 1×, and add 5 mL of serum and antibiotic-free DMEM.
4. Add the mixed DNA–Fugene® 6 solution dropwise on the cells.
5. Incubate at 37 °C in a 5 % CO_2 incubator for 6 h.
6. Replace with fresh complete media.
7. Incubate at 37 °C in a 5 % CO_2 incubator for 24 h.

3.2 Cell Lysis

1. Collect cells by trypsinization and wash them in cold PBS containing SUMO and ubiquitin protease inhibitor: 5 mM *N*-ethylmaleimide (NEM) (*see* **Note 5**).
2. Lyse cells under denaturing condition using an SDS containing lysis buffer composed of one volume of buffer I and three volumes of buffer II and containing protease inhibitors and NEM.
3. Incubate lysate on ice for 15 min.
4. Sonicate each sample for 20 s (*see* **Note 6**).
5. Centrifuge in microcentrifuge for 10 min at 10,000 ×*g* at 4°C and transfer supernatants to clean tubes.
6. Use a protein quantitation assay to determine the protein concentration of each sample (*see* **Note 7**).
7. Dilute sample using equivalent amount of proteins in 5× loading buffer (*see* **Note 8**).

3.3 Immunoblotting

1. Prepare a 8% polyacrylamide gel composed by a resolving portion and a stacking portion (*see* **Notes 9** and **10**).
2. Heat samples at 95 °C for 5 min.
3. Load sample on the gel together with a prestained molecular weight standard (*see* **Note 11**).
4. Run gel in running buffer at 100 mV until samples has entered the resolving gel, and then continue at 180 mM until the dye front has come out from the gel.
5. Transfer protein from the gel to a PVDF membrane at 250 mA for 1 h in transfer buffer (*see* **Note 12**).
6. Incubate membrane in blocking solution for 1 h at RT.
7. Incubate with primary antibody diluted in 5% milk/TBST overnight at 4 °C.
8. Wash membrane three times with TBST.
9. Incubate with secondary antibody diluted in 5% milk/TBST for 1 h at RT.
10. Wash membrane three times with TBST.
11. Develop the membrane using ECL according to manufacturer's instructions (*see* **Note 13**). The results of the immunoblotting analysis are shown in Figs. 1 and 2 (*see* **Note 14**).

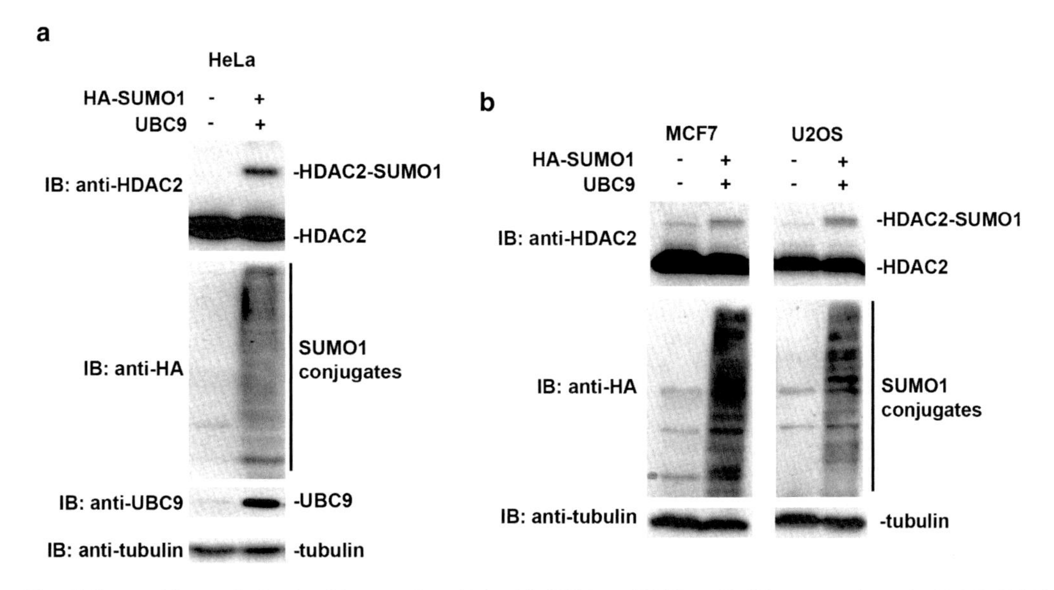

Fig. 1 Western blot analysis of cell lysates from HeLa (**a**), U2OS and MCF7 cells (**b**) co-transfected with HA-SUMO1 and UBC9 or empty vector. Cells were lysed in SDS lysis buffer and lysates were resolved on denaturing polyacrylamide gel. The expression level of HDAC2 and sumoylated-HDAC2 proteins was determined by immunoblotting (IB) using an anti-HDAC2 antibody. Tubulin was used as loading control and overexpression of HA-SUMO1 and UBC9 was checked by immunoblotting (IB) with anti-HA and anti-UBC9 antibodies respectively

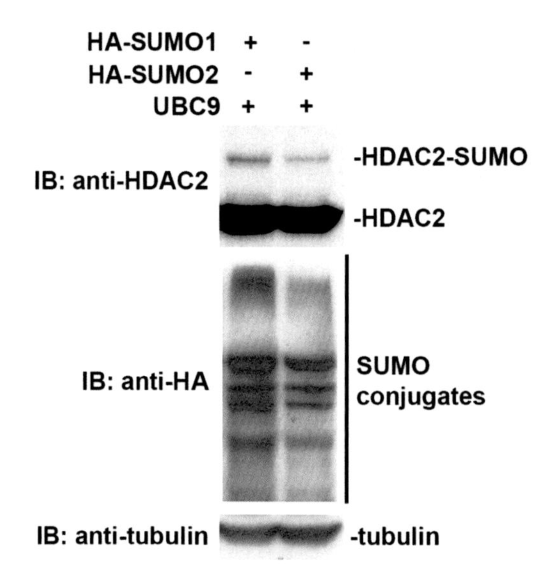

Fig. 2 Western blot analysis of cell lysates from HeLa cells co-transfected with HA-SUMO1 or HA-SUMO2 and UBC9. Cells were lysed in SDS lysis buffer and lysates were resolved on denaturing polyacrylamide gel. The expression level of HDAC2 and sumoylated-HDAC2 proteins was determined by immunoblotting (IB) using an anti-HDAC2 antibody. Tubulin was used as loading control and overexpression of HA-SUMO1 and HA-SUMO2 was checked by immunoblotting with anti-HA antibody

4 Notes

1. The inhibitors are freshly added to the buffers. A protease inhibitor cocktail for mammalian cell extract commercially available can be used instead.
2. 0.5 M NEM is prepared in absolute ethanol, stored at −20 °C and freshly added to the buffers.
3. APS can be dissolved in water and aliquots can be stored at −20 °C.
4. HeLa cells can be efficiently transfected using also calcium phosphate transfection procedure. MCF7 and U2OS can be also transfected using Lipofectamine® 2000 (Life Technologies), but we noticed an increased cell death using this reagent compared to Fugene® 6.
5. NEM is a SUMO and ubiquitin protease inhibitor used in the buffers to avoid deconjugation of sumoylated proteins.
6. Due to the high concentration of SDS in the lysis buffer, lysates appear very viscous.
7. Cell lysates can be also processed for anti-HDAC2 Immunoprecipitation. Samples have to be diluted at least 1:4 in a non-denaturing lysis buffer to dilute the amount of SDS before incubating them with the antibody.

8. β-mercaptoethanol can be replaced by dithiothreitol (DTT), used at a final concentration of 0.04 M in loading buffer 1×.
9. HDAC2 molecular weight is around 55 kDa, the addition of one SUMO protein covalently bound to HDAC2 increases the size of the protein in a range of 15–17 kDa. HDAC2 seems to have only one lysine modified by SUMO, but in the case of a protein with multiple target sites, multiples of this size increase are expected.
10. UBC9's molecular weight is around 18 kDa, and thus, a 15% polyacrylamide gel is necessary to detect it.
11. A consistent amount of proteins in necessary to detect SUMO-modified HDAC2; at least 80 μg of proteins need to be loaded on the gel for each sample.
12. PVDF membranes are preferred compared to nitrocellulose membranes. PVDF has to be activated by soaking it in pure methanol. After 1 min or less of incubation, methanol is removed and the membrane is washed once in TBST. Do not let the membrane dry. It is preferred not to stain the membrane with Ponceau after transfer.
13. Detection can be done by preferred method.
14. After the incubation and detection with the first antibody (anti-HDAC2), the membrane can be stripped using a blot stripping solution and probed with other antibodies (e.g., anti-HA and anti-tubulin, as shown in Figs. 1 and 2).

Acknowledgments

This work was supported by grants from Associazione Italiana per la Ricerca sul Cancro to Susanna Chiocca (AIRC IG5732, AIRC IG12075). SCi was supported by a fellowship from Fondazione Umberto Veronesi (FUV).

References

1. Yang WM, Inouye C, Zeng Y et al (1996) Transcriptional repression by YY1 is mediated by interaction with a mammalian homolog of the yeast global regulator RPD3. Proc Natl Acad Sci U S A 93:12845–12850
2. Zeng Y, Tang CM, Yao YL et al (1998) Cloning and characterization of the mouse histone deacetylase-2 gene. J Biol Chem 273: 28921–28930
3. Segre CV, Chiocca S (2011) Regulating the regulators: the post-translational code of class I HDAC1 and HDAC2. J Biomed Biotechnol 2011:690848
4. Tsai SC, Seto E (2002) Regulation of histone deacetylase 2 by protein kinase CK2. J Biol Chem 277:31826–31833
5. Adenuga D, Rahman I (2010) Protein kinase CK2-mediated phosphorylation of HDAC2 regulates co-repressor formation, deacetylase activity and acetylation of HDAC2 by cigarette smoke and aldehydes. Arch Biochem Biophys 498:62–73
6. Kramer OH, Zhu P, Ostendorff HP et al (2003) The histone deacetylase inhibitor valproic acid selectively induces proteasomal degradation of HDAC2. EMBO J 22:3411–3420

7. Li Y, Li X, Guo B (2010) Chemopreventive agent 3,3′-diindolylmethane selectively induces proteasomal degradation of class I histone deacetylases. Cancer Res 70:646–654
8. Yang SR, Chida AS, Bauter MR et al (2006) Cigarette smoke induces proinflammatory cytokine release by activation of NF-kappaB and posttranslational modifications of histone deacetylase in macrophages. Am J Physiol Lung Cell Mol Physiol 291:L46–L57
9. Nott A, Watson PM, Robinson JD et al (2008) S-Nitrosylation of histone deacetylase 2 induces chromatin remodelling in neurons. Nature 455:411–415
10. Doyle K, Fitzpatrick FA (2010) Redox signaling, alkylation (carbonylation) of conserved cysteines inactivates class I histone deacetylases 1, 2, and 3 and antagonizes their transcriptional repressor function. J Biol Chem 285:17417–17424
11. Citro S, Chiocca S (2013) Sumo paralogs: redundancy and divergencies. Front Biosci 5:544–553
12. Rodriguez MS, Dargemont C, Hay RT (2001) SUMO-1 conjugation in vivo requires both a consensus modification motif and nuclear targeting. J Biol Chem 276:12654–12659
13. Bossis G, Melchior F (2006) SUMO: regulating the regulator. Cell Div 1:13
14. Hay RT (2005) SUMO: a history of modification. Mol Cell 18:1–12
15. Brandl A, Wagner T, Uhlig KM et al (2012) Dynamically regulated sumoylation of HDAC2 controls p53 deacetylation and restricts apoptosis following genotoxic stress. J Mol Cell Biol 4:284–293
16. Xu X, Vatsyayan J, Gao C et al (2010) HDAC2 promotes eIF4E sumoylation and activates mRNA translation gene specifically. J Biol Chem 285:18139–18143
17. Citro S, Jaffray E, Hay RT et al (2013) A role for paralog-specific sumoylation in histone deacetylase 1 stability. J Mol Cell Biol 5: 416–427
18. Desterro JM, Rodriguez MS, Hay RT (1998) SUMO-1 modification of IkappaBalpha inhibits NF-kappaB activation. Mol Cell 2: 233–239
19. Colombo R, Boggio R, Seiser C et al (2002) The adenovirus protein Gam1 interferes with sumoylation of histone deacetylase 1. EMBO Rep 3:1062–1068
20. Tatham MH, Jaffray E, Vaughan OA et al (2001) Polymeric chains of SUMO-2 and SUMO-3 are conjugated to protein substrates by SAE1/SAE2 and Ubc9. J Biol Chem 276:35368–35374

Chapter 3

Analysis of Epigenetic Regulation of Hypoxia-Induced Epithelial–Mesenchymal Transition in Cancer Cells by Quantitative Chromatin Immunoprecipitation of Histone Deacetylase 3 (HDAC3)

Jian-Qiu Wang, Min-Zu Wu, and Kou-Juey Wu

Abstract

Epigenetics plays a key role in gene expression control. Histone modifications including acetylation/deacetylation or methylation/demethylation are major epigenetic mechanisms known to regulate epithelial–mesenchymal transition (EMT)-associated gene expression during hypoxia-induced cancer metastasis. Chromatin immunoprecipitation (ChIP) assay is a powerful tool for investigation of histone modification patterns of genes of interest. In this chapter, we describe a protocol that uses chromatin immunoprecipitation (ChIP) to analyze the epigenetic regulation of EMT marker genes by deacetylation of acetylated Histone 3 Lys 4 (H3K4Ac) under hypoxia in a head and neck cancer cell line FaDu cells. Not only a method of ChIP coupled by real-time quantitative PCR but also the detailed conditions are provided based on our previously published studies.

Key words Chromatin immunoprecipitation, Real-time qPCR, Histone modification, Epithelial–mesenchymal transition, Hypoxia, HDAC3

1 Introduction

Epigenetic regulation by histone acetylation/deacetylation or methylation/demethylation mediated by various histone modifiers are implicated in many human diseases, especially cancer [1, 2]. The amino-terminal lysine acetylation and deacetylation of histones are catalyzed by two groups of enzymes: histone acetyltransferase (HAT) and histone deacetylase (HDAC) [3]. Generally, acetylation of the lysine residues within the N-terminal of histones removes the positive charge on the histones, thereby decreasing their affinity with the negatively charged DNA, resulting in a more relaxed chromatin structure which increases accessibility for gene transcription [4]. This relaxation can be reversed by histone deacetylation [5]. Indeed, it is quite possible that histone deacetylation may have

Sibaji Sarkar (ed.), *Histone Deacetylases: Methods and Protocols*, Methods in Molecular Biology, vol. 1436,
DOI 10.1007/978-1-4939-3667-0_3, © Springer Science+Business Media New York 2016

either positive or negative regulation of gene expression, depending on the gene context involved [6]. In one of our studies, we discovered that hypoxia-inducible factor-1 (HIF-1)-induced histone deacetylase 3 (HDAC3) is essential for hypoxia-induced EMT and metastatic phenotypes [7]. On the one hand, HDAC3 interacts with hypoxia-induced WDR5 and recruits the HMT complex to increase H3K4 methylation levels and promote mesenchymal genes (e.g., N-cadherin) activation. On the other hand, HDAC3 also decreases acetylated H3K4 (H3K4Ac) levels and represses epithelial genes (e.g., E-cadherin) expression [7].

Chromatin immunoprecipitation (ChIP) assay is nearly the most powerful tool for investigating histone modification patterns of genes of interest and has been extensively developed in recent decades [8, 9]. This assay determines whether a certain protein–DNA interaction is present at a given location, condition, and time point. Briefly, the chromatin DNA and associated proteins (e.g., chromatin modifier such as HDAC3, or histone mark such as H3K4Ac) are temporarily cross-linked to form DNA–protein complex using cross-linking reagent such as formaldehyde. The chromatin DNA is then sheared to smaller fragments of ~500 bp by sonication or nuclease digestion. Then the DNA fragments bound with the protein of interest are selectively immunoprecipitated by a specified antibody against the protein. Prior to immunoprecipitation, an aliquot of the input DNA should be saved as a reference sample. The cross-links between DNA–protein complexes from input control and immunoprecipitated samples are reversed using high concentration of NaCl. After purification the DNA can be analyzed using various techniques, such as quantitative PCR, microarray or high-throughput sequencing [10–12].

Although ChIP is a widely used technique to investigate epigenetic histone marks in genomic DNA, there are several conditions to be optimized by an experimenter empirically. It is intricate and time consuming to initially setup ChIP procedures. In this article, we describe a method of ChIP for investigation of epigenetic regulation of two classical EMT marker genes, E-cadherin and N-cadherin, by deacetylation of H3K4Ac under hypoxia, coupled by real-time qPCR. The protocol described here also refers to several published protocols listed in reference [13–15], with a number of modifications to fit the aims of our study. We also describe detailed conditions for ChIP optimization based on our previous publications.

2 Materials

2.1 Chromatin Immunoprecipitation Components

1. 37% formaldehyde.
2. 1 M glycine.

3. SDS buffer: 50 mM Tris–HCl, pH 8.0, 0.5% SDS, 100 mM NaCl, 5 mM EDTA. Filter-sterilize (0.2 μm) and store at room temperature. Add protease inhibitors immediately before use.
4. IP buffer: 100 mM Tris–HCl, pH 8.6, 0.3% SDS, 1.7% Triton X-100, 5 mM EDTA. Filter-sterilize (0.2 μm) and store at 4 °C. Add protease inhibitors immediately before use.
5. Wash buffer 1: 20 mM Tris–HCl, pH 8.0, 150 mM NaCl, 5 mM EDTA, 5% sucrose, 1% Triton X-100, 0.2% SDS. Filter-sterilize (0.2 μm) and store at room temperature.
6. Wash buffer 2: 50 mM HEPES–KOH, pH 7.5, 0.1% deoxycholic acid, 500 mM NaCl, 1 mM EDTA, 1% Triton X-100. Filter-sterilize (0.2 μm) and store at room temperature.
7. Wash buffer 3: 10 mM Tris–HCl, pH 8.0, 0.5% deoxycholic acid, 250 mM LiCl, 1 mM EDTA, 0.5% NP-40. Filter-sterilize (0.2 μm) and store at room temperature.
8. Wash buffer 4: 10 mM Tris–HCl, pH 7.5, 1 mM EDTA, pH 8.0. Filter-sterilize (0.2 μm) and store at room temperature.
9. Elution buffer: 10 mM EDTA, 1% SDS, 50 mM Tris–HCl, pH 8.0. Filter-sterilize (0.2 μm) elution buffer and store it at room temperature.
10. Blocked protein A beads.

Centrifuge to collect the pre-blocked protein A beads. Aspirate the supernatant and add 1 ml blocking buffer containing 0.5 mg/ml BSA. Incubate the beads with constant rotation at 4 °C for 1 h. Centrifuge and discard the supernatant. Make 50% protein A beads slurry by adding the equal amount of IP buffer. Store it at 4 °C.

2.2 DNA Extraction

1. 25:24:1 phenol–chloroform–isoamyl alcohol.
2. 4 M LiCl.
3. 100% ethanol.
4. 3 M sodium acetate, pH 5.2.
5. 70% ethanol.

3 Methods

3.1 Chromatin Immunoprecipitation Protocol

1. Cross-link protein–DNA complex by incubating cells with 37% formaldehyde directly added to culture media to a 1% final concentration and incubate cells on a shaking device for 15 min at room temperature (*see* **Note 1**).
2. Stop the cross linking reaction by adding 1 M glycine diluted to a final concentration of 0.125 M. Incubate at room temperature for 5 min, pellet the cells and discard the supernatant. Wash the cells with PBS three times.

3. Resuspend the cells in 5 ml of SDS buffer with protease inhibitors per 10^7 cells. Incubate on ice for 10 min, and harvest the nuclei by centrifuge at 1500 rpm ($650 \times g$) for 10 min at 4 °C.
4. Resuspend the nuclei with 2 ml of IP buffer containing protease inhibitors. Pipette up and down thoroughly, and incubate on ice for 15 min (*see* **Note 2**).
5. Sonicate chromatin to an average length of about 500 bp. Optimization for sonicator setting is required. While sonicating, avoid introducing bubbles to samples in order to obtain the best shearing efficiency (*see* **Note 3**).
6. Centrifuge the lysates at 12,000 rpm ($11{,}000 \times g$) for 15 min at 4 °C. Carefully collect the supernatant in a clean tube and discard the pellet. Samples can be stored at −80 °C for several months or subjected to the IP reaction immediately.
7. Divide or dilute samples with IP buffer to a total reaction volume of 1 ml. Add 5–10 μl of a specific antibody or control IgG to samples. Be sure to retain 10% of sample as your “input.” Incubate samples on a rotation device at 4 °C for 1–2 h.
8. Add 50 μl of blocked 50% protein A beads slurry into samples and incubate at 4 °C overnight.
9. Collect beads by centrifuging at 12,000 rpm ($11{,}000 \times g$) for 1 min at 4 °C.
10. Wash samples with IP buffer one time. Perform four washes starting with wash buffer 1 to wash buffer 4 (*see* **Note 4**).
11. Add 250 μl of elution buffer. Vortex samples vigorously at room temperature for 20 min.
12. Centrifuge samples and transfer the supernatant to a new tube.
13. Add elution buffer to samples including “input” to a final volume of 500 μl, and incubate at 65 °C for at least 5 h or overnight.
14. Add 55 μl of 4 M LiCl and 250 μl of 25:24:1 phenol/chloroform/isoamyl alcohol. Vortex vigorously for 30 s, and centrifuge at full speed for 20 min at room temperature.
15. Carefully remove the top (aqueous) phase containing the DNA and transfer to a new tube. Add 1 ml of 100% ethanol and 50 μl of 3 M sodium acetate, pH 5.2. Mix by vortexing briefly, and precipitate DNA at −80 °C for 1 h or −20 °C overnight (*see* **Note 5**).
16. Centrifuge 20 min at high speed and discard the supernatant.
17. Wash the pellets with 1 ml of 70% ethanol. Allow to air-dry for 20 min.
18. Resuspend DNA pellet in 20 μl of H_2O. Store DNA samples at −20 °C.

Table 1
Reaction components for real-time PCR

Component	Amount per reaction (μl)	Final concentration
2× SYBR Green mix	2.5	
Primer mix	0.5	1 μM each
DNA	2	20× diluted

Table 2
The cycling program for real-time PCR

	Step 1	Step 2	Step 3
Temperature	95 °C	95 °C	60 °C
Time	10 min	15 s	30 s
Cycles	1	40	

3.2 qPCR Protocol

1. Design PCR primers for amplification length in a range of 50–150 bases.
2. Set up real-time PCR reactions with a diluted input and precipitated DNA, as shown in Table 1.
3. Set up the cycling parameters for real-time PCR reaction as shown in Table 2. The melting curve analysis should be performed to evaluate the quality of primer set.
4. Calculate data by the CT method and plotted as percent (%) input DNA. qChIP values are calculated by the following formula: percent (%) input recovery = [100/(input fold dilution/bound fold dilution)] $\times 2^{(\text{input CT} - \text{bound CT})}$.

4 Notes

1. The duration of the fixation and the final concentration of the formaldehyde can affect the efficiency of the procedure. Longer incubation may cause cells to form into a cross-linked aggregate that might reduce the shearing efficiency. However, a shorter incubation may affect the yield of precipitated DNA.
2. Make sure the equalization of cells subjected to the ChIP procedure.

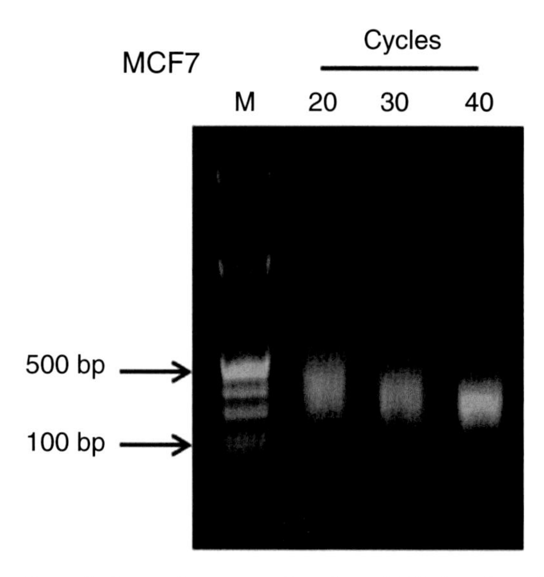

Fig. 1 Sonication of chromatin preparation from MCF-7 cells using different cycles showed the different sizes of DNA, which can be used as an indicator to optimize the cycle number so the chromatin will be suitable for subsequent chromatin immunoprecipitation. M: DNA marker

3. The sonicator settings need to be optimized based on cell type, cell number, and sonicator. Figure 1 shows the size of DNA of breast cancer cell-MCF7 upon DNA shearing in different cycles. Insufficient sonication may lead to nonspecific binding. The larger or smaller size of DNA may affect the ChIP efficiency. Generating too much heat during sonication may lead to protein degradation, resulting in low yield of precipitated DNA. The sonication step should be done at 4 °C or in ice-cold state.
4. Insufficient wash may result in high noise ratio. Extension of wash duration or steps may improve the quality of signal.
5. While recovering DNA, adding inert carrier such as glycogen can significantly increase the recovery of precipitated DNA.

Acknowledgment

This work was supported in part to J.Q.W. by National Natural Science Foundation of China [81301850]; Educational Commission of Zhejiang Province of China [Y201328812]; and to K.J.W. by Ministry of Science and Technology Summit grant [MOST 103-2745-B-039-001-ASP]; National Science Council Frontier grant [NSC102-2321-B-010-001]; center of excellence for cancer research at Taipei Veterans General Hospital [MOHW104-TDU-B-211-124-001]; and National Health Research Institutes [NHRI-EX104-10230SI].

References

1. Esteller M (2007) Cancer epigenomics: DNA methylomes and histone-modification maps. Nat Rev Genet 8(4):286–298
2. Wang JQ, Wu KJ (2015) Epigenetic regulation of epithelial-mesenchymal transition by hypoxia in cancer: targets and therapy. Curr Pharm Des 21(10):1272–1278
3. Barneda-Zahonero B, Parra M (2012) Histone deacetylases and cancer. Mol Oncol 6(6): 579–589
4. Grunstein M (1997) Histone acetylation in chromatin structure and transcription. Nature 389(6649):349–352
5. Gray SG, Teh BT (2001) Histone acetylation/ deacetylation and cancer: an "open" and "shut" case? Curr Mol Med 1(4):401–429
6. Zupkovitz G, Tischler J, Posch M et al (2006) Negative and positive regulation of gene expression by mouse histone deacetylase 1. Mol Cell Biol 26(21):7913–7928
7. Wu MZ, Tsai YP, Yang MH et al (2011) Interplay between HDAC3 and WDR5 is essential for hypoxia-induced epithelial-mesenchymal transition. Mol Cell 43(5):811–822
8. Costlow N, Lis JT (1984) High-resolution mapping of DNase I-hypersensitive sites of Drosophila heat shock genes in Drosophila melanogaster and Saccharomyces cerevisiae. Mol Cell Biol 4(9):1853–1863
9. Gilmour DS, Lis JT (1984) Detecting protein-DNA interactions in vivo: distribution of RNA polymerase on specific bacterial genes. Proc Natl Acad Sci U S A 81(14):4275–4279
10. Mukhopadhyay A, Deplancke B, Walhout AJ, Tissenbaum HA (2008) Chromatin immunoprecipitation (ChIP) coupled to detection by quantitative real-time PCR to study transcription factor binding to DNA in Caenorhabditis elegans. Nat Protoc 3(4):698–709
11. Qin J, Li MJ, Wang P, Zhang MQ, Wang J (2011) ChIP-Array: combinatory analysis of ChIP-seq/chip and microarray gene expression data to discover direct/indirect targets of a transcription factor. Nucleic Acids Res 39:W430–W436, Web Server issue
12. Schmidt D, Wilson MD, Spyrou C, Brown GD, Hadfield J, Odom DT (2009) ChIP-seq: using high-throughput sequencing to discover protein-DNA interactions. Methods 48(3): 240–248
13. Carey MF, Peterson CL, Smale ST (2009) Chromatin immunoprecipitation (ChIP). Cold Spring Harb Protoc 2009(9), pdb prot5279
14. Nelson JD, Denisenko O, Bomsztyk K (2006) Protocol for the fast chromatin immunoprecipitation (ChIP) method. Nat Protoc 1(1):179–185
15. Schoppee Bortz PD, Wamhoff BR (2011) Chromatin immunoprecipitation (ChIP): revisiting the efficacy of sample preparation, sonication, quantification of sheared DNA, and analysis via PCR. PLoS One 6(10):e26015

Chapter 4

Molecular and Functional Characterization of Histone Deacetylase 4 (HDAC4)

Lin Li and Xiang-Jiao Yang

Abstract

Histone deacetylases (HDACs) regulate various nuclear and cytoplasmic processes. In mammals, these enzymes are divided into four classes, with class II further divided into two subclasses: IIa (HDAC4, HDAC5, HDAC7, HDAC9) and IIb (HDAC6 and HDAC10). While HDAC6 is mainly cytoplasmic and HDAC10 is pancellular, class IIa HDACs are dynamically shuttled between the nucleus and cytoplasm in a signal-dependent manner, indicating that they are unique signal transducers able to transduce signals from the cytoplasm to chromatin in the nucleus. Once inside the nucleus, class IIa HDACs interact with MEF2 and other transcription factors, mainly acting as transcriptional corepressors. Although class IIa HDACs share many molecular properties in vitro, they play quite distinct roles in vivo. This chapter lists methods that we have used for molecular and biochemical characterization of HDAC4, including development of regular and phospho-specific antibodies, deacetylase activity determination, reporter gene assays, analysis of subcellular localization, and determination of interaction with 14-3-3 and MEF2. Although described specifically for HDAC4, the protocols should be adaptable for analysis to the other three class IIa members, HDAC5, HDAC7, and HDAC9, as well as for other proteins with related properties.

Key words Lysine acetylation, Histone deacetylase, HDAC4, 14-3-3, MEF2, Transcriptional repression, Nucleocytoplasmic trafficking

1 Introduction

Lysine acetylation has emerged as a major posttranslational modification for histones and nonhistone proteins [1, 2]. This is a reversible modification process that is controlled by lysine acetyltransferases and deacetylases. Traditionally, the deacetylases have been referred to as histone deacetylases (HDACs), even though it is now known that they also target nonhistone proteins. In mammals, 18 HDACs have been identified. Based on sequence homology to the yeast deacetylases Rpd3, HDA1, and Sir2, the 18 mammalian HDACs are divided into four classes, namely class I (HDAC1, -2, -3, -8), class II (HDAC4, -5, -6, -7, -9, and -10), class III (containing seven sirtuins), and class IV (HDAC11) [3, 4]. Class II HDACs have been further classified into class IIa

Sibaji Sarkar (ed.), *Histone Deacetylases: Methods and Protocols*, Methods in Molecular Biology, vol. 1436,
DOI 10.1007/978-1-4939-3667-0_4,

(HDAC4, -5, -7, and -9) and class IIb (HDAC6 and 10) [4]. In the deacetylase superfamily, class IIa HDACs are unique due to their roles as novel signal transducers able to transmit signal information from the cytoplasm to the nucleus [4]. For serving in this role, they contain intrinsic nuclear import and export signals for dynamic shuttling between the cytoplasm and nucleus (Fig. 1) [5]. Moreover, this dynamic trafficking is actively regulated by phosphorylation-dependent interaction with 14-3-3 proteins, as well as by association with other proteins such as the MEF2 family of transcription factors (Fig. 1) [5]. Different kinases then phosphorylate class IIa HDACs at three or four sites, to promote 14-3-3 binding and subsequent retention of the deacetylases in the cytoplasm [4]. Once in the nucleus, class IIa HDACs interact with transcription factors such as MEF2 to repress transcription [5]. Although class IIa HDACs share many molecular properties, they play quite distinct roles in vivo. HDAC4 and HDAC7 are important for brain and vasculature development, respectively, whereas HDAC5 and HDAC9 play critical roles in regulating skeletal and cardiac muscle development [6]. Here we list methods used for molecular and biochemical analyses of HDAC4. These methods are also applicable to characterization of HDAC5, -7, and -9, as well as for other proteins with related properties.

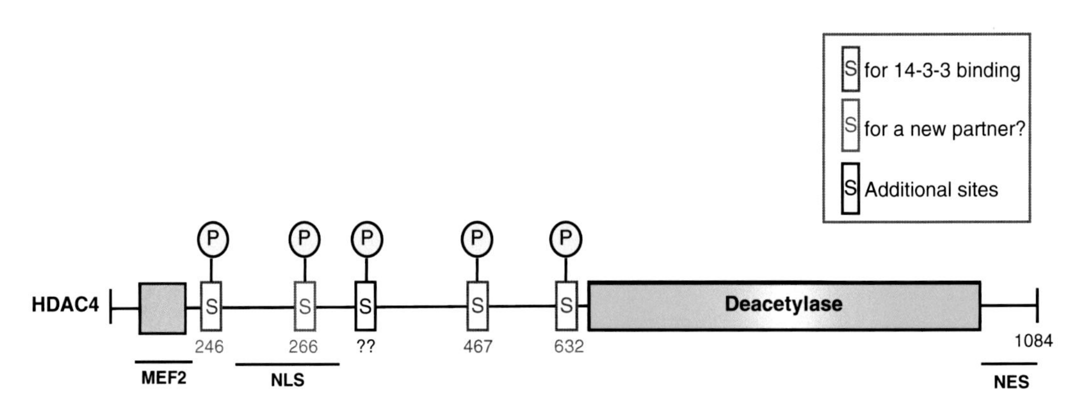

Fig. 1 Schematic illustration of HDAC4. This deacetylase contains a nuclear localization signal (NLS) in the N-terminal part, a highly conserved deacetylase domain, and a nuclear export signal (NES) at the C-terminal end [26]. Moreover, the N-terminal part of HDAC4 possesses not only an MEF2-binding site conserved from *C. elegans* to humans, but also three 14-3-3-binding sites, S246, S467, and S632, the first of which is conserved from *C. elegans* to humans. These three sites need to be phosphorylated for mediating interaction with 14-3-3 proteins. S266 is a fourth phosphorylation site and is located within the NLS, so its phosphorylation impedes nuclear localization of HDAC4 [8]. There are other phosphorylation sites [27–29], but their functions are less clear. HDAC5, HDAC7, and HDAC9 have very similar domain organizations [5]. Not illustrated here is an N-terminal PxLPxI motif conserved among HDAC4, HDAC5, and HDAC9, but not HDAC7. This motif is important for interaction with the ankyrin-repeat protein ANKRA2 [28, 30–32]. In addition, HDAC4 possesses an efficient sumoylation site within the N-terminal part [33] and is able to promote sumoylation of binding partners such as MEF2 [34–36]. HDAC4 is also subject to ubiquitination, degradation, and specific cleavage [37, 38]

2 Materials

2.1 Cell Lines and Animals

1. NIH3T3, SKN, HEK293, and 293T cells, which are cultured in DMEM media supplemented with 10 % FBS and 100 U penicillin and streptomycin in a 37 °C incubator at 5 % CO_2.
2. Rabbits (young rabbits 10–16 weeks of age, 2.5–3.0 kg).

2.2 Reagents and Materials

1. Peptides and proteins: For generation of a regular polyclonal HDAC4 antibody, an N-terminal fragment of human HDAC4 protein was expressed and affinity-purified as an MBP (maltose-binding protein) fusion protein; for producing an anti-phospho-Ser-246 HDAC4 antibody, the peptide LRKTApSEPNLKC was synthesized; FLAG peptide (DYKDDDDK); bovine serum albumin (BSA).
2. Plasmids: pBluescript KSII (+) and mammalian expression vectors for GFP-HDAC4, HA-14-3-3β, the yeast transcription factor Gal4 (residues 1-147), Gal4 (resides 1-147)-HDAC4, and β-galactosidase, all under the control of the CMV promoter.
3. Antibodies: Anti-HDAC4 polyclonal antibodies (anti-rabbit, produced in our lab), anti-HA monoclonal antibody, anti-14-3-3 antibodies.
4. Freund's adjuvant, complete and incomplete.
5. Anesthetic and disinfectant: Acepromazine, butorphanol, EMLA cream, and chlorhexidine.
6. Affinity resins and beads: CNBr-activated Sepharose 4B, anti-Flag M2 agarose beads, immobilized protein A/G beads.
7. Disposable column, tips, and Eppendorf tubes.
8. Sulfhydryl Immobilization Kit (which includes Sulfhydryl coupling resin (6 % cross-linked beaded agarose supplied as a 50 % slurry in the storage buffer (10 mM EDTA–Na, 0.05 % NaN_3, 50 % glycerol)); sample preparation buffer: 0.1 M sodium phosphate, 5 mM EDTA–Na, pH 6.0; coupling buffer: 50 mM Tris, 5 mM EDTA–Na, pH 8.5; wash solution: 1.0 M NaCl and 0.05 % NaN_3; L-cysteine·HCl; polyethylene porous discs: five discs, to add to the top to the gel bed after coupling to prevent the column from running dry).
9. mcKLH (mariculture keyhole limpet hemocyanin).
10. Cell culture dishes and plates.
11. Transfection reagents.
12. Drugs: 5 μg/ml leptomycin B in ethanol, 3 μM trichostatin A in DMSO.

13. Chemicals: [^{3}H] acetate (2.4 Ci/mmol), [^{3}H] acetylcoenzyme A (4.7 Ci/mmol), D-(−)-luciferin, acetone.

2.3 Solutions and Buffers

1. Phosphate-buffered saline (PBS): 137 mM NaCl, 2.7 mM KCl, 10 mM Na_2HPO_4 and 2 mM KH_2PO_4 (pH 7.4 or pH 8.5 adjusted with NaOH for the latter).
2. Buffer A: 50 mM Tris–HCl pH 8.0, 10 % glycerol, 1 mM DTT, 1 mM PMSF, 0.1 mM EDTA, and 10 mM sodium butyrate.
3. Buffer B: 20 mM Tris–HCl pH 8.0, 10 % glycerol, 5 mM MgCl2, 0.1 % NP-40, protease inhibitors, and KCl (0.15 or 0.5 M).
4. Buffer N: 10 mM Tris–HCl [pH 8.0], 250 mM sucrose, 2 mM $MgCl_2$, 1 mM $CaCl_2$, 1 % Triton X-100, 1 mM phenylmethylsulfonyl fluoride (PMSF), and protease inhibitors.
5. Buffer H: 50 mM Tris–HCl pH 8.0, 10 % glycerol, 1 mM dithiothreitol, 0.1 mM EDTA, and 1 mM PMSF.
6. Hypotonic lysis buffer: 20 mM HEPES pH 7.6, 20 % glycerol, 10 mM NaCl, 1.5 mM $MgCl_2$, 0.2 mM EDTA, 0.1 % Triton X-100, 25 mM NaF, 25 mM β-glycerophosphate, 1 mM dithiothreitol, and protease inhibitors.
7. 1 and 2 mM HCl.
8. 0.1 M Ethanolamine in PBS.
9. 6 M Guanidine in PBS.
10. 0.1 M Glycine–HCl (pH 2.5).
11. 0.1 M Tris–HCl (pH 8.0).
12. Elution buffer (0.1 M glycine–HCl pH 2.7).
13. 3× Sodium dodecyl sulfate (SDS) sample buffer: 150 mM Tris–HCl pH 6.8, 30 % glycerol, 6 % SDS, 0.3 % bromophenol blue, and 300 mM DTT.
14. 0.4 N H_2SO_4.
15. 5 M NaCl.
16. 0.1 M HCl-0.16 M acetic acid.

2.4 Equipment

1. Sterile needles (22 G) and syringes.
2. Dialyzer.
3. Flow-through UV monitor.
4. Pipettes.
5. Cell scrapper.
6. Fluorescence microscope.
7. Luminometer plate reader.
8. Benchtop centrifuge.

3 Methods

3.1 Development of Antibodies Specific to Regular and Phosphorylated HDAC4

To study the function and regulation of HDAC4, regular and phospho-specific antibodies were developed as described [7–9]. For the regular antibody, an HDAC4 fragment was expressed as a MBP-tagged fusion protein (Subheadings 3.1.1 and 3.1.2). Protocols for bacterial expression and protein purification can be found elsewhere [10, 11]. For the phospho-specific antibody, a phosphopeptide (LRKTApSEPNLKC), corresponding to the sequence of Ser-246 and its surrounding residues of human HDAC4, was synthesized commercially (Subheading 3.1.3).

3.1.1 Rabbit Immunization for Antiserum Production

Rabbit immunization and antiserum production were carried out at McGill Animal Resource Center (https://www.mcgill.ca/cmarc/home), and the following protocol was modified from a standard operating procedure (#406) developed by the Center.

1. Choose two young adult female rabbits (2–2.5 kg; specific pathogen free) for immunization.
2. Tranquilize the rabbits by injecting intramuscularly with mixed acepromazine (0.5 mg/kg) and butorphanol (0.2 mg/kg).
3. Apply EMLA cream over the ear at the blood collection site 15 min before puncture to collect 1 ml of blood using a 22 G needle from the ear artery. Centrifuge and collect the supernatant as the preimmune serum. Aliquot and keep it at −80 °C.
4. For primary immunization, combine 0.5 ml antigen (200–500 μg of MBP-HDAC4 in 500 μl of sterile PBS) with 0.5 ml of Freud's complete adjuvant (FCA) and emulsify (*see* **Note 1**).
5. Rinse the injection site with chlorhexidine. Inject the 1 ml sample subcutaneously into ten sites, 0.1 ml per site, bilaterally along the thoracic-lumbar region of the spine. Before removing the needle, withdraw on plunger slightly to prevent the leakage of adjuvant into the dermal layer (*see* **Note 2**).
6. Wait for 3–4 weeks for the rabbits to build up a primary immunological response.
7. For secondary immunization, inject in the vicinity of the initial sites as described except that Freud's incomplete adjuvant (FIA) is used (*see* **Note 3**).
8. Wait for another 3–4 weeks after secondary immunization.
9. Collect 1 ml blood sample and determine the titer by indirect ELISA if the titer is not sufficient; repeat **steps 8** and **9**. In most cases, the antibody titer reaches an acceptable level after two boosters.
10. When the antibody titer is sufficient, inject 0.5 mg of antigen (in 0.5 ml sterile PBS, without any adjuvant) into the ear artery.

11. If the rabbits are read for bleeding out, exsanguinate via cardiac puncture under general anesthesia; otherwise, collect 8.5 ml/kg of blood every 4 weeks. Do not exceed 6 months after initial immunization.
12. About 50 ml of antiserum can be collected per rabbit. The antiserum can be used directly or for affinity purification to enrich the antibody.

3.1.2 Antibody Affinity Purification on CNBr-Activated Sepharose

This procedure is adapted from a published method [12].

1. Dialyze the antigen at least twice with PBS (pH8.5) to remove any trace amount of Tris (*see* **Note 4**).
2. Centrifuge and collect the supernatant (if there are precipitates).
3. Determine the protein concentration (0.5–2 mg is needed).
4. Swell 0.17 g of CNBr-activated Sepharose 4B in a chromatography column (0.8 × 4 cm) containing 12 ml HCl (1 mM) for 10 min at room temperature.
5. Wash the resin three times quickly with 1 mM HCl (~12 ml each).
6. Wash the resin once with 5 ml PBS (pH 8.5).
7. Add the antigen prepared above (**step 2**) to the resin and rotate the suspension at room temperature for at least 2 h.
8. Check protein concentration of the supernatant to determine the cross-linking efficiency.
9. Wash the resin three times with regular PBS if the coupling efficiency is good (i.e., more than a half of the input has disappeared from the supernatant).
10. Incubate the resin in 10 ml PBS/0.1 M ethanolamine for 1 h at room temperature to inactivate the resin.
11. Wash the resin sequentially with PBS/6 M guanidine, PBS, and 0.1 M glycine–HCl pH 2.5 (10 ml/each).
12. Repeat the wash cycle once.
13. Wash the resin once with 10 ml PBS/6 M guanidine and twice with 10 ml PBS.
14. Suspend the resin in 0.5 ml PBS and store it at 4 °C.
15. Mix 0.5 ml resin prepared above with 10 ml serum in a disposable plastic column such as poly-prep chromatography columns (0.8 × 4 cm from Bio-Rad, with caps at both ends). Cap the column at both ends, put it inside a 50 ml Falcon tube (to prevent the cap from falling off), and rotate the column at room temperature for at least 2 h.
16. Centrifuge the column for ~2 min inside the 50 ml Falcon tube at ~3,000 × *g*, suspend the resin in 10 ml PBS, and incubate at room temperature for ~2 min.

17. Repeat step 16 four times.
18. Add 0.5 ml 0.1 M glycine–HCl pH 2.5.
19. Incubate at room temperature for 5 min (important: no longer than this) and transfer the column quickly to a 15 ml Falcon tube containing 0.5 ml 0.1 M Tris–HCl pH 8.0 so that the eluate drains directly to the Tris buffer for immediate neutralization.
20. Add 0.5 ml 0.1 M Tris–HCl pH 8.0 to resin and allow the eluate to drain directly to the eluate from the above step. The resulting mixture contains the affinity-purified antibody. Procced immediately to step 21. The resin is washed later with PBS and stored at 4 °C for further purification of additional antisera.
21. Take a small drop to estimate the pH value with a pH paper. It should be 7.0–8.0; if not, add more 0.1 M Tris–HCl pH 8.0 to neutralize the solution. Flash-freeze the purified antibody in aliquots on dry ice and keep it at −80 °C.

3.1.3 Antibody Affinity Purification on Sulfhydryl Coupling Resin

A rabbit polyclonal antibody specific to phospho-Ser246 of HDAC4 was prepared by use of a phosphopeptide antigen as the affinity tag. For the anti-phospho-Ser246 HDAC4 antibody, 10–20 mg of the phosphopeptide LRKTApSEPNLKC (where pS is phospho-serine, corresponding to residues 241-251 of HDAC4) was synthesized and HPLC-purified to 85 % purity in a commercial vendor; this amount was sufficient for both immunization of two rabbits and subsequent affinity purification. Two rabbits were immunized with the phosphopeptide [9]. Similarly, an anti-phospho-Ser266 antibody was developed [8]. The C-terminal Cys is not from HDAC4, but was added for antigen preparation by conjugation of the peptide to mariculture keyhole limpet hemocyanin (mcKLH), according to the manufacturer's instruction. The conjugated antigens were used for serum production as described in Subheading 3.1.1. After serum production, the same phosphopeptides were used for conjugation to sulfhydryl coupling resin, via the extra Cys residue, to prepare affinity gel for purification of the antibody. The following protocol is adapted from the manufacturer's manual (Pierce Manual 20402). It is based on a gel bed volume of 1 ml; for column with other gel bed volumes, adjust all solution (e.g., sample, wash, elution) volumes accordingly.

1. Equilibrate the sulfhydryl coupling resin to room temperature and add an appropriate volume of gel slurry to a disposable poly-prep column (*see* Subheading 3.1.1). For example, add 2 ml of gel slurry to obtain a gel bed of 1 ml.
2. Equilibrate the resin with four column volumes of coupling buffer (*see* **Note 5**).

3. Dissolve 2 mg of sulfhydryl-containing peptide in 1 ml coupling buffer and transfer it to the column prepared from the above step. Retain a small aliquot of the peptide solution for determination of the coupling efficiency later.
4. Cap the column at both ends, put it inside a 50 ml Falcon tube (to prevent the caps from falling off), and rotate the column at room temperature by end-over-end mixing at room temperature for 15 min.
5. Incubate the column at room temperature for additional 30 min without mixing.
6. Sequentially remove top and bottom column caps and allow the solution to drain from the column into a clean tube.
7. Place the column over a new collection tube and wash column with 3 column volumes of coupling buffer.
8. Determine the coupling efficiency by measuring and comparing absorbance (at 280 nm) of the unbound fraction (**step 6**) and the initial peptide solution (**step 3**).
9. Install the bottom cap onto the column (*see* **Note 6**).
10. Prepare 50 mM L-cysteine·HCl in coupling buffer.
11. Apply 1 ml of 50 mM L-cysteine to the column and install the top cap.
12. Put the capped column inside a 50 ml Falcon tube (to prevent the cap from falling off) and rotate the column at room temperature by end-over-end mixing at room temperature for 15 min. Then incubate for additional 30 min, without mixing.
13. Sequentially remove the top and bottom caps, allow the buffer to drain from the column, and discard the drainage.
14. Wash the column with at least six column volumes of wash solution. The resin can be used for antibody purification in the same manner as described in **steps 15–21** of Subheading 3.1.2.
15. If not used immediately, wash the column with two column volumes of the storage buffer. Install the bottom cap and add 2 ml storage buffer. The column can be capped and kept at 4 °C for storage (*see* **Note 7**).

3.2 HDAC4 Shuttling Between the Cytoplasm and Nucleus

3.2.1 Analysis of Subcellular Localization of HDAC4 by Fractionation

1. Seed NIH3T3 cells in a 6-well plate (3×10^5 cells per well).
2. Wash cells twice with PBS when they reach ~90 % confluency.
3. Add 1 ml of ice-cold hypotonic lysis buffer. Put the plate on ice for 5 min, with occasional shaking.
4. Harvest the cells with a scrapper and transfer the cell suspension into a 1.5 ml tube.

5. Centrifuge at 1,300 ×*g* and 4 °C for 5 min. Transfer the supernatant into a clean tube and leave the pellet in the original tube.
6. Centrifuge the supernatant at 16,000 ×*g* and 4 °C for 10 min, and collect the supernatant as the cytoplasmic fraction.
7. Suspend pellet from **step 5** in 0.2 ml of the hypotonic lysis buffer containing 0.5 M NaCl, and rotate the lysate at 4 °C for 20 min.
8. Centrifuge the nuclear lysate at 16,000 ×*g* and 4 °C for 10 min, and collect the supernatant as the nuclear extract.
9. Analyze cytoplasmic and nuclear extracts by Western blotting with regular and phospho-specific anti-HDAC4 antibodies.

Endogenous HDAC4 in NIH3T3 cells is mainly localized to the cytoplasm, but there is also a small fraction in the nucleus [7]. Similarly, ectopic expression of GFP-HDAC4 in NIH3T3 showed that GFP-HDAC4 mainly resides in the cytoplasm. For HEK293 cells, GFP-HDAC4 exists in the cytoplasm in a majority of them, but in a very small proportion, GFP-HDAC4 forms dot-like structure in the nucleus [7].

3.2.2 Analysis of Subcellular Localization of HDAC4 by Nuclear Export Inhibition

In NIH3T3, HDAC4 is mainly in cytoplasm, but in other cell lines such as HeLa, HDAC4 and its paralogs are mainly in the nucleus [13–15]. Also, an HDAC4 mutant lacking the N-terminal 117 residues was first reported to be actively exported to the cytoplasm [16], implying that nucleocytoplasmic shuttling is a regulatory mechanism for HDAC4. To study whether HDAC4 is actively shuttled between the cytoplasm and the nucleus, we treated cells with leptomycin B [7], a specific inhibitor of CRM1-mediated nuclear export [17, 18]. The following protocol is modified from the published studies [17, 18].

1. Seed NIH3T3 and other cells onto three wells of a 12-well plate.
2. On the next day, 0.1 μg of GFP-HDAC4-expressing plasmid is transfected into the cells by use of a transfection reagent such as lipofactamine 2000 and Superfect according to the manufacturer's instructions.
3. About 24 h later, examine live GFP fluorescence under a fluorescence microscope.
4. To the three wells, add leptomycin B to a final concentration of 10 ng/ml at three different time points, 0, 15, and 40 min. Alternatively, use one well and examine live GFP fluorescence under a fluorescence microscope at different time points.
5. Examine live GFP fluorescence under a fluorescence microscope. As shown [7], GFP-HDAC4 is relocated from the cytoplasm to the nucleus.

3.3 Interaction of 14-3-3 Proteins with HDAC4

3.3.1 Co-immunoprecipitation of HDAC4 with 14-3-3

HDAC4 is actively exported to the cytoplasm in a CRM1-dependent manner [7]. Protein shuttling often involves anchor proteins. 14-3-3 proteins are such anchors. For example, 14-3-3 proteins inhibit the translocation of the forkhead transcription factor FKHRL1 and the cell cycle-regulating phosphatase CDC25C from the nucleus to the cytoplasm [19]. To study whether HDAC4 interacts with 14-3-3β in cells, we carried out co-immunoprecipitation as described [7]. The following protocol is adapted from our published study [7].

1. Seed HEK293 cells into four 6 cm dishes (~10^6 each dish).
2. When cells reach appropriate confluency (~95 %), 3 μg of Flag-HDAC4 expression plasmid is co-transfected into the cells with or without an expression plasmid for hemagglutinin (HA)-tagged human 14-3-3β (3 μg) by use of a transfection reagent such as lipofectamine 2000 and Superfect according to the manufacturer's instructions.
3. Incubate the cells in a 37 °C incubator (5 % CO_2), and replace the transfection medium with fresh complete DMEM medium 3 h later.
4. About 48 h after transfection, wash the cells twice with PBS.
5. If the cells will not be lysed immediately, remove the PBS completely and keep the dishes at −80 °C. Otherwise, add 0.5 ml buffer B containing 0.5 M KCl, scrap the cells, and collect the suspension into a 1.5 ml Eppendorf tubes.
6. Centrifuge at 16,000 × *g* and 4 °C for 5 min.
7. Incubate the supernatant with 30 μl M2 agarose beads, or with the mouse anti-HA monoclonal antibody bound to protein G beads, overnight at 4 °C.
8. Wash the beads four times with buffer B supplemented with 0.15 M KCl (*see* **Note 8**).
9. Add 30 μl buffer B and 1.5 μl FLAG peptide (4 mg/ml) to M2 beads, rotate the suspension at 4 °C for 60 min, centrifuge briefly at 5,000 × *g*, transfer 20 μl of the supernatant for mixing with 10 μl 3×SDS sample buffer, and boil for 5 min. For protein G beads, add 30 μl 1× SDS sample buffer to the beads and boil for 5 min.
10. Use boiled samples for Western blotting with anti-Flag or anti-HA antibody.

3.3.2 Mapping 14-3-3-Binding Sites on HDAC4

14-3-3 proteins bind to two types of consensus sites: R-(S/Ar)-(+/S)-pS-(L/E/A/M)-P and R-X-(Ar/S)-(+)-pS-(L/E/A/M)-P, where Ar is an aromatic amino acid, pS is phosphoserine, + is a basic amino acid, and X is any amino acid [20, 21]. There are five potential 14-3-3-binding sites in HDAC4: 242- RKTASEP-248, 464-RTQSAP-469, 516-RQPESHP-522, 629-RAQSSP-632, and 703-RGRKATL-709 (conserved residues are underlined) [7]. Various HDAC4 mutants were used to determine which parts of HDAC4 are responsible for binding to 14-3-3 proteins. Deletion analyses showed that residues 531–1084 of HDAC4 contain one actual 14-3-3-binding site [7]. Two of the predicted motifs are within this range, namely 629-RAQSSP-632 and 703-RGRKATL-709. To test if S632 is essential, S632 was replaced with alanine. This mutant was unable to bind to 14-3-3 proteins [7], indicating that S632 but not T708 is important for 14-3-3 binding. To investigate whether S246 is important for 14-3-3 binding, S246 was replaced with alanine, and the resulting mutant was unable to bind to 14-3-3. Similarly, S467 was shown to be important for binding as well. Thus, S246, S467, and S632 of HDAC4 mediate the binding of 14-3-3 proteins (Fig. 1) [7]. Western blotting with an antibody against phosphor-S246 showed that in wild-type HDAC4, S246 is phosphorylated [9]. In agreement with this, no phosphorylation signal was detectable in the S246A mutant [9].

3.4 Determination of HDAC4 Deacetylase Activity

The following protocols are adapted from our published paper [22]. Although not described here, there are alternatives [23]. For example, another method is to perform deacetylation assays with hyperacetylated histones and subsequent immunoblotting analyses with anti-acetylated histone antibodies that are commercially available from different vendors.

3.4.1 Preparation of [^{3}H] Acetyl-Histones

1. Seed HeLa cells in a 6-well plate (0.3×10^6 cells per dish).
2. Replace culture medium containing 50 μCi of [^{3}H]acetate (2.4 Ci/mmol) per ml and 3 μM trichostatin A (TSA) when the cells reach 95 % confluency (~10^6 cells).
3. After incubation for 2–6 h, wash the cells once with ice-cold PBS.
4. Lyse HeLa cells with 500 μl buffer N and rotate the lysate at 4 °C for 1 h.
5. Centrifuge at 10,000 × *g* and 4 °C for 1 min to collect the nuclei. Wash the pellet three times with buffer N.
6. Discard the supernatant, add 0.1 ml of 0.4 N H_2SO_4, and resuspend the pellet completely by vortexing. Then add 9 volumes of acetone for precipitation of histones and incubate the mixture on ice for at least 1 h.
7. Centrifuge at 15,000 × *g* and at 4 °C for 10 min to collect histones and discard the supernatant.
8. Dissolve the precipitated histone pellet in 0.1 ml of 100 mM Tris–HCl (pH 8.0) and precipitate histones with cold acetone. Repeat **steps 6–8** for 3–4 times.
9. Air-dry the histones and dissolve in 100 μl 2 mM HCl. Determine protein concentration using Bradford methods.

10. Levels of histone acetylation are verified by using triton–acetic acid-urea gels.

Alternatively, [^{3}H] acetyl-histones could also be prepared by in vitro labeling [22].

1. Incubate 50 μg histones with 50 pmol of [^{3}H]acetyl-coenzyme A (4.7 Ci/mmol) and 0.5 μg of Flag-PCAF (p300/CBP-associated factor) in 100 μl of Buffer A at 30 °C for 30 min.
2. Add 2 μl of 5 M NaCl, 1 ml of cold acetone, and 65 μg of BSA to precipitate histones (*see* **Note 9**).
3. Leave the tube on dry ice for 2 h.
4. Centrifuge at 15,000 × *g* for 5 min at 4 °C.
5. Wash the resulting pellet with 1 ml of cold acetone, air-dry, and dissolve in 100 μl of 2 mM HCl.

3.4.2 Determination of Deacetylase Activity of HDAC4

1. Reactions are carried out in 0.2 ml of buffer H, with purified Flag-HDAC4.
2. Allow the reaction to proceed at 37 °C for 90 min.
3. Stop the reaction by addition of 0.1 ml of 0.1 M HCl–0.16 M acetic acid.
4. To extract the released [^{3}H]acetate, add 0.9 ml of ethyl acetate.
5. After vigorous vortexing, take 0.6 ml of the upper organic phase for quantification by liquid scintillation counting.

3.5 Transcriptional Repression by HDAC4

The MEF2 family of transcription factors plays important roles in muscle, bone, heart, brain, hematopoiesis and other developmental processes [24]. In 1999, we and others identified MEF2 proteins as interaction partners of HDAC4 [16, 22, 25]. The binding site was mapped to a small region in the N-terminal part of HDAC4 (Fig. 1) [16, 22, 25]. Similarly, MEF2 interacts with HDAC5, HDAC7, and HDAC9 [14, 26]. Co-immunoprecipitation procedures are similar as those described in Subheading 3.3.1. HDAC4 interacts with MEF2 to repress transcription [22].

1. Plate 293T or NIH3T3 cells onto 6 cm dishes (1×10^6).
2. On the next day, when the cells reach 95 % confluency, start transfection.
3. The luciferase report plasmid Gal4-tk-Luc (50–200 ng) is co-transfected with a vector for expressing HDAC4 as a protein fused to the DNA-binding domain of the yeast Gal4 transcription factor (residues 1–147) or for the DNA-binding domain alone (50–200 ng) with transfection reagents. pBluescript KSII (+) s used to normalize the total amount of plasmids used in each transfection, and pCMV-β-Gal (50 ng) s cotransfected for normalization of transfection efficiency (*see* **Note 10**).

4. About 32 h after transfection, expose the transfected cells to 0.3 μM trichostatin A for 16 h.
5. Then the cells are lysed in situ with a lysis buffer and the luciferase reporter activity is determined by use of D-(−)-luciferin as the substrate.
6. The chemiluminescence from activated luciferin or Galacto-Light Plus is measured on a Luminometer plate reader.

4 Notes

1. Combine the antigen and the adjuvant using two syringes and locking connector (for instance, three-way stopcock) and emulsify until it no longer separates.
2. Injection sites must be sufficiently distant to prevent coalescence of the local inflammatory response.
3. If primary injection, use FCA. Then use FIA or other alternative adjuvant.
4. Extensive dialysis is necessary as any amine-based buffers (such as Tris and EDTA) inactivate CNBr-Sepharose and interfere antigen cross-linking.
5. Throughout the entire procedure, it is important not to allow the gel bed to run dry. Add sufficient solution or install the bottom cap on the column whenever no buffer is above the top of the gel bed.
6. **Steps 9–16** are to block nonspecific binding sites on the affinity gel.
7. All buffers to be run through a column can be degassed to avoid air bubbles.
8. For washing, it is better to use low-salt buffer in order to reduce the nonspecific background.
9. This step is to remove unincorporated [^{3}H]acetyl-coenzyme A.
10. Because the transfection efficiency varies in different wells, the expression plasmid CMV-β-Gal is co-transfected as an internal control for transfection efficiency normalization. The luciferase activities are normalized based on β-galactosidase activity.

References

1. Kim GW, Yang XJ (2011) Comprehensive lysine acetylomes emerging from bacteria to humans. Trends Biochem Sci 36.211–220
2. Choudhary C, Weinert BT, Nishida Y, Verdin E, Mann M (2014) The growing landscape of lysine acetylation links metabolism and cell signalling. Nat Rev Mol Cell Biol 15: 536–550
3. Khochbin S, Verdel A, Lemercier C, Seigneurin-Berny D (2001) Functional significance of histone deacetylase diversity. Curr Opin Genet Dev 11:162–166
4. Yang XJ, Seto E (2008) The Rpd3/Hda1 family of lysine deacetylases: from bacteria and yeast to mice and men. Nat Rev Mol Cell Biol 9:206–218

5. Yang XJ, Grégoire S (2005) Class II histone deacetylases: from sequence to function, regulation and clinical implication. Mol Cell Biol 25:2873–2884
6. Haberland M, Montgomery RL, Olson EN (2009) The many roles of histone deacetylases in development and physiology: implications for disease and therapy. Nat Rev Genet 10:32–42
7. Wang AH, Kruhlak MJ, Wu J, Bertos NR, Vezmar M, Posner BI, Bazett-Jones DP, Yang XJ (2000) Regulation of histone deacetylase 4 by binding of 14-3-3 proteins. Mol Cell Biol 20:6904–6912
8. Walkinshaw DR, Weist R, Xiao L, Yan K, Kim GW, Yang XJ (2013) Dephosphorylation at a conserved SP motif governs cAMP sensitivity and nuclear localization of class IIa histone deacetylases. J Biol Chem 288:5591–5605
9. Walkinshaw DR, Weist R, Kim GW, You L, Xiao L, Nie J, Li CS, Zhao S, Xu M, Yang XJ (2013) The tumor suppressor kinase LKB1 activates the downstream kinases SIK2 and SIK3 to stimulate nuclear export of class IIa histone deacetylases. J Biol Chem 288:9345–9362
10. Pelletier N, Champagne N, Lim H, Yang XJ (2003) Expression, purification, and analysis of MOZ and MORF histone acetyltransferases. Methods 31:24–32
11. Yan K, Wu CJ, Pelletier N, Yang XJ (2012) Reconstitution of active and stoichiometric multisubunit lysine acetyltransferase complexes in insect cells. Methods Mol Biol 809:445–464
12. Harlow E, Lane D (1999) Using antibodies: a laboratory manual. Cold Spring Harbor Laboratory Press, Cold Spring Harbor, NY
13. Fischle W, Emiliani S, Hendzel MJ, Nagase T, Nomura N, Voelter W, Verdin E (1999) A new family of human histone deacetylases related to Saccharomyces cerevisiae HDA1p. J Biol Chem 274:11713–11720
14. Lemercier C, Verdel A, Galloo B, Curtet S, Brocard MP, Khochbin S (2000) mHDA1/HDAC5 histone deacetylase interacts with and represses MEF2A transcriptional activity. J Biol Chem 275:15594–15599
15. Kao HY, Downes M, Ordentlich P, Evans RM (2000) Isolation of a novel histone deacetylase reveals that class I and class II deacetylases promote SMRT-mediated repression. Genes Dev 14:55–66
16. Miska EA, Karlsson C, Langley E, Nielsen SJ, Pines J, Kouzarides T (1999) HDAC4 deacetylase associates with and represses the MEF2 transcription factor. EMBO J 18:5099–5107
17. Kudo N, Matsumori N, Taoka H, Fujiwara D, Schreiner EP, Wolff B, Yoshida M, Horinouchi S (1999) Leptomycin B inactivates CRM1/exportin 1 by covalent modification at a cysteine residue in the central conserved region. Proc Natl Acad Sci U S A 96:9112–9117
18. Fornerod M, Ohno M, Yoshida M, Mattaj IW (1997) CRM1 is an export receptor for leucine-rich nuclear export signals. Cell 90:1051–1060
19. Brunet A, Bonni A, Zigmond MJ, Lin MZ, Juo P, Hu LS, Anderson MJ, Arden KC, Blenis J, Greenberg ME (1999) Akt promotes cell survival by phosphorylating and inhibiting a Forkhead transcription factor. Cell 96:857–868
20. Yaffe MB, Rittinger K, Volinia S, Caron PR, Aitken A, Leffers H, Gamblin SJ, Smerdon SJ, Cantley LC (1997) The structural basis for 14-3-3:phosphopeptide binding specificity. Cell 91:961–971
21. Rittinger K, Budman J, Xu J, Volinia S, Cantley LC, Smerdon SJ, Gamblin SJ, Yaffe MB (1999) Structural analysis of 14-3-3 phosphopeptide complexes identifies a dual role for the nuclear export signal of 14-3-3 in ligand binding. Mol Cell 4:153–166
22. Wang AH, Bertos NR, Vezmar M, Pelletier N, Crosato M, Heng HH, Th'ng J, Han J, Yang XJ (1999) HDAC4, a human histone deacetylase related to yeast HDA1, is a transcriptional corepressor. Mol Cell Biol 19:7816–7827
23. Yuan Z, Rezai-Zadeh N, Zhang X, Seto E (2009) Histone deacetylase activity assay. Methods Mol Biol 523:279–293
24. Potthoff MJ, Olson EN (2007) MEF2: a central regulator of diverse developmental programs. Development 134:4131–4140
25. Sparrow DB, Miska EA, Langley E, Reynaud-Deonauth S, Kotecha S, Towers N, Spohr G, Kouzarides T, Mohun TJ (1999) MEF-2 function is modified by a novel co-repressor, MITR. EMBO J 18:5085–5098
26. Wang AH, Yang XJ (2001) Histone deacetylase 4 possesses intrinsic nuclear import and export signals. Mol Cell Biol 21:5992–6005
27. Mathias RA, Guise AJ, Cristea IM (2015) Post-translational modifications regulate class IIa histone deacetylase (HDAC) function in health and disease. Mol Cell Proteomics 14:456–470
28. Xu C, Jin J, Bian C, Lam R, Tian R, Weist R, You L, Nie J, Bochkarev A, Tempel W et al (2012) Sequence-specific recognition of a PxLPxI/L motif by an ankyrin repeat tumbler lock. Sci Signal 5:ra39

29. Paroni G, Cernotta N, Dello Russo C, Gallinari P, Pallaoro M, Foti C, Talamo F, Orsatti L, Steinkuhler C, Brancolini C (2008) PP2A regulates HDAC4 nuclear import. Mol Biol Cell 19:655–667
30. Wang AH, Gregoire S, Zika E, Xiao L, Li CS, Li H, Wright KL, Ting JP, Yang XJ (2005) Identification of the ankyrin repeat proteins ANKRA and RFXANK as novel partners of class IIa histone deacetylases. J Biol Chem 280:29117–29127
31. McKinsey TA, Kuwahara K, Bezprozvannaya S, Olson EN (2006) Class II histone deacetylases confer signal responsiveness to the ankyrin-repeat proteins ANKRA2 and RFXANK. Mol Biol Cell 17:438–447
32. Nie J, Xu C, Jin J, Aka JA, Tempel W, Nguyen V, You L, Weist R, Min J, Pawson T et al (2015) Ankyrin repeats of ANKRA2 recognize a PxLPxL motif on the 3M syndrome protein CCDC8. Structure 23:700–712
33. Kirsh O, Seeler JS, Pichler A, Gast A, Muller S, Miska E, Mathieu M, Harel-Bellan A, Kouzarides T, Melchior F et al (2002) The SUMO E3 ligase RanBP2 promotes modification of the HDAC4 deacetylase. EMBO J 21:2682–2691
34. Grégoire S, Tremblay AM, Xiao L, Yang Q, Ma K, Nie J, Mao Z, Wu Z, Giguere V, Yang XJ (2006) Control of MEF2 transcriptional activity by coordinated phosphorylation and sumoylation. J Biol Chem 281:4423–4433
35. Grégoire S, Yang XJ (2005) Association with class IIa histone deacetylases upregulates the sumoylation of MEF2 transcription factors. Mol Cell Biol 25:2273–2282
36. Hietakangas V, Anckar J, Blomster HA, Fujimoto M, Palvimo JJ, Nakai A, Sistonen L (2006) PDSM, a motif for phosphorylation-dependent SUMO modification. Proc Natl Acad Sci U S A 103:45–50
37. Cernotta N, Clocchiatti A, Florean C, Brancolini C (2011) Ubiquitin-dependent degradation of HDAC4, a new regulator of random cell motility. Mol Biol Cell 22:278–289
38. Backs J, Worst BC, Lehmann LH, Patrick DM, Jebessa Z, Kreusser MM, Sun Q, Chen L, Heft C, Katus HA et al (2011) Selective repression of MEF2 activity by PKA-dependent proteolysis of HDAC4. J Cell Biol 195:403–415

Chapter 5

Approaches for Studying the Subcellular Localization, Interactions, and Regulation of Histone Deacetylase 5 (HDAC5)

Amanda J. Guise and Ileana M. Cristea

Abstract

As a member of the class IIa family of histone deacetylases, the histone deacetylase 5 (HDAC5) is known to undergo nuclear–cytoplasmic shuttling and to be a critical transcriptional regulator. Its misregulation has been linked to prominent human diseases, including cardiac diseases and tumorigenesis. In this chapter, we describe several experimental methods that have proven effective for studying the functions and regulatory features of HDAC5. We present methods for assessing the subcellular localization, protein interactions, posttranslational modifications (PTMs), and activity of HDAC5 from the standpoint of investigating either the endogenous protein or tagged protein forms in human cells. Specifically, given that at the heart of HDAC5 regulation lie its dynamic localization, interactions, and PTMs, we present methods for assessing HDAC5 localization in fixed and live cells, for isolating HDAC5-containing protein complexes to identify its interactions and modifications, and for determining how these PTMs map to predicted HDAC5 structural motifs. Lastly, we provide examples of approaches for studying HDAC5 functions with a focus on its regulation during cell-cycle progression. These methods can readily be adapted for the study of other HDACs or non-HDAC-proteins of interest. Individually, these techniques capture temporal and spatial snapshots of HDAC5 functions; yet together, these approaches provide powerful tools for investigating both the regulation and regulatory roles of HDAC5 in different cell contexts relevant to health and disease.

Key words HDAC5, Protein complexes, Immunoaffinity purification, Phosphorylation, Protein interactions, Immunofluorescence microscopy, Cell cycle, Kinase inhibition, siRNA knockdown, In-gel digestion

1 Introduction

Histone deacetylase 5 (HDAC5) is a member of the class IIa family of HDACs. Being implicated in cardiac development, the misregulation of HDAC5 is known to be associated with cardiac hypertrophy and myopathies [1, 2]. Misregulation of HDAC5 is further linked to other human diseases, including tumorigenesis [3], viral infection-induced diseases [4], and drug abuse-induced epigenetic alterations [5–7] (Fig. 1a).

Sibaji Sarkar (ed.), *Histone Deacetylases: Methods and Protocols*, Methods in Molecular Biology, vol. 1436,
DOI 10.1007/978-1-4939-3667-0_5,

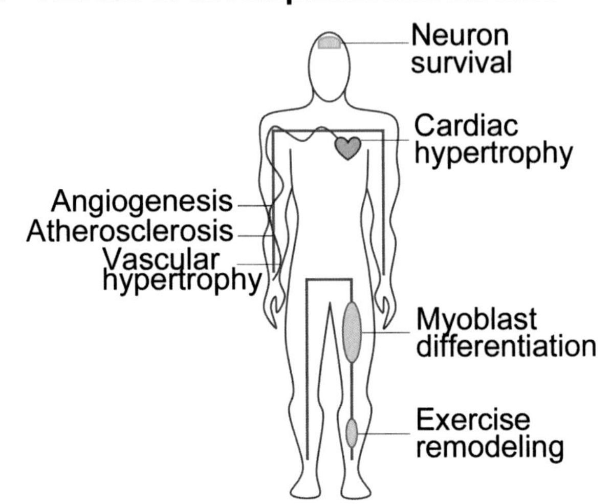

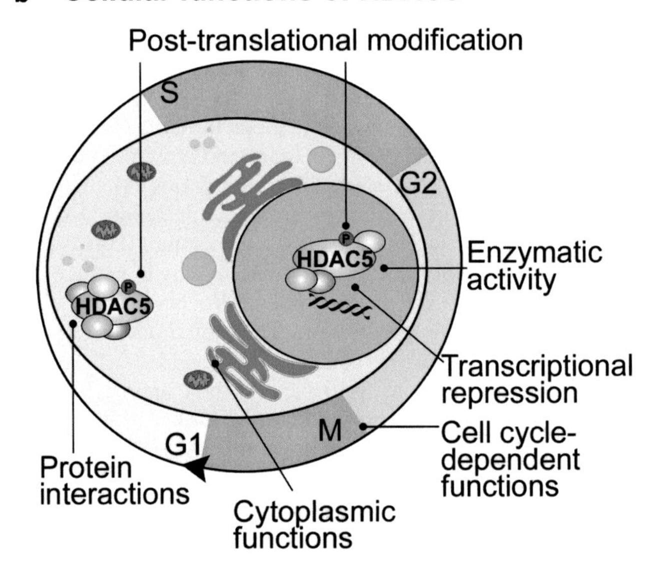

Fig. 1 Exploring the broad impact of HDAC5 on human health and disease and its essential intracellular functions. (**a**) HDAC5 is implicated in muscle development, cardiac disease, and neuron survival (adapted from Ref. [68]) (**b**) Within the cell, HDAC5 shuttles between the nucleus and the cytoplasm and has critical roles in transcriptional control, cell cycle regulation, and localization-dependent protein interactions

Class IIa HDACs (HDAC4, HDAC5, HDAC7, and HDAC9) are known to have limited intrinsic enzymatic activity, owing to an amino acid substitution within their catalytic domains when compared to the prominent class I deacetylases (e.g., HDAC1 and HDAC2) [8]. Thus, rather than being established as regulators of substrate acetylation levels, the main functions of HDAC5 and the other class IIa enzymes are connected to their critical ability to act as transcriptional repressors through participation in multi-protein complexes [9, 10] (Fig. 1b). In the nucleus, HDAC5 associates with co-repressor complexes, including the nuclear co-repressor complex (NCOR) and the class I enzyme HDAC3 to repress transcriptional targets [9]. HDAC5 is also capable of repressing transcriptional targets independently of these co-repressor complexes through association with transcription factors, such as MEF2 [11]. The HDAC5-MEF2 association is mediated through the N-terminal region of class IIa HDACs that contains transcription factor binding sites, as well as a nuclear localization signal (NLS) (Fig. 2). The C-terminus of HDAC5 contains the deacetylation domain (DAC) and nuclear export sequence (NES) [12]. The domain features of HDAC5 were established through the generation of

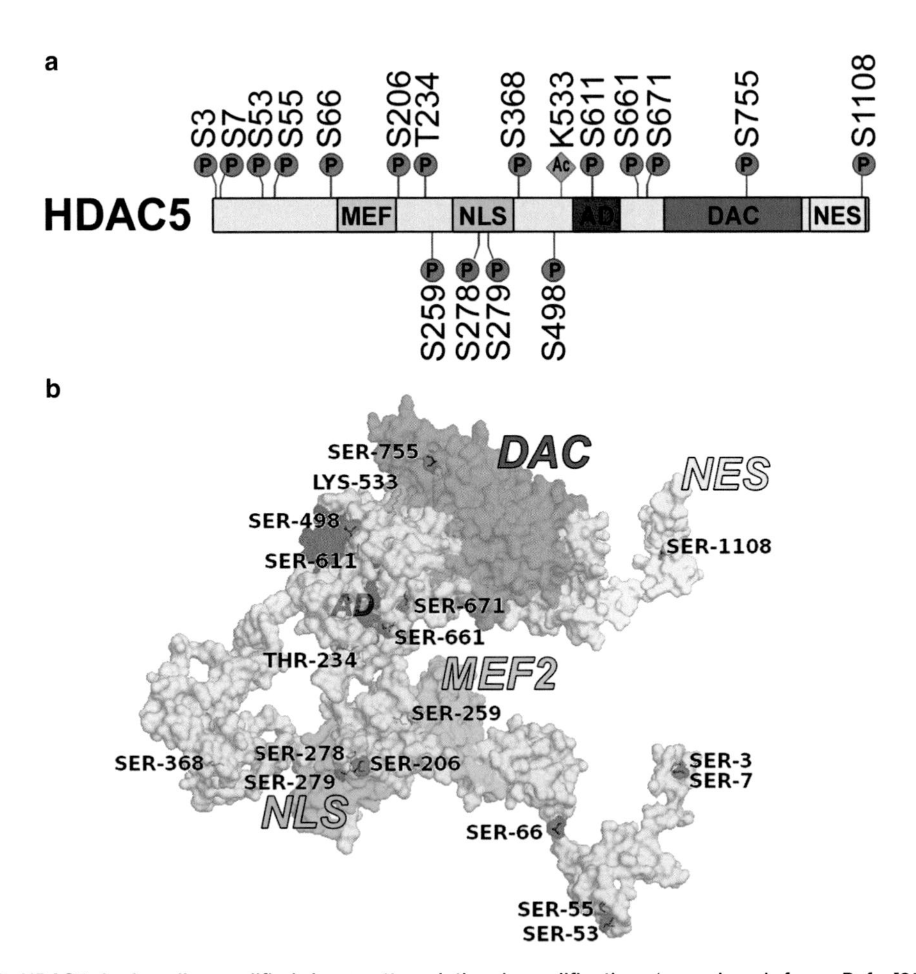

Fig. 2 HDAC5 is heavily modified by posttranslational modification (reproduced from Ref. [35]). (**a**) Phosphorylation and acetylation of HDAC5 spans the protein sequence. HDAC5 domains are indicated: *MEF* mef2 binding domain (*blue*); *NLS* nuclear localization signal (*green*); *AD* acidic domain (*red*); *DAC* deacetylation domain (*purple*); *NES* nuclear export sequence (*yellow*). (**b**) Numerous posttranslational modifications of HDAC5 map to the external surface of the predicted protein structure

mutated HDAC5 constructs and subsequent observation of the impact on localization patterns and enzymatic activity, as well as through comparison with other class IIa HDACs [13–18]. Moreover, putative localization signals were fused to GFP to assess their impact on localization, allowing the definition of NLS and NES motifs [19].

Since its discovery, HDAC5 has been the focus of intensive study. Numerous signaling pathways have been shown to converge on phosphorylation sites within HDAC5, including CaMK family members, protein kinases A, C, and D, and Aurora B [2, 20–30]. In particular, HDAC5 has been the target of investigation for its roles in heart development and disease [26, 31–33], especially with respect to its change in localization during hypertrophic conditions. An important feature of class IIa HDACs is their ability to shuttle in and out of the nucleus. In conjunction with the NLS and NES, HDAC5 subcellular localization is regulated by site-specific phosphorylation. Phosphorylation of two sites (Ser259 and Ser498) flanking the NLS promotes binding of 14–3–3 chaperone proteins and nuclear export [11, 19, 34], while phosphorylation of Ser279 within the NLS is important for nuclear localization of HDAC5 [16, 29] (Fig. 2). The kinases responsible for individual phosphorylations have been identified by examining consensus kinase target motifs, and confirmed by activation/inhibition studies and site-directed mutagenesis of target residues [7, 19, 29, 30]. While the nuclear roles of HDAC5 in regulating transcription are the best understood, HDAC5 roles in the cytoplasm and outside of transcriptional control remain poorly defined. A recent investigation of HDAC5 has revealed a cell cycle-dependent regulation of class IIa HDACs, including HDAC5, HDAC4, and HDAC9, by Aurora B-mediated phosphorylation [30]. Phosphorylation of HDAC5 within the NLS is accompanied by midzone and midbody localization of HDAC5 within a phosphorylation gradient during mitosis [30]. Using immunoaffinity purification of EGFP-tagged HDAC5, numerous phosphorylation sites, including those detailed above, were identified [16] (Fig. 2). Site-specific mutations of HDAC5 has provided further insight into the functional roles of individual phosphorylations, for example, identifying a new point of regulation of its nuclear localization [16, 30, 35]; however, numerous sites await further characterization. The impact of phosphorylation on protein function can be assessed through the effect of phosphorylation site mutation on subcellular localization, protein interactions, and, possibly, transcriptional repression functions or enzymatic activity.

This chapter describes several methods that have proven valuable for studying the subcellular localization, protein interactions, posttranslational modifications, and activity of HDAC5 (Fig. 1b). We start by discussing the value and the challenges associated with studying either endogenous or tagged versions of HDAC5, and by describing a method for generating stable cell lines expressing tagged HDAC5. We next describe a frequently used microscopy approach for assessing the sub-cellular localization of HDAC5 in fixed and live cells. Given that two of the most critical regulators of HDAC5 functions are its protein interactions and posttranslational modifications, we dedicate the next two sections to methods for isolating HDAC5 protein complexes and identifying its interactions and modifications, and for predicting associated HDAC5 structural motifs. Lastly, we provide examples of methods for studying HDAC5 functions and activity, by focusing on its regulation during the cell-cycle progression and its assessment of enzymatic activity.

2 Materials and Equipment

2.1 Studying Endogenous and Tagged HDAC5 in Cell Culture

2.1.1 Generating Cell Lines Expressing GFP-Tagged HDAC5

1. HDAC5-EGFP construct in retroviral vector (e.g., pLXSN).
2. Phoenix cells (production cell line).
3. 6-well tissue culture plates or T75 flasks.
4. Tissue culture dishes (10 or 15 cm) or flasks (T75 or T175).
5. Incubator set at 37 °C, 5 % CO_2.
6. DMEM supplemented with 10 % FBS (DMEM+) or RMPI supplemented with 10 % FBS (RPMI+).
7. DPBS.
8. Wild-type HEK293 cells (or other recipient cell type).
9. FuGene transfection reagent.
10. Centrifuge and rotor capable of spinning at 1000 × g for 5 min.
11. Serological pipettes.
12. Sterile 15 mL conical tubes.
13. Centrifuge adaptors for 15 mL conical tubes.
14. 0.45 μm low-binding membrane filter (e.g., cellulose acetate or polysulfonic membrane filters).
15. Polybrene.
16. G418.
17. Materials for FACS analysis (*see* Subheading 2.5.2).
18. Materials for immunofluorescence microscopy (*see* Subheading 2.2).

2.1.2 Cell Culture for Studies of Endogenous or Tagged HDAC5

1. Sterile DPBS (Dulbecco's phosphate-buffered saline (1×), liquid) (DPBS).
2. DMEM++ (500 mL DMEM (Dulbecco's modified Eagle media (1×), liquid) + 50 mL fetal bovine serum + 5 mL penicillin–streptomycin).

3. 1× trypsin (5 mL trypsin Stock + 45 mL sterile DPBS).
4. Cell stock (typically frozen stock in an 85 % FBS/10 % DMSO solution, stored at −144 °C).
5. Water bath set at 37 °C.
6. Incubator set at 37 °C, 5 % CO_2.
7. Sterile tissue culture hood with UV sterilization capability.
8. Centrifuge with swinging bucket rotor capable of spins at 254 × *g* at room temperature.
9. Conical tube adaptors for centrifuge rotor.
10. 15 mL conical tubes.
11. 50 mL conical tubes.
12. Serological pipettes and pipettor.
13. Tissue culture dishes (100 mm or 150 mm) or flasks (T25 or T75).
14. Light microscope for visualizing cells (optional).
15. Hemocytometer (optional).

2.2 Assessing Subcellular Localization of HDAC5

2.2.1 Fixed Cell Imaging

1. Starting cell culture (ca. 1×10^6 cells).
2. Poly-D-lysine.
3. Glass coverslips.
4. Glass microscopy slides or glass-bottom dishes.
5. Phosphate buffered saline (DPBS).
6. 2 % paraformaldehyde in DPBS.
7. 0.1 M glycine in DPBS (0.375 g glycine in 50 mL DPBS).
8. 0.1 % Triton X-100 in DPBS (500 μL 10 % Triton X-100 in 49.5 mL DPBS).
9. 0.2 % Tween in DPBS (100 μl 100 % Tween in 49.9 mL DPBS).
10. BSAT solution: 2 % BSA, 0.2 % Tween in DPBS (1 g BSA and 100 μL 100 % Tween in 49.9 mL DPBS).
11. Primary antibodies.
12. Fluorescent secondary antibodies.
13. DAPI or To-Pro3 dye (if staining DNA).
14. Anti-fade solution/mounting media (e.g., Aqua PolyMount).
15. Nail Polish.
16. Aluminum Foil.

2.2.2 Live Cell Imaging

1. Starting cell culture (ca. 1×10^6 cells) expressing fluorescently tagged proteins.
2. Glass-bottom dishes (35 mm glass-bottom dishes).

3. DMEM++ (*see* Subheading 2.1.2)
4. Hank's balanced saline solution (HBSS) or phosphate buffered saline (DPBS).
   (a) HBSS is available commercially or can be made to the following specifications: 0.137 M NaCl, 5.4 mM KCl, 0.25 mM Na_2HPO_4, 0.1 g glucose, 0.44 mM KH_2PO_4, 1.3 mM $CaCl_2$, 1.0 mM $MgSO_4$, 4.2 mM $NaHCO_3$.
5. Live-cell compatible fluorescent dyes (e.g., ER-Tracker, MitoTracker, LysoTracker, Hoechst) dissolved in DMSO according to manufacturer's instructions.
6. Opti-MEM or other dye-free media for imaging.
7. Aluminum Foil.
8. Fluorescence microscope capable of live cell imaging and laser excitation at wavelengths corresponding to fluorescent tags.

2.3 Defining HDAC5 Protein Interactions

2.3.1 Harvesting Cells

1. Sterile DPBS (*see* Subheading 2.1.2).
2. 1× trypsin (*see* Subheading 2.1.2).
3. DMEM++ or standard culture media (*see* Subheading 2.1.2).
4. Protease inhibitor cocktail.
5. Freezing buffer (20 mM Na-HEPES/1.2% PVP (w/v), pH = 7.4).

   For 50 mL, use 0.2383 g HEPES (MW = 238.3 g/mol) and 0.6 g polyvinylpyrrolidone. Adjust the pH to 7.4 with NaOH. Adjust volume to 50 mL with dH_2O. Sterile filter solution and store at room temperature.

2.3.2 Cryogenic Cell Lysis

1. Ball mill-type homogenizer (mixer mill) capable of mechanically disrupting frozen tissue and cells within removable grinding chambers, such as a Retsch Mixer Mill.
2. Grinding chamber (Stainless steel or tungsten carbide).
3. Grinding ball (Stainless steel or tungsten carbide).
4. Hot H_2O (tap water is sufficient).
5. Windex.
6. 10% bleach solution.
7. Ultrapure H_2O.
8. Methanol.
9. Metal spatula.
10. Liquid nitrogen bath.
11. 50 mL conical tubes.
12. Long-handled tongs.

2.3.3 Conjugation of Magnetic Beads to Antibodies

1. Superparamagnetic beads with surface epoxy groups (e.g., this protocol was specifically optimized for M-270 Epoxy Dynabeads) (Store at 4 °C).
2. Affinity purified antibody against tag or protein of interest (e.g., GFP or FLAG or HDAC5).
3. 0.1 M Na-phosphate buffer, pH 7.4 (Na-phosphate buffer). Prepare solution as 19 mM NaH_2PO_4, 81 mM Na_2HPO_4 and adjust pH to 7.4. Filter-sterilize and store at 4 °C.
4. 3 M ammonium sulfate ($AmSO_4$). Prepare solution in 0.1 M sodium phosphate buffer, pH 7.4. Filter-sterilize and store at 4 °C.
5. 100 mM glycine–HCl, pH 2.5. Prepare in H_2O and adjust to pH 2.5 with HCl. Filter-sterilize and store at 4 °C.
6. 10 mM Tris, pH 8.8. Prepare in water and adjust to pH 8.8 with HCl. Filter-sterilize and store at 4 °C.
7. 100 mM triethylamine. Prepare a fresh solution in water on the day of bead washing steps. **CAUTION**: Triethylamine must be handled in a chemical hood and be disposed of as hazardous chemical waste. Triethylamine is toxic and flammable.
8. DPBS, pH 7.4 (*see* Subheading 2.1.2).
9. DPBS containing 0.5 % Triton X-100.
10. DPBS containing 0.02 % sodium azide (NaN_3). **CAUTION**: Sodium azide must be handled in a chemical hood and disposed appropriately. Sodium azide is a toxic solid compound.
11. Tube rotator kept at 30 °C.
12. Magnetic tube rack separator.
13. TOMY micro tube mixer.
14. 2 mL round bottom tubes.
15. Ultrapure water (e.g., from a Milli-Q Integral Water Purification System).

2.3.4 Immunoaffinity Purifications of Endogenous or Tagged HDAC5

1. Frozen cell powder stored at −80 °C prior to use.
2. 3 M NaCl, filter-sterilized.
3. 10 % Triton X-100 in Milli-Q H_2O, filter-sterilized.
4. 10× TBT with salts, filter-sterilized. Store at 4 °C.

   To prepare 100 mL of 10× TBT:

   (a) 4.766 g HEPES (MW 238.31).

   (b) 10.797 g potassium acetate (MW 98.15).

   (c) 0.407 g magnesium chloride hexahydrate (MW 203.30).

   (d) 2.0 μL 0.5 M $ZnCl_2$.

   (e) 2.0 μL 0.5 M $CaCl_2$.

   (f) Adjust pH to 7.4 with 5 M KOH.

(g) Add 1.0 mL Tween 20.

(h) Add Milli-Q H_2O up to 100 mL.

(i) Sterile filter and store at 4 °C.

5. Mammalian protease inhibitor cocktail containing inhibitors with broad specificity against serine proteases, cysteine proteases, acid proteases, and aminopeptidases.
6. Phosphatase inhibitor cocktail containing sodium orthovanadate, sodium molybdate, sodium tartrate, and imidazole.
7. Phosphatase inhibitor cocktail containing cantharidin, (-)-*p*-Bromolevamisole, and calyculin A.
8. DNase I (1 mg/mL stock).
9. Milli-Q H_2O or other Ultrapure H_2O.
10. Optimized lysis buffer (prepared immediately before isolations). Store at 4 °C.
11. Magnetic beads conjugated with antibodies. Store at 4 °C.
12. 50 mL conical tubes.
13. Immersion tissue homogenizer/emulsifier (e.g., Polytron from Kinematica) or vortex for smaller-scale isolations.
14. Centrifuge and rotor capable of spinning at 8000 ×*g*, set at 4 °C with 50 mL conical tube adaptors.
15. Tube rotator kept at 4 °C.
16. 2 mL round-bottom tubes.
17. 1.5 mL tubes.
18. Bar magnets (for conical tubes) and magnetic separation rack (for micro tubes).
19. 4× LDS elution buffer. For 10 mL solution: dissolve 0.666 g of Tris–HCl, 0.682 g of Tris base, 0.8 g of LDS, and 0.006 g of EDTA (free acid) in ultrapure H_2O. Store at –20 °C.
20. 10× reducing agent (e.g., 1 M DTT).
21. 10× alkylating agent (e.g., 1 M iodoacetamide, light-sensitive, store at –20 °C).
22. Heating block or bath set at 70 °C.

2.3.5 In-Gel Digestion of Isolated Protein Complexes

1. Milli-Q water.
2. SimplyBlue SafeStain.
3. HPLC grade water.
4. 100 % acetonitrile (ACN).
5. 0.1 M ammonium bicarbonate (ABC).
6. 0.1 % formic acid.
7. 0.1 % formic acid–50 % ACN.
8. 0.5 μg/μL trypsin stock.

9. 10% TFA stock.
10. Nu-PAGE gel containing separated immunoaffinity isolated samples.
11. Low-protein-binding or low-retention pipet tips.
12. Gel slicer capable of generating 1 mm slices.
13. Razor blade for gel lane excision.
14. Metal spatula or pick for separating gel slices.
15. Ceramic surface for separating gel slices.
16. Optional: multichannel pipettor and reagent reservoirs.

2.3.6 Peptide Extraction and Sample Desalting (StageTips)

1. 1.5 mL tubes.
2. Low-protein-binding or low-retention pipet tips.
3. HPLC grade water.
4. 100% acetonitrile (ACN).
5. 0.1 M ammonium bicarbonate (ABC).
6. 0.1% formic acid.
7. 0.1% formic acid–50% ACN.
8. 10% TFA stock.
9. 0.2% TFA (wash buffer).
10. 0.10 M ammonium formate–0.5% formic acid–40% ACN (hereafter referred to as Elution #1).
11. 0.15 M ammonium formate–0.5% formic acid–60% ACN (hereafter referred to as Elution #2).
12. 5% ammonium hydroxide–80% ACN (hereafter referred to as Elution #3).
13. 1% FA/4% ACN (Dilution Solution).
14. 16 G Needle and plunger.
15. Hydrophilic sulfonated poly(styrenedivinylbenzene) copolymer disks (e.g., SDB-RPS Empore disks).
16. SpeedVac or vacuum concentrator.

2.4 Investigating Regulation of HDAC5 by Posttranslational Modification

2.4.1 Identification of Posttranslational Modifications

All reagents from Subheadings 2.3.4, 2.3.5, and 2.3.6.

2.4.2 Characterizing PTM Functions

1. FASTA sequence of interest.
2. Computer with internet access.
3. E-mail address to retrieve results (Optional).

2.4.3 Structure Predictions

1. FASTA sequence for protein of interest.
2. List of PDB accession numbers for structures to be included/excluded.
3. Computer with internet access.
4. E-mail address to retrieve results (Optional).

2.5 Cell Cycle Dependent Functions of HDAC5

2.5.1 G1/S, G2/M, and Cytokinesis Arrests

1. Cell stocks or cells growing in culture, plus standard culturing materials (*see* Subheading 2.1.2).
2. DMSO (negative control).
3. Hydroxyurea suspended in DMSO.
4. Nocodazole suspended in DMSO.
5. Hesperadin suspended in DMSO.
6. DPBS (*see* Subheading 2.1.2).
7. Trypsin (*see* Subheading 2.1.2).
8. DMEM++ (*see* Subheading 2.1.2).

2.5.2 Flow Cytometry

1. Cell stocks or cells growing in culture, plus standard culturing materials (*see* Subheading 2.1.2).
2. DPBS (*see* Subheading 2.1.2).
3. 1× Trypsin (*see* Subheading 2.1.2).
4. DPBS supplemented with 2–5 % FBS.
5. 15 mL conical tubes.
6. 95 % Ethanol.
7. Propidium iodide (1 mg/mL).
8. RNase A (10 mg/mL).
9. Barrier pipet tips.
10. Flow cytometry-compatible culture tube.
11. Ice bucket.
12. Aluminum foil.
13. Microcentrifuge capable of rotation at 500 ×*g*, set at room temperature.
14. Vortex.
15. Flow cytometer (e.g., BD LSRII FACS) and software (e.g., BD FACS Diva).

2.5.3 Kinase Inhibition Studies

Small-Molecule Kinase Inhibitors

1. Cell stocks or cells growing in culture, plus standard culturing materials (*see* Subheading 2.1.1).
2. DMSO (negative control).
3. Hydroxyurea suspended in DMSO.
4. Nocodazole suspended in DMSO.
5. Hesperadin suspended in DMSO.

siRNA-Mediated Knockdown of Kinases

1. Cell stocks or cells growing in culture, plus standard culturing materials (*see* Subheading 2.1.1).
2. Lipofectamine RNAiMAX Reagent or similar transfection reagent.
3. Opti-MEM medium (Life Technologies).
4. siRNA complex targeting kinase.
5. RNase-free barrier pipet tips.
6. RNase-free 1.7 mL tubes.
7. Incubator set at 37 °C, 5 % CO_2.
8. Sterile tissue culture hood with UV sterilization capability.

2.6 HDAC5 Enzymatic Activity

1. All reagents from the Subheading 2.3 *Immunoaffinity isolations*.
2. Fluorometric HDAC activity assay kit. The following protocol has been optimized for the Fluor-de-Lys® HDAC fluorometric activity assay kit, but may be adapted for other fluorometric kits.
3. 96-well plate with optical bottom.
4. 1.5 mL tubes.
5. Microplate-reading fluorimeter capable of excitation at 350–380 nm and detection at 450–480 nm.

3 Methods

3.1 Studying Endogenous and Tagged HDAC5 in Cell Culture

Depending on the biological questions of interest, experiments can be performed by studying either endogenous or tagged HDAC5. The studies on endogenous HDAC5 most frequently rely on the use of specific anti-HDAC5 antibodies. For example, the protein expression levels of endogenous HDAC5 can be studied across cell types and under different stimuli, treatments, or disease conditions using anti-HDAC5 antibodies and western blotting. The protein levels can be compared to HDAC5 mRNA levels, which can be assessed by qRT-PCR. Anti-HDAC5 antibodies also allow visualization of the subcellular localization of HDAC5 (e.g., [19]) (*see* Subheading 3.2) and have the potential to allow the identification of HDAC5 protein interactions via immunoaffinity purification assays (*see* Subheading 3.3). Anti-HDAC5 antibodies can also be designed to recognize a certain modified state of HDAC5 (Fig. 2), for example via the generation of antibodies against site-specific phosphorylations (i.e., phosphoSer259 [36, 37], phosphoSer278 [30], phosphoSer498 [38]). Custom antibody generation can be achieved through the synthesis of a peptide corresponding to the site of modification and a sufficiently long flanking region to impart specificity. Noteworthy, although this has not been yet used extensively in HDAC5 studies and, therefore, it is not covered in this chapter, the endogenous HDAC5, unmodified or modified, can also be detected by targeted mass spectrometry analyses (as in Refs. [35, 39, 40]). Efficient digestion of HDAC5 can be achieved with trypsin, and the use of additional enzymes can further expand the protein sequence coverage and aid in the identification of PTMs (see the in-gel digestion method in Subheading 3.3.5).

Making use of tagged HDAC5 constructs can be beneficial in numerous studies. One valuable application is in live cell studies, in which the dynamic localization of the enzyme can be monitored in real time using fluorescently tagged-HDAC5 constructs. Such studies have been used effectively to visualize the nuclear–cytoplasmic shuttling of HDAC5 (e.g., [16]). The tagging of HDAC5 has also proved valuable when studying its protein interactions. Antibodies against a tag, such as green fluorescent protein (GFP) [41], tend to provide effective isolations of the protein of interest, usually by having higher affinity and specificity than the antibodies against the endogenous protein. Importantly, tagged constructs provide a mean to study HDAC5 mutants that assess the functions of posttranslational modifications. HDAC5 mutants that either mimic or abolish a modification state can be used to study the functions of individual modification sites (e.g., [16, 30, 35]).

When working with a tagged protein construct, it is critical that the functionality of the tagged protein is assessed and validated. It may be practically impossible to rule out all potential effects of tagging, as these studies are limited by the current knowledge regarding the protein functions and the assays available. However, as an example, the functionality of HDAC5-EGFP was assessed by visualization of its subcellular localization by immunofluorescence microscopy, by confirming that the tagged HDAC5 retained its interaction with HDAC3 by immunoaffinity purification, and thereby its in vitro enzymatic activity via a fluorometric assay. Several important parameters have to be considered when generating tagged constructs, including the selection of the promoter driving the expression of the tagged proteins. Greater overexpression of HDAC5 can be achieved using the pCMV promoter in certain cell types [42, 43], while a moderate stable expression can be achieved using retrovirus or lentivirus transduction, such as with pLXSN [44]. Avoiding a substantial overexpression of a protein can be critical when constructing cell lines, as protein accumulation can toxic to the cell and can contribute to the formation of protein aggregates. In this chapter, we provide a protocol for generating stable cell lines that express tagged-HDAC5. This protocol has been successfully used in HDAC5 studies, as well as in studies of the other ten HDACs and of the seven sirtuins [10, 16, 39, 45–47]. As our laboratory has used this protocol for studying HDAC5 in both plated (HEK293) and suspension (CEM T) cells, differences in the respective protocols are indicated.

3.1.1 HDAC5 Cell Line Generation

Generation of Stable HDAC5-Flag-EGFP Cell Lines via Retroviral Transduction

1. DAY 1: plate Phoenix cells in a 6-well plate (3×10^5 cells in 2 mL media per well). Plating is best performed in the afternoon to allow cells to grow overnight (ca. 24 h). Incubate at 37 °C, 5 % CO_2 overnight.
2. DAY 2: transfect Phoenix cells in afternoon with 1 μg pLXSN-HDAC5-Flag_EGFP plasmid + 3uL FuGene transfection reagent. Incubate at 37 °C, 5 % CO_2 overnight.
3. In the morning of Day 3, wash cells with fresh DMEM+ media.
4. DAY 3: plate HEK293 cells in a 6-well plate (3×10^5 cells in 2 mL media per well). Culture cells overnight (ca. 24 h). For suspension cells (e.g., CEM T) wait to split cells until Day 5. Incubate all cell cultures at 37 °C, 5 % CO_2.
5. DAY 4: collect the supernatant from the cell culture containing released viral particles in a 15 mL conical tube.
6. Centrifuge collected supernatant from **step 5** at $1000 \times g$ for 5 min and reserve resulting supernatant.
7. Filter the supernatant from **step 6** using a 0.45 μm low-binding membrane filter (e.g., cellulose acetate or polysulfonic). The filtered fraction is the Viral Media.
8. Aspirate the media from HEK293 cells (from **step 4**) and add filtered Viral Media such that the volume is ca. 2 mL. Add polybrene (1:1000). Incubate at 37 °C, 5 % CO_2 overnight. Alternatively, dilute a confluent T25 suspension culture 1:10 and add 1 mL of culture to each well. Add 1 mL of Viral Media and 2 μL of polybrene and incubate at 37 °C, 5 % CO_2 overnight.
9. DAY 5: change the media, or if the cells are sufficiently confluent transfer adherent cells to a 10 cm dish (or T25 flask for suspension cells). Incubate at 37 °C, 5 % CO_2 overnight.
10. DAY 6: add G418 (300 μg/mL) to the media. Incubate at 37 °C, 5 % CO_2.
11. For subsequent expansions on days 7–12, split cells following standard procedures, but continue to add G418 to the media (at 300 μg/mL).
12. For cells intended to stably express a fluorescently tagged protein, examine fluorescent signal by immunofluorescence microscopy to assess incorporation.
13. DAY 13: or after ca. 1 week following the initial G418 selection, collect ca. 1×10^5 cells for fluorescence activated cell sorting (FACS) to isolate and retain the fluorescent population.

3.1.2 Cell Culture

The following protocol is described for use of adherent cells. For suspension cells, such as CEM T cells, the trypsin steps should be omitted, and RPMI 1640 media should be substituted for

DMEM. Both adherent and suspension cells can be cultured in filter-cap flasks or cell culture dishes. All cell culture work is to be done under the hood. Spray anything that will go into the hood with 70 % ethanol. Latex or nitrile gloves and a lab coat should be worn for all tissue culture and cell culture work. Spills under the hood can be cleaned using bleach.

1. Prepare sterile solutions for cell culture.
   (a) DPBS.
   (b) DMEM++ (500 mL DMEM Media + 50 mL fetal bovine serum + 5 mL penicillin–streptomycin).
   (c) 1× trypsin (5 mL trypsin stock + 45 mL sterile DPBS).
2. One hour prior to working with cells, warm media, DPBS, and trypsin solution in a 37 °C water bath.
3. Thaw frozen aliquot of cell stock. Cell stocks are typically frozen in an 85 % FBS, 10 % DMSO solution and are stored long-term at −144 °C.
4. Pipet 5 mL of DMEM++ into a sterile 15 mL tube using a serological pipet.
5. Pipet cell stock solution into the same tube as in **step 4**, mixing the stock solution with the 5 mL of media. Pipet up and down gently to suspend cells in media.
6. Centrifuge 15 mL tubes at ca. 254 × *g* for 5 min at room temperature in a swinging bucket rotor.
7. Add 8 mL fresh DMEM++ to every plate to which cells will be added.
8. When the spin in **step 6** is finished, aspirate the supernatant and discard. Aspiration steps can be performed using an aspirating pipet and tubing connected to a vacuum collection flask. **CAUTION**: Be careful not to disturb the cell pellet at the bottom of the tube or aspirate cell material when removing the supernatant.
9. To divide the cells among plates, add 2 mL of fresh DMEM++ for each plate (e.g., for three plates, pipet 6 mL of fresh DMEM++ into the tube) and mix up and down by pipetting to suspend the cell pellet.
10. Add 2 mL of cell suspension to each plate. Rock the plate back and forth a few times to distribute the cells across the surface of the plate.
11. Examine plates under the light microscope to ensure cell transfer after plating.
12. Place plates in the 37 °C incubator and allow to adhere and grow.

13. Check plates daily to assess confluence and media color. Cells should be ready to split when confluent in 3–4 days. Once >70 % confluent, split cells 1:3 or 1:4.
14. Aspirate old media.
15. Wash cells on plate with 2 mL DPBS. Aspirate DBPS.
16. Add 2 mL 1× Trypsin to plate and tip plate to spread solution over the surface. Allow to sit for 1–2 min. Tap the edges of the plate firmly to dislodge cells from surface of the plate. While incubating the plates with trypsin, prepare fresh plates (3–4 new plates per original plate) and add 8 mL of DMEM++ to each new plate.
17. Once cells are no longer adhering to the surface of the original plate, add an appropriate amount of fresh DMEM++ to the plate and pipet up and down to mix. For example, add 4 mL of DM++ for a total volume of 6 mL (including Trypsin solution) for distribution to three new plates.
18. Add 2 mL of this cell suspension to each fresh plate and rock plate to distribute cells.
19. Return cells to incubator.

3.2 Assessing the Subcellular Localization of HDAC5

HDAC5 localization can be examined either in fixed cells to capture snapshots of HDAC5 localization [10, 16, 29, 30, 35], which is the most commonly used technique and is described in this section, or by live cell imaging to assess dynamic localizations of HDAC5 in different contexts (e.g., altered phosphorylation status, varying cell cycle stages) [16, 38] (*see* Subheading 3.2.1).

Fixation of cells allows imaging the endogenous HDAC5 using commercial antibodies and assessing its co-localization with other proteins of interest. Antibodies raised against specific phosphorylated sites can further provide spatial and temporal information about the posttranslational status of HDAC5. Localization studies can be further aided by the use of cells expressing fluorescently tagged HDAC5. Such methods allow for visualizing the protein via direct fluorescence, which can prevent potential cross-specificity of antibodies used for immunostaining. Alternatively, antibodies against the fluorescent tag can also be used for visualization. The protocol for examining the localization of tagged HDAC5 in fixed cells is identical to that of endogenous HDAC5, with the exception of the antibodies used. The visualization of fluorescently tagged HDAC5 (either by direct fluorescence or antibodies against the tag) can be performed in tandem with the visualization of other proteins using antibody staining procedures to assess their co-localization. The protocols detailed in the next two sections describe the preparation of cells for studies of HDAC5 localization in fixed or live cells and includes instructions on how to prepare cells expressing fluorescently tagged HDAC5 (e.g., HDAC5-EGFP) for live cells imaging, and how to incorporate in these studies staining of cellular organelles, such as trackers for ER (indicated below) or mitochondria.

3.2.1 Preparing Cells for Visualization of Fixed Cells

1. DAY 0: Coat coverslips with poly-D-lysine solution (diluted a 1 mg/mL stock 1:100) and incubate overnight (or at least 8 h) at room temperature.
2. DAY 1: Plate 5×10^5 cells on cover slips in a 6-well plate. Incubate at 37 °C, 5 % CO2 overnight. Cells may be plated on glass-bottom dishes or EZ-Slides as an alternative to coverslips.
3. DAY 2: Check to ensure that cells are ca. 80–90 % confluent.
4. Wash coverslips 1× with 1 mL DPBS. Dispense DPBS with a pipet onto the side of the well rather than directly over top of the coverslip to avoid disturbing the cell monolayer.
5. Fix cells with 1 mL 2 % Paraformaldehyde for 15 min at 40 °C and incubate on a plate shaker with gentle rocking.
6. Wash 3 × 5 min with 1 mL 0.1 M glycine in DPBS by gentle rocking on a shaker.
7. Permeabilize cells with 1 mL 0.1 % Triton X-100 in DPBS for 15 min, while gently rocking on plate shaker.
8. Wash 3 × 5 min with 1 mL 0.2 % Tween in DPBS. Incubate on plate shaker (gentle).
9. Block cells with 1 mL 2 % BSA, 0.2 % Tween in DPBS, 60 min, on plate shaker (gentle). Alternatively, cells can be blocked at 4 °C overnight.
10. Aspirate blocking solution and add 1 mL of the primary antibody solution. For example, for visualizing HDAC5-EGFP by indirect fluorescence, Roche anti-GFP antibodies can be diluted at 1:5000 in 2 % BSA, 0.2 % Tween in DPBS to make the primary antibody solution. EGFP-tagged HDAC5 can also be visualized using direct fluorescence, in which case the antibody incubation steps can be omitted. Aurora B visualization can be achieved by preparation of a 1:1000 anti-Aurora B antibody solution in 2 % BSA, 0.2 % Tween in DPBS. Wrap the plate in aluminum foil to protect from light. Incubate for 30–60 min at room temperature on a plate shaker (gentle).
11. Wash cells 3 × 5 min with 1 mL 0.2 % Tween in DPBS on a plate shaker (gentle).
12. Add 1 mL secondary antibody solution. Wrap the plate in foil to protect from light and incubate for 30–60 min at room temperature on a plate shaker (gentle).
13. Wash cells 3 × 5 min with 1 mL 0.2 % Tween in DPBS, Plate shaker (gentle).

14. Stain cells in 1 mL 1 μM DAPI or To-Pro3. Wrap the plate in foil to protect from light and incubate for 30–60 min at room temperature on a plate shaker (gentle).
15. Wash cells 3×5 min with 1 mL 0.2% Tween in DPBS on a plate shaker (gentle). Keep cells in the last wash.
16. To mount coverslips, drop ca. 15 μL of anti-fade solution onto the slide.
17. Place cover slip on top of the slide such that the cell monolayer and anti-fade solution are sandwiched between the slide and the coverslip.
18. Wipe off the excess anti-fade solution and seal the edges of slide/coverslip with nail polish.
19. Store the slide at 4 °C in the dark prior to imaging. Slides can be stored for 1–2 weeks at 4 °C in the dark before signal deterioration.

3.2.2 Preparing Cells for Live Cell Imaging by Immunofluorescence Microscopy

Immunofluorescence Microscopy by Live Cell Imaging

1. DAY 0: plate cells on glass-bottom dishes (e.g., 35 mm glass-bottom dishes) in DMEM++
2. DAY 1: Dilute ER-Tracker solution to 500 nM in HBSS Solution and warm to 37 °C.
3. Aspirate DMEM ++ and wash cells with 1 mL HBSS.
4. Aspirate HBSS wash and add 1 mL of 500 nM ER-Tracker/HBSS staining solution to dish.
5. Wrap dish in foil and incubate cells in staining solution at 37 °C, 5% CO_2 for 30 min.
6. Aspirate staining solution and add 1 mL of Opti-MEM media to dish.
7. Wrap dish in foil and proceed to fluorescence imaging.

3.3 Defining HDAC5 Protein Interactions

Given the nuclear–cytoplasmic shuttling of HDAC5, many of its functions are modulated through the formation of temporal and spatial protein interactions (Fig. 1b). Therefore, similar to the studies of other histone deacetylases, there has been a great interest in characterizing these HDAC5 interactions and their regulation by posttranslational modifications. Immunoaffinity purification in conjunction with mass spectrometry has become a robust and routine approach for identifying HDAC5 protein interactions in different biological context in an effective and unbiased manner. Although important insights have been gained from such studies, it is evident that further analyses will be needed to expand this knowledge and characterize tissue-, cell type-, and subcellular localization-specific HDAC5 protein interactions in the context of health or disease states. The protocol presented in this section has been successfully implemented in HDAC5 protein interactions studies [10, 16, 30, 35], and contains as main steps:

(1) cell harvesting, (2) cryogenic cell lysis, (3) conjugation of magnetic beads with antibody, (4) immunoaffinity purification of HDAC5, and (5) analysis of interacting proteins using in-gel digestion with trypsin and mass spectrometry. This workflow can be used to study the protein interactions of either endogenous or tagged HDAC5. The isolation of endogenous HDAC5 would require the use of an anti-HDAC5 antibody with validated affinity and specificity. The isolation of tagged HDAC5 requires the use of antibodies against the tag. For example, we have effectively isolated HDAC5-EGFP using magnetic beads conjugated with anti-EGFP antibodies [10, 16, 30, 35, 41]. It is important to note that the antibodies used for these workflows, regardless if these are against a protein or a tag, should be purified. Otherwise, all the other proteins present in the antibody solution would also bind to the magnetic beads, leading to the presence of numerous non-specific associations. The protocol described below is suitable for both adherent and suspension cells (*see* **Note 1**). Additional protocols for immunoaffinity purification of proteins from other tissue types or from studies incorporating metabolic labeling have been described in detail elsewhere [48].

3.3.1 Harvesting and Freezing Cells

1. Aspirate media from cell culture.
2. Wash with 8 mL DPBS to remove dead cells and wash away serum.
3. Aspirate DBPS wash.
4. Add 2 mL 1× trypsin–EDTA in DPBS and place the dish on a rocker. Tap gently to loosen cells.
5. Quench trypsin digest with 6 mL DMEM/10% FBS/1% (Pen/Strep) and transfer to conical tube (or larger autoclaved cell harvesting centrifuge tubes if the volumes are large).
6. Spin the samples at ~900 × *g* for 5 min at 4 °C to pellet cells.
7. Aspirate media.
8. Wash cell pellet in 50 mL DPBS.
9. Re-pellet cells (900 × *g* for 5 min at 4 °C).
10. Wash the cell pellet again in 50 mL DPBS.
11. Re-pellet cells (900 × *g* for 5 min at 4 °C).
12. Record the weight of the final cell pellet (*see* **Note 2**).
13. Suspend pelleted cells in freezing buffer (20 mM Na-HEPES/1.2% PVP (w/v), pH = 7.4), using ~100 μL for 0.5 g of cell powder.
14. Poke 5–10 small holes in the cap of a fresh 50 mL falcon tube with a syringe needle. Place the tube in a liquid nitrogen bath and allow to cool, and fill half of the tube with liquid nitrogen (*see* **Note 3**).

15. Using a pipet, allow the suspended cell solution to fall dropwise into the falcon tube from **step 14**. The cell droplets will flash freeze into small spheres upon contact with the liquid nitrogen. Continue this process until the whole cell solution is frozen into small spheres.
16. Re-cap the falcon tube with the cap with holes and agitate the tube to allow the residual liquid nitrogen to escape. Replace the lid with a fresh cap that has no holes.
17. Store frozen cell spheres at −80 °C until ready to proceed with cryo-disruption.

3.3.2 Cryogenic Cell Lysis

1. Wash mill chambers, grinding ball, plastic washer, and metal spatula in the following order with:
   (a) Hot H_2O.
   (b) Windex.
   (c) Bleach.
   (d) Windex.
   (e) Methanol.
2. Allow mill chamber and ball to dry in the hood for ca. 15 min.
3. Once dried, prechill the mill chamber and the ball together in liquid nitrogen until the boiling appearance stops (ca. 3–5 min).
4. Using long-handled tongs, remove the mill chamber and ball from the liquid nitrogen.
5. Add cell or tissue material to be ground to the deeper half of chamber.
6. Place the ball on top of sample.
7. Tightly screw on the other half of the mill chamber and then submerge the entire chamber containing cell material and grinding ball in the liquid nitrogen bath for 5 min.
8. Using the tongs, remove the mill chamber from the nitrogen back and lock the chamber into the pulverizer and press start (setting: 30 Hz, 2.5 min/cycle). Ensure that the machine is balanced.
9. Re-freeze chamber.
10. Repeat **steps** 8–9 seven times for mammalian cells or 11 times for bacterial cells or 20 times for yeast cells.
11. Label a fresh 50 mL conical tube and place upright in a stand in the liquid nitrogen bath to keep tube cool.
12. Unscrew the mill chamber taking care not to spill any cell/tissue powder. Scrape sample into 50 mL conical using a prechilled metal spatula. Transfer the ball between sides of the mill chamber and scrape to maximize yield.

13. Clean mill chamber, ball, washer, and spatula as in **step 1** and allow to dry in hood.

3.3.3 Conjugation of Magnetic Beads

The protocol described below has been optimized for the conjugation of high-affinity antibodies to magnetic M-270 Epoxy Dynabeads; however, this protocol can be adapted for the use of different bead diameters or conjugation chemistries (e.g., carboxylic acid, tosylated, streptavidin, amine) for isolation of other targets (e.g., glycoproteins or biotinylated proteins). This protocol is compatible with in-house purified antibodies, as well as many commercially available antibodies against both affinity tags and endogenous proteins. Therefore, the protocol can be used for studying endogenous and tagged HDAC5. For example, for tagged HDAC5, the magnetic beads can be conjugated with IgG for Protein A-tagged HDAC5, with anti-GFP antibodies for HDAC5-EGFP, and with anti-FLAG for isolating HDAC5-FLAG.

1. DAY 0, afternoon: Weigh the necessary amount (start with ca. 4–5 mg beads/isolation) of magnetic beads (M-270 Epoxy Dynabeads. in a round-bottom 2 mL tube using clean metal spatula and a sensitive balance (*see* **Note 4**).
2. Wash beads with 1 mL sodium phosphate buffer; vortex for 30 s; seal cap with Parafilm.
3. Mix for 15 min on TOMY shaker (speed #7, room temperature).
4. Place tube against magnetic rack and remove the ferrous buffer once the beads have separated from the buffer solution.
5. Wash beads with 1 mL Na-phosphate buffer; vortex 30 s.
6. Place tube against magnetic rack and remove buffer.
7. Prepare antibody solution to resuspend beads. Use 5 μg Ab/1 mg beads for purified, high-affinity, antibodies. To calculate the amount of antibody, $AmSO_4$, and Na-phosphate buffer required for antibody solution, assume a total volume of 20 μL per mg of beads. The final concentration of $AmSO_4$ should be 1 M (i.e., one-third of the total volume if starting with a 3 M stock). The remaining volume can be made up with Na-phosphate buffer (i.e., $Volume_T - Volume_{Ab} - Volume_{AmSO4} = Volume_{Na\text{-}phosphate}$). Add the 3 M $AmSO_4$ last to limit protein precipitation.
8. Carry out the conjugation overnight on a rotor at 30 °C. Use Parafilm to seal the top of the tube and rotate on *y*-axis.
9. DAY 1, morning: place the tube on a magnet. Remove supernatant and wash beads with:
   (a) 1 mL NaPhosphate buffer.
   (b) 1 mL 100 mM glycine–HCl (fast wash) (*see* **Note 5**).

(c) 1 mL 10 mM Tris.

(d) 1 mL freshly prepared 100 mM triethylamine solution (fast wash) (140 μL + 10 mL Milli-Q H_2O in hood) (*see* **Note 5**).

(e) 1 mL DPBS (wash 4×).

(f) 1 mL DPBS + 0.5 % Triton X-100 for 15 min.

(g) 1 mL DPBS

10. Resuspend beads for storage at 4 °C in DPBS–0.02 % NaN_3 and use within 2–3 weeks (*see* **Note 6**). Store at C_f ca. 15 μL/mg beads.

3.3.4 Immunoaffinity Purification of HDAC5 With Interacting Partners

1. Resuspend frozen cell powder in an appropriate volume of lysis buffer (LB), e.g., 10 mL LB/g cells. It is advisable to optimize the lysis buffer composition for each isolation of HDAC5, as the efficiency of isolation can be cell-, tissue-, and stimulus-type dependent. However, as a starting point, the buffer detailed below provides an example of a previously optimized lysis condition that was successfully used to isolate HDAC5 from HEK293, U2OS, and CEM T cells [10, 16, 30, 35]. The buffer composition is: 20 mM HEPES–KOH, pH 7.4, 0.1 M potassium acetate, 2 mM $MgCl_2$, 0.1 % Tween 20, 1 μM $ZnCl_2$, 1 μM $CaCl_2$, 0.5 % Triton X-100, 250 mM NaCl, 4 μg/mL DNase I, 1/100 (v/v) protease, and phosphatase inhibitor cocktails. Therefore, 10 mL of this lysis buffer can be prepared by mixing the following reagents:
   (a) 833 μL 3M NaCl.
   (b) 500 μL 10 % Triton X-100.
   (c) 1 mL 10× TBT with salts.
   (d) 100 μL protease inhibitor cocktail (for mammalian cells).
   (e) 100 μL phosphatase inhibitor cocktail (*see* Subheading 2.3.4, **item 6**).
   (f) 100 μL phosphatase inhibitor cocktail (*see* Subheading 2.3.4, **item 7**).
   (g) 40 μL DNase I (1 mg/mL stock).
   (h) 7.33 mL Milli-Q H_2O.
2. Mix with Polytron for 20 s; or vortex for 30 s for small volumes or viral samples.
3. Incubate for 5 min with agitation at room temp for DNaseI activity (*see* **Note 7**).
4. Centrifuge at 4 °C for 10 min at 8000 × *g*.
5. Place conjugated Dynabeads on magnet and remove buffer.
6. Wash 3× with 1 mL of WB and resuspend in 100 μL of WB.

7. Transfer lysate S/N to clean container (conical or culture tube)—make sure no particles are present, as these can clog the magnetic beads and lead to non-specific associations. Save pellet for Western blot analysis.
8. Add washed beads to cell lysate S/N.
9. Incubate at 4 °C with gentle rotation (5 min to 3 h (*see* **Note 8**)).
10. Place tube on magnet.
11. Transfer flow-through to another tube and store at 4 °C for Western blot analysis.
12. Wash beads 4× with 1 mL WB (during first and fourth washes, transfer beads to fresh tubes) and 2× with 1 mL DPBS.
13. Elute proteins with 30 μL 1× LDS sample buffer (EB) or 1× (LDS SB/Reducing Agent).
14. Agitate on TOMY shaker (setting #7) for 10 min and place sample at 70 °C for 10 min. At this point, samples can be kept at 4 °C (if running gel within next 24 h) or at −20 °C for longer. Keep 10% for Western.
15. Carry on either with **steps 16–22** for determining the efficiency of HDAC5 isolation or with Subheading 3.3.5 for preparing the samples for mass spectrometry analysis.
16. Perform Western blotting to check IP efficiency using the following samples:
    (a) Whole cell lysate (WCL): Precipitate protein with four volumes of 100% acetone (ice cold) and place at −20 °C for at least 30 min (can leave this overnight). Spin precipitated protein solution at 3000 × *g* for 10 min at 4 °C. Wash pellet 2× with 75% acetone. Use 1% of precipitated FT protein (instead of 10% because there should be a lot of protein present) to prepare samples for Western blotting.
    (b) Pellet: suspend pellet in water; boil and vortex if necessary; dissolve pellet as well as possible. Use 10% of pellet to prepare samples for Western blotting.
    (c) Flow-through (FT): Take 10% of flow-through volume. Precipitate protein with four volumes of 100% acetone (ice cold) and place at −20 °C for at least 30 min (can leave this overnight). Spin precipitated protein solution at 3000 × *g* for 10 min at 4 °C. Wash pellet 2× with 75% acetone. Allow to dry for 15 min (*see* **Note 9**). Add LDS SB and prepare samples for Western blotting.
    (d) Magnetic Beads: Resuspend beads in LDS SB and use 10% of beads to prepare samples for Western blotting. Eluate: Take 10% of IP eluate and prepare samples for Western blotting.

17. Prepare samples in LDS sample buffer (4×) such that final sample concentration is 1× LDS sample buffer.
18. Add NuPAGE reducing agent or DTT (10× stocks) to a final concentration of 1×.
19. Heat at 70 °C for 10 min.
20. Add iodoacetamide (final = 10 %) to sample and incubate in dark for 20–30 min at room temperature.
21. Separate samples on gel, transfer to a PVDF or nitrocellulose membrane, incubate with antibody against the bait protein or tag to determine efficiency of isolation.
22. Upon confirmation of efficient isolation of bait protein by Western blotting, separate remaining eluate by 1D SDS-PAGE.
    (a) Add reducing agent (e.g., DTT) to final concentration of 100 mM.
    (b) Heat sample at 70 °C for 10 min.
    (c) Add iodoacetamide at 10 % (from 1 M stock, final [IAA] = 100 mM); incubate in dark for 20 min on ice.

3.3.5 In-gel Digestion of HDAC5 Co-isolated Proteins (See **Note 10***)*

Proteins co-isolated with HDAC5 can be analyzed by mass spectrometry or by western blotting. This section discusses in-gel digestion techniques for proteomic-based mass spectrometry analysis; however, in-solution digestion can provide a less time-consuming alternative to in-gel methods. The advantages of using in-gel digestion lie in the ability of gel resolution to remove detergents from IP samples, as well as to resolve complex protein samples [49]. Use of in-solution digestion [48] can minimize sample loss and aid in reproducibility for higher-throughput analyses [10, 50, 51]. The protocol described in this section includes enzymatic digestion using trypsin; however, additional and/or alternative enzymes may be appropriate for the generation of specific target peptides or for the analysis of other proteins of interest. When selecting an enzyme for protein digestion, *in silico* digestions of the protein sequences can be performed to determine the expected cleavage products and assess enzyme suitability, which can be achieved using online platforms such as the MS-Digest utility (http://prospector.ucsf.edu/prospector/cgi-bin/msform.cgi?form=msdigest) of Protein Prospector (Baker, P.R. and Clauser, K.R. http://prospector.ucsf.edu).

1. To prepare remaining eluates for mass spectrometry analysis, separate samples on a pre-cast NuPAGE Gradient 4–12 % using a MOPS SDS or MES SDS running buffer. Use the following settings to separate proteins: 100 V for 5 min; 200 V for ca. 20 min. Allow the dye front to progress ca. one-third of the way down the gel.

2. After removing the gel from the pre-cast cassette, wash 3 × 5 min in Milli-Q dH_2O.
3. After third wash, stain gel using Simply Blue SafeStain for 30–60 min.
4. Destain gel in Milli-Q dH_2O and store gel in Milli-Q dH_2O at 4 °C until ready to process for digestion.
5. In-gel digestion can be performed in either individual 1.7 mL tubes or in wells of a 96-well plate with sealing mat (useful when more than four samples are processed at a time). Rinse the 96-well plate with Milli-Q water and shake/tap out excess liquid. Each well/tube is considered a single sample (*see* **Note 11**).
6. Prepare destaining solution on the day of digestion by mixing 2.0 mL of 100 % ACN and 2.0 mL of 0.1 M ABC in a solution basin (*see* **Note 12**).
7. Excise a single gel lane from the gel and cut the gel lane into 1 mm slices. When using a 96-well format, do not put more than four 1 mm slices into each well.
8. Add 90 μL of destaining solution to each well and agitate at 4 °C for 10 min. Aspirate and discard (*see* **Note 13**).
9. Add 90 μL of fresh destaining solution to each well and agitate at 4 °C for 10 min.
10. While destaining aliquot the following amounts of the following solutions into separate basins to allow use of the multichannel pipette for dehydration and rehydration steps:
    (a) 5 mL of 100 % ACN for dehydration steps.
    (b) 5 mL of 50 mM ABC for rehydration steps.
11. Aspirate the destaining solution from wells and add 90 μL ACN to each well. Invert the plate several times and incubate until gel pieces are white (ca. 1 min).
12. Remove ACN and add 90 μL of 50 mM ABC to each well. Invert the plate several times and shake at 4 °C until the gel pieces swell and become translucent (ca. 5 min).
13. Aspirate ABC and add 90 μl of 100 % ACN to dehydrate, and incubate as above.
14. Remove ACN and add 50 mM ABC to samples and let rehydrate as above.
15. Aspirate ABC and perform a final (third) dehydration with 90 μL of 100 % ACN.
16. Aspirate ACN. Allow the residual ACN in wells to evaporate during preparation of the trypsin solution (*see* **step 13**).
17. Mix 331.5 μL of 50 mM ABC and 8.5 μL of 0.5 μg/μL trypsin in an 1.7 mL tube.

18. Add 20 μL of trypsin–ABC solution to the dehydrated gel pieces. Allow gel pieces to rehydrate and swell (ca. 5 min).
19. Add 30–50 μL of 50 mM ABC to gel pieces, using a sufficient volume to fully cover the gel pieces. Seal the plate and incubate for 10 min at 37 °C. Check gel pieces after 10 min of incubation to ensure they are still fully covered by ABC, then incubate overnight at 37 °C.

3.3.6 Peptide Extraction and Sample Desalting (StageTips)

Following peptide extraction, samples can be desalted using either StageTips (as described in this section) [50, 52] or just prior to liquid chromatographic separation on the analytical column through the use of online trap columns.

1. Add an equal volume (relative to amount of ABC added on Day 2, **step 15**) of 1 % formic acid to each sample and incubate for 4 h at room temperature.
2. Transfer the extracted peptides to a clean 1.7 mL tube. Avoid transferring large gel pieces. Use a different tip for each well; however, the tip can be reused for the same well in **step 4** below. Adjacent samples can be pooled at this step, if desired, for samples corresponding to gel regions with lighter staining/ lower estimated protein content.
3. Add an equal volume (*see* **step 1**) of 0.5 % formic acid–50 % ACN solution to the gel pieces and incubate for an additional 2 h at room temperature.
4. After 2 h, combine the second extraction (**step 3**) with the first peptide extraction (**step 2**).
5. SpeedVac samples to ~20 μL to remove ACN. During SpeedVac concentration, assemble StageTips. Assemble one StageTip for each sample.
   (a) Cut Empore SDB disk out using the 16 gauge needle and use the plunger to pack the disk into a P200 tip.
   (b) Tap the disk down into place until it is approximately ¾ of the way down from the lowest graduated ring of the P200 tip.
   (c) Each disk binds ca. 15 μg of peptides. Additional disks can be layered or a larger disk diameter can be selected if needed (e.g., for 25–30 μg, use 14 gauge needle).
6. Dilute each sample to approximately 40 μL with 0.5 % TFA.
7. Adjust samples to 1 % TFA (*see* **Note 14**).
8. Pipet samples over the top of the StageTip. Hold StageTip and flick wrist to remove excess air between the sample and the top of the SBD disk.
9. Place each StageTip into a collection tube and centrifuge tips at 1000–2000 × *g* until the entire sample has passed through

the disk (*see* **Note 15**). Flow-through can be saved if desired for analysis of the efficiency of isolation.

10. Pipet 100 μL wash buffer over the top of the StageTip and centrifuge as above until all wash solution has passed through the disk (*see* **Note 16**).
11. If fractionating sample, continue to **step 13**. If performing a single elution, apply 50 μL of the Elution #3 solution to the top of each StageTip. Proceed to **step 14**.
12. Elute peptides from the disk into autosampler vials by applying air pressure with the syringe. Each elution should take approximately 15 s.
13. For fractionation, perform stepwise elutions into separate autosampler vials using Elution #1, #2, and #3 solutions. Proceed to **step 14**.
14. Concentrate the samples to approximately 1 μL in the SpeedVac (*see* **Note 17**).
15. Add Dilution Solution (1% FA/4% ACN) to achieve the desired peptide concentrations (e.g., 9 μL for routine LC-MS/MS analysis).

3.4 Investigating Regulation of HDAC5 by Posttranslational Modification

3.4.1 Identification of Posttranslational Modifications (PTMs)

Immunoaffinity approaches coupled with mass spectrometry analysis can also be used for the identification of posttranslational modifications, including phosphorylation, acetylation, sumoylation, and ubiquitination, among others. Enrichment for the isolation of specific modifications can be achieved through the use of antibody-based enrichments (e.g., using pan-anti-acetyl or anti-phospho antibodies) [53–55] or through the use of affinity chromatography methods that make use of metal-based resins with affinity for phosphorylated peptides such as IMAC (Ni-based) or TiO_2 enrichments [56–59].

3.4.2 Characterizing PTM Functions

Amino Acid Substitutions

The impact of posttranslational modifications on HDAC5 localization, activity, and function can be assessed by generation of site-specific null or mimetic mutants at identified sites of modification (Fig. 2). For example, a phosphorylated serine residue can be mutated to alanine to simulate an unmodified state, while mutation of the residue to aspartic acid can mimic constitutive phosphorylation. Specifically, mutation of Ser259 and Ser298 to alanine results in a loss of association with 14–3–3 proteins and increased nuclear localization of HDAC5 [15, 60]. Alternatively, mutation of Ser279 to alanine results in cytoplasmic localization of HDAC5 [16, 29]. Similarly, phosphorylated threonine or tyrosine residues can be mutated to glutamic acid to generate phosphomimetics. Acetylation of lysine can be mimicked with replacement of lysine with glutamine, while mutation of lysine to arginine would eliminate potential acetylation of the site while retaining the charge state of the amino acid. These mutations can be achieved by site-directed mutagenesis using a variety of available kits and cloning techniques.

3.4.3 Motif Searching

Determining potential enzymes that can act on PTM sites can also provide insight into the regulation and function of individual modifications. Algorithms such as those found at gps.biocuckoo.org or expasy.org that search amino acid sequences for motifs matching established consensus sequences can be employed for prediction of multiple modifications.

1. Obtain FASTA sequence of interest.
2. Navigate to http://www.cbs.dtu.dk/services/NetPhosK/
3. Paste FASTA sequence into "Submission" window or upload FASTA submission using the "Choose file" radio button.
4. Select filtering type (for rapid prediction, select "Prediction without filtering").
5. Click submit to begin the search.

3.4.4 Structure Predictions

While a crystal structure for HDAC5 has not yet been published, the complete structure of another deacetylase, HDAC8, and of the deacetylase domain of HDAC4 have been investigated by X-ray crystallography and modeling [61–63]. Due to the high degree of sequence homology among HDACs, especially within the enzymatic domain, these known structures provide a useful platform for further predicting HDAC5 structural elements. These structural predictions can be informative when investigating how certain posttranslational modifications may impact HDAC5 functions.

Predictions of HDAC5 structural conformation can be generated based on the primary amino acid sequence of HDAC5 and its similarity in sequence and motifs to other previously crystallized structures. Predictions for HDAC5 can thus be developed based on PDB structures currently available in the RCSD PDB databank. Online structural prediction platforms, such as I-TASSER (http://zhanglab.ccmb.med.umich.edu/I-TASSER/), allow for searching of protein FASTA sequences and also allow addition of customizable restraints to the search space through the addition of inclusion and exclusion lists for similar proteins and protein motifs with solved crystal structures [64–66]. Such user-defined restrains and curated structure libraries can further refine the algorithm for individual proteins.

1. Obtain FASTA sequence of HDAC5.
2. Navigate to http://zhanglab.ccmb.med.umich.edu/I-TASSER in the web browser.

3. Copy and paste FASTA sequence into the box under the heading "Copy and paste your sequence here."
4. Optional: enter an e-mail address in the corresponding box to retrieve I-TASSER prediction results.
5. If desired, select additional constraints for structural modeling by selecting the appropriate filter by clicking on the black triangle next to each "Option" category. (i.e., Option I: Assign additional restraints & templates to guide I-TASSER modeling, Option II: Exclude some templates from I-TASSER template library, Option III: Specify secondary structure for specific residues)
6. Click the "Run I-TASSER" button to execute the modeling algorithm.

3.5 Cell Cycle-Dependent Functions of HDAC5

Accumulating evidence has demonstrated that histone deacetylases are regulated during the progression of the cell cycle. For example, class IIa HDACs (HDAC4, HDAC5, and HDAC9) were shown to be substrates of the mitotic kinase Aurora B, and protein interactions and phosphorylations of HDAC5 were reported to change during the cell cycle. These reports demonstrate the value and the need for future studies that characterize the cell cycle-dependent functions of HDACs. Cell cycle arrests in conjunction with protein immunoaffinity purifications, imaging techniques, and examination of posttranslational modifications can provide important insights into HDAC5 functions (Fig. 1b). The protocols presented in this chapter can be used to assess diverse biological functions of HDAC5 during cell cycle progression, and have been selected given their successful use in previous HDAC5 studies [30]. The protocols describe the use of HEK293 cells, but similar methods can be used for studies in other cell types.

3.5.1 Arresting Cells in G1/S, G2/M, and Cytokinesis

1. Seed HEK293 cells expressing HDAC5-EGFP and culture to 30% confluence.
2. To arrest cells in G1/S: add hydroxyurea to standard media DMEM++ to a final concentration of 4.0 mM and incubate for 40 h.
3. To arrest cells in G2/M: add nocodozole to standard media DMEM++ to a final concentration of 100 ng/mL and incubate for 20 h.
4. To block the completion of cytokinesis: add hesperadin to DMEM++ to a final concentration of 20–60 nM and incubate for 24 h.
5. Harvest cells by trypsinization and centrifugation.
6. Lyse cells and analyze lysates by SDS-PAGE and Western blotting.

3.5.2 Analysis of Cell Cycle Stage by Flow Cytometry

1. Wash cells with 5 mL DPBS.
2. Perform trypsin digestion.
3. Transfer into 5 mL DPBS + 2–5 % FBS serum in a 15 mL conical (to quench trypsin reaction).
4. Spin at 250 × *g* for 5 min.
5. Aspirate supernatant.
6. Vortex pellet and add 1.5 mL of cold DPBS.
7. Slowly add 3 mL of cold 95 % EtOH in drop wise fashion.
8. Store overnight at 4 °C (can be stored up to 2 weeks).
9. Spin cells at 1800 × *g* for 5 min.
10. Pour off supernatant and resuspend in 5 mL DPBS.
11. Spin at 1800 × *g* for 5 min.
12. Pour off supernatant and resuspend in 5 mL DPBS.
13. Spin at 1800 × *g* for 5 min.
14. Pour off supernatant and use a barrier pipette tip to resuspend in 500 μL propidium iodide (PI)–RNase mixture (400 μL PI (1 mg/mL) + 200 μL RnaseA (10 mg/mL)) (final [PI] = 40 μg/mL).
15. Transfer cell solution to a flow cytometry compatible tube.
16. Cover tube in foil and incubate at room temperature for 1 h in the dark.
17. Vortex tube immediately prior to flow cytometry analysis to limit cell clumping.
18. Acquire and record PI signal using a BD LSRII FACS instrument and the BD FACS Diva program.

3.5.3 Kinase Inhibition Studies

As HDAC5 is known to be heavily modified by phosphorylation at multiple sites, kinases play critical roles in regulating its functions. Many studies have aimed to define the kinases that modulate HDAC5 phosphorylations and the downstream impact of these modifications [2, 6, 16, 24, 29, 30, 67] and reviewed in Ref. [68]. While not all the used protocols can be covered here, one commonly used approach is to inhibit kinases and study the impact of that inhibition on HDAC5 localization, interactions, and transcriptional regulatory functions. Inhibition of specific kinases can be achieved at a global level using Staurosporine or at a more specific level using selective kinase inhibitors (e.g., hesperadin to inhibit Aurora B). Small molecule inhibitors should be selected based on high selectivity toward targets and low cytotoxicity. Selective kinase inhibition can also be achieved through siRNA mediated knockdown of target kinases.

3.5.4 Small-Molecule Kinase Inhibitors

1. Seed HEK293 cells expressing HDAC5-EGFP and culture to 30 % confluence.

2. Add hesperadin (or other selective kinase inhibitors) to DMEM++ to a final concentration of 20–60 nM and incubate for 24 h.
3. Harvest cells by trypsinization and centrifugation.
4. Lyse cells and analyze lysates by Western blotting, room temperature-PCR, or immunofluorescence microscopy.

3.5.5 siRNA-Mediated Knockdown of Kinases

1. Seed HEK293 cells and culture to 60–80 % confluence.
2. Dilute Lipofectamine RNAiMAX reagent in Opti-MEM medium. For a 6-well dish, dilute 9 μL Lipofectamine RNAiMAX in 150 μL Opti-MEM.
3. Dilute siRNA in Opti-MEM Medium. For a 6-well dish, dilute 30 pmol of siRNA complex in 150 μL Opti-MEM.
4. Mix diluted Lipofectamine solution (**step 3**) and siRNA solution (**step 4**) 1:1.
5. Incubate the combined solution for 5 min at room temperature.
6. Add combined solution (siRNA–lipid complex) to plated cells.
7. Incubate cells at 37 °C for 24–72 h.
8. Analyze transfected cells for efficient knockdown by Western blotting, reverse transcription-qPCR, or immunofluorescence microscopy.

3.6 Enzymatic Activity of HDAC5

The innate deacetylase activity of HDAC5 against acetylated lysine is limited in vivo due to the absence of a catalytic tyrosine residue present in the enzymatic domains of other HDACs. In HDAC5 and the other class IIa HDACs (4, 7, and 9), this tyrosine residue is replaced with histidine [8]. However, HDAC5 seems to retain activity toward a trifluoroacetylated lysine substrate [8], which can be used to assess enzymatic activity immunoaffinity purified class IIa HDACs. It is important to consider that HDAC5 closely associates with the class I enzyme HDAC3 in vivo, resulting in ambiguity in analysis of immunoaffinity purified HDAC5. Therefore, to determine HDAC5 activity, the co-isolation of HDAC3 must also be assessed. Alternatively, cell lines expressing shHDAC3 may be suitable for determining the individual contributions of HDAC5 to enzymatic activity. Assessment of deacetylation activity most commonly involves the use of a fluorometric assay. Following immunoaffinity isolation of HDAC5 (and HDAC3) using magnetic beads, the fluorometric enzymatic assay can be performed while the proteins themselves are bound to the beads. Following the fluorometric assay, protein samples can be eluted from the beads for additional analysis of protein expression levels and co-isolating proteins by Western blotting.

3.6.1 Fluorometric Deacetylase Assay

1. Start with 0.2 g cells/enzymatic assay. Use 2 mg beads per 0.2 g cells and 1 mL of lysis buffer per 0.2 g cells.

2. Follow the protocol above for immunoaffinity isolation of protein complexes (*see* Subheading 3.3.4 *Immunoaffinity purification of HDAC5 with interacting partners* **steps 1–12(b)**).
3. Wash 3× with buffer used for enzymatic activity assay (e.g., phosphatase buffer or assay buffer).
4. After the final wash, resuspend the sample in a small volume buffer (300 μl) and split it to 5 round bottom tubes (60 μl in each). The remaining tubes are for individual assay conditions as indicated below.
   (a) Tube 1 is reserved for Western blotting. Proceed to elution in 20 μl 1× TEL buffer.
   (b) Tube 2: Time point 1 (e.g., time = 0 min), No inhibitor (-TSA).
   (c) Tube 3: Time point 1 (e.g., time = 0 min), +Inhibitor (+TSA) (*see* **Note 18**).
   (d) Tube 4: Time point 2 (e.g., time = 90 min), No inhibitor (-TSA).
   (e) Tube 5: Time point 2 (e.g., time = 90 min), +Inhibitor (+TSA).
5. Re-split the sample from each tube (**step 4b-e (Tubes 2-5)**) into three round bottom tubes for three replicates.
6. Place all tubes on magnetic rack, discard supernatant. Set up the enzymatic reactions on the magnetic beads, taking into account the following pre-steps:
   (a) Prepare master mixes with 400 μM substrate ± 2 μM TSA in assay buffer and pre-warm to 37 °C.
   (b) Add 20 μL of the appropriate master mix solutions to Tubes 4 and 5 and mix well.
   (c) Allow Tubes 2 and 3 to stand at 37 °C. Do not add master mixes to these tubes at this time.
7. Allow the reactions in Tubes 4 and 5 to proceed for 90 min at 37 °C, flick the tubes to mix intermittently during the incubation.
8. Quench the deacetylation reaction by adding 40 μl of 1× Developer in assay buffer. Also add the appropriate master mix solutions to Tubes 2 and 3, followed by an immediate addition of the developer.
9. After the development step, briefly spin, place tubes on magnetic rack, and transfer all supernatants to individual wells on 96-well plates for fluorescence measurements. Save the beads for **step 12**.
10. Read samples in a microplate-reading fluorimeter capable of excitation at a wavelength in the range 350–380 nm and detection of emitted light in the range 450–480 nm (e.g.,

Synergy Mx fluorometer). Completion of signal development can be assessed by taking fluorescence readings at 5 min intervals. The Developer reaction is complete when the fluorescence readings reach a maximum and plateau. Fluorescence signal remains stable for at least 60 min. after reaching the maximum plateau.

11. Elute the protein from the magnetic beads with 20 μl 1× TEL buffer to assess isolation by Western blotting.

4 Notes

1. The volumes provided for reference are for adherent cells in culture plates with 100 mm diameter; adjust volumes accordingly for larger dishes. For cells in suspension, trypsinization steps should be omitted.
2. The mass of the wet cell pellet can be most easily calculated by recording the mass of the empty tube and the final mass of the tube including cell pellet after harvesting. To avoid overestimation of the mass of the pellet, try to remove as much of the DPBS was as possible from the pellet without aspirating cell material.
3. For freezing cell material in the liquid nitrogen bath, it is helpful to secure a Styrofoam tube rack in the bottom of another Styrofoam box to hold the falcon tube upright. This can be easily achieved by using pipet tips to pin the rack down in the nitrogen bath. Alternatively, a metal tube rack can be carefully placed in the liquid nitrogen bath to hold the conical tubes.
4. Bead conjugation is easier to perform with bead amounts of >10 mg. Therefore, it is advisable to conjugate the beads for the bait and control and replicates at the same time.
5. In the "fast wash" steps, the beads should not be left to sit in the wash solution to minimize prolonged exposure of the antibody-conjugated bead solution to the highly acidic (pH = 2.5) glycine solution or to highly basic triethylamine solution.
6. The isolation efficiency after 1 month of storage at 4 °C decreases to ca. 60% of the original performance. Therefore, use the magnetic beads within 2 weeks after conjugation.
7. Addition of DNAse is important for the analysis of proteins that are known to bind to DNA in order to determine the dependency of co-isolating proteins on the presence of DNA in immunoaffinity purifications. In lieu of DNase I, benzonase can also be used for efficient degradation of DNA in samples [69].
8. Incubation time should be optimized for efficient recovery of the bait protein while minimizing non-specific background proteins. It is best to use as short an incubation time as possi-

ble, as extended times lead to the accumulation of nonspecific associations and the loss of weak interactions [41]. For HDAC5 isolations, incubations of 30–60 min are sufficient to capture the protein and its interacting partners.

9. Allowing the precipitated protein to dry for too long will make the pellet hard to resuspend. Insufficient drying retains traces of acetone, which will interfere with the running of the gel. The dried pellet should be dull white and opaque in appearance. A translucent pellet indicates that the pellet is not dry. A bright white pellet indicates that the pellet may be overly dry.
10. Preparation for in-gel digestion can also be performed on same day as digestion to shorten protocol. For very complex samples or if loading > 50 μg of protein per lane, full gel resolution can be performed to distribute sample complexity across the gel length. However, for the majority of protein complexes, ideal gel resolution is achieved when the dye-front is between a third and half way down the full gel length.
11. Working solution volumes in this protocol have been calculated for 16 samples, which corresponds to the average number of wells needed to process two gel lanes in which samples have been resolved one-third of the way down the gel (ca. 8 fractions per lane), assuming that 90 μL of solution are used per well, unless noted otherwise.
12. If using 96-well plates, all solutions can be added/removed from wells by a multichannel pipettor, keeping same set of tips for each sample set for all steps performed on this day.
13. For all aspirations, ensure that gel pieces remain in the wells and do not stick to the pipet tips. To minimize loss of gel pieces, avoid pushing the pipet tip down to the bottom of well and limit the number of gel pieces in each well.
14. The final volume of the sample loaded over the StageTip should be between 50 and 150 μL. The final concentration of ACN should be <10 %.
15. Choose a centrifuge speed that gives a flow rate over the disk of approximately 50 μL sample/min.
16. Empty collection tubes if necessary.
17. For fractionated samples, the three elution solutions will concentrate at different rates. The concentration speed of Elution #3 is the fastest, then Elution #2, then Elution #1. For reference, concentration of Elution #3 takes 8–12 min in our SpeedVac (on the no-heat setting).
18. TSA (trichostatin A) treatment of samples is included to inhibit HDAC activity and serve as a control to demonstrate that measured deacetylation activity in parallel experiments is dependent on the presence of HDAC activity.

Acknowledgments

We are grateful to the current and past members of the Cristea Lab for development and refinement of the techniques discussed in this chapter. This work was supported by grants from the National Institutes of Health (DP1 DA026192 and R01 HL127640) to IMC and an NSF Graduate Research Fellowship to AJG.

References

1. Zhang CL, McKinsey TA, Chang S, Antos CL, Hill JA, Olson EN (2002) Class II histone deacetylases act as signal-responsive repressors of cardiac hypertrophy. Cell 110(4):479–488
2. Vega RB, Harrison BC, Meadows E, Roberts CR, Papst PJ, Olson EN, McKinsey TA (2004) Protein kinases C and D mediate agonist-dependent cardiac hypertrophy through nuclear export of histone deacetylase 5. Mol Cell Biol 24(19):8374–8385
3. Huang Y, Tan M, Gosink M, Wang KK, Sun Y (2002) Histone deacetylase 5 is not a p53 target gene, but its overexpression inhibits tumor cell growth and induces apoptosis. Cancer Res 62(10):2913–2922
4. Lomonte P, Thomas J, Texier P, Caron C, Khochbin S, Epstein AL (2004) Functional interaction between class II histone deacetylases and ICP0 of herpes simplex virus type 1. J Virol 78(13):6744–6757
5. Renthal W, Maze I, Krishnan V, Covington HE 3rd, Xiao G, Kumar A, Russo SJ, Graham A, Tsankova N, Kippin TE, Kerstetter KA, Neve RL, Haggarty SJ, McKinsey TA, Bassel-Duby R, Olson EN, Nestler EJ (2007) Histone deacetylase 5 epigenetically controls behavioral adaptations to chronic emotional stimuli. Neuron 56(3):517–529
6. Dietrich JB, Takemori H, Grosch-Dirrig S, Bertorello A, Zwiller J (2012) Cocaine induces the expression of MEF2C transcription factor in rat striatum through activation of SIK1 and phosphorylation of the histone deacetylase HDAC5. Synapse 66(1):61–70. doi:10.1002/syn.20988
7. Taniguchi M, Carreira MB, Smith LN, Zirlin BC, Neve RL, Cowan CW (2012) Histone deacetylase 5 limits cocaine reward through cAMP-induced nuclear import. Neuron 73(1):108–120. doi:10.1016/j.neuron.2011.10.032
8. Lahm A, Paolini C, Pallaoro M, Nardi MC, Jones P, Neddermann P, Sambucini S, Bottomley MJ, Lo Surdo P, Carfi A, Koch U, De Francesco R, Steinkuhler C, Gallinari P (2007) Unraveling the hidden catalytic activity of vertebrate class IIa histone deacetylases. Proc Natl Acad Sci U S A 104(44):17335–17340
9. Fischle W, Dequiedt F, Hendzel MJ, Guenther MG, Lazar MA, Voelter W, Verdin E (2002) Enzymatic activity associated with class II HDACs is dependent on a multiprotein complex containing HDAC3 and SMRT/N-CoR. Mol Cell 9(1):45–57
10. Joshi P, Greco TM, Guise AJ, Luo Y, Yu F, Nesvizhskii AI, Cristea IM (2013) The functional interactome landscape of the human histone deacetylase family. Mol Syst Biol 9:672. doi:10.1038/msb.2013.26
11. Lemercier C, Verdel A, Galloo B, Curtet S, Brocard MP, Khochbin S (2000) mHDA1/HDAC5 histone deacetylase interacts with and represses MEF2A transcriptional activity. J Biol Chem 275(20):15594–15599
12. Bertos NR, Wang AH, Yang XJ (2001) Class II histone deacetylases: structure, function, and regulation. Biochem Cell Biol 79(3):243–252
13. Verdel A, Khochbin S (1999) Identification of a new family of higher eukaryotic histone deacetylases – coordinate expression of differentiation-dependent chromatin modifiers. J Biol Chem 274(4):2440–2445
14. Downes M, Ordentlich P, Kao HY, Alvarez JG, Evans RM (2000) Identification of a nuclear domain with deacetylase activity. Proc Natl Acad Sci U S A 97(19):10330–10335
15. McKinsey TA, Zhang CL, Olson EN (2001) Identification of a signal-responsive nuclear export sequence in class II histone deacetylases. Mol Cell Biol 21(18):6312–6321
16. Greco TM, Yu F, Guise AJ, Cristea IM (2011) Nuclear Import of Histone Deacetylase 5 by Requisite Nuclear Localization Signal Phosphorylation. Molecular & Cellular Proteomics 10 (2). doi: 10.1074/mcp.M110.004317
17. Borghi S, Molinari S, Razzini G, Parise F, Battini R, Ferrari S (2001) The nuclear local-

ization domain of the MEF2 family of transcription factors shows member-specific features and mediates the nuclear import of histone deacetylase 4. J Cell Sci 114(Pt 24):4477–4483
18. Zhang CL, McKinsey TA, Lu JR, Olson EN (2001) Association of COOH-terminal-binding protein (CtBP) and MEF2-interacting transcription repressor (MITR) contributes to transcriptional repression of the MEF2 transcription factor. J Biol Chem 276(1):35–39
19. McKinsey TA, Zhang CL, Lu J, Olson EN (2000) Signal-dependent nuclear export of a histone deacetylase regulates muscle differentiation. Nature 408(6808):106–111
20. Harrison BC, Kim MS, van Rooij E, Plato CF, Papst PJ, Vega RB, McAnally JA, Richardson JA, Bassel-Duby R, Olson EN, McKinsey TA (2006) Regulation of cardiac stress signaling by protein kinase D1. Mol Cell Biol 26(10):3875–3888. doi:10.1128/mcb.26.10.3875-3888.2006
21. Huynh QK, McKinsey TA (2006) Protein kinase D directly phosphorylates histone deacetylase 5 via a random sequential kinetic mechanism. Arch Biochem Biophys 450(2):141–148
22. Matthews SA, Liu P, Spitaler M, Olson EN, McKinsey TA, Cantrell DA, Scharenberg AM (2006) Essential role for protein kinase D family kinases in the regulation of class II histone deacetylases in B lymphocytes. Mol Cell Biol 26(4):1569–1577. doi:10.1128/MCB.26.4.1569-1577.2006
23. Sucharov CC, Langer S, Bristow M, Leinwand L (2006) Shuttling of HDAC5 in H9C2 cells regulates YY1 function through CaMKIV/PKD and PP2A. Am J Physiol Cell Physiol 291(5):C1029–C1037
24. Backs J, Backs T, Bezprozvannaya S, McKinsey TA, Olson EN (2008) Histone deacetylase 5 acquires calcium/calmodulin-dependent kinase II responsiveness by oligomerization with histone deacetylase 4. Mol Cell Biol 28(10):3437–3445
25. Ha CH, Wang W, Jhun BS, Wong C, Hausser A, Pfizenmaier K, McKinsey TA, Olson EN, Jin ZG (2008) Protein kinase D-dependent phosphorylation and nuclear export of histone deacetylase 5 mediates vascular endothelial growth factor-induced gene expression and angiogenesis. J Biol Chem 283(21):14590–14599
26. Backs J, Backs T, Neef S, Kreusser MM, Lehmann LH, Patrick DM, Grueter CE, Qi X, Richardson JA, Hill JA, Katus HA, Bassel-Duby R, Maier LS, Olson EN (2009) The delta isoform of CaM kinase II is required for pathological cardiac hypertrophy and remodeling after pressure overload. Proc Natl Acad Sci U S A 106(7):2342–2347. doi:10.1073/pnas.0813013106
27. Liu Y, Contreras M, Shen T, Randall WR, Schneider MF (2009) Alpha-adrenergic signalling activates protein kinase D and causes nuclear efflux of the transcriptional repressor HDAC5 in cultured adult mouse soleus skeletal muscle fibres. J Physiol 587(Pt 5):1101–1115
28. Peng Y, Lambert AA, Papst P, Pitts KR (2009) Agonist-induced nuclear export of GFP-HDAC5 in isolated adult rat ventricular myocytes. J Pharmacol Toxicol Methods 59(3):135–140
29. Ha CH, Kim JY, Zhao JJ, Wang WY, Jhun BS, Wong C, Jin ZG (2010) PKA phosphorylates histone deacetylase 5 and prevents its nuclear export, leading to the inhibition of gene transcription and cardiomyocyte hypertrophy. Proc Natl Acad Sci U S A 107(35):15467–15472. doi:10.1073/pnas.1000462107
30. Guise AJ, Greco TM, Zhang IY, Yu F, Cristea IM (2012) Aurora B-dependent regulation of class IIa histone deacetylases by mitotic nuclear localization signal phosphorylation. Mol Cell Proteomics 11(11):1220–1229. doi:10.1074/mcp.M112.021030, Pii: M112.021030
31. Berger I, Bieniossek C, Schaffitzel C, Hassler M, Santelli E, Richmond TJ (2003) Direct interaction of Ca2+/calmodulin inhibits histone deacetylase 5 repressor core binding to myocyte enhancer factor 2. J Biol Chem 278(20):17625–17635
32. Ago T, Liu T, Zhai P, Chen W, Li H, Molkentin JD, Vatner SF, Sadoshima J (2008) A redox-dependent pathway for regulating class II HDACs and cardiac hypertrophy. Cell 133(6):978–993
33. Calalb MB, McKinsey TA, Newkirk S, Huynh K, Sucharov CC, Bristow MR (2009) Increased phosphorylation-dependent nuclear export of class ii histone deacetylases in failing human heart. Clin Transl Sci 2(5):325–332. doi:10.1111/j.1752-8062.2009.00141.x
34. McKinsey TA, Zhang CL, Olson EN (2000) Activation of the myocyte enhancer factor-2 transcription factor by calcium/calmodulin-dependent protein kinase-stimulated binding of 14-3-3 to histone deacetylase 5. Proc Natl Acad Sci U S A 97(26):14400–14405
35. Guise AJ, Mathias RA, Rowland EA, Yu F, Cristea IM (2014) Probing phosphorylation-dependent protein interactions within functional domains of histone deacetylase 5 (HDAC5). Proteomics 14(19):2156–2166. doi:10.1002/pmic.201400092

36. Zhang L, Jin M, Margariti A, Wang G, Luo Z, Zampetaki A, Zeng L, Ye S, Zhu J, Xiao Q (2010) Sp1-dependent activation of HDAC7 is required for platelet-derived growth factor-BB-induced smooth muscle cell differentiation from stem cells. J Biol Chem 285(49):38463–38472. doi:10.1074/jbc.M110.153999
37. Mihaylova MM, Vasquez DS, Ravnskjaer K, Denechaud PD, Yu RT, Alvarez JG, Downes M, Evans RM, Montminy M, Shaw RJ (2011) Class IIa histone deacetylases are hormone-activated regulators of FOXO and mammalian glucose homeostasis. Cell 145(4):607–621. doi:10.1016/j.cell.2011.03.043
38. Cho Y, Cavalli V (2012) HDAC5 is a novel injury-regulated tubulin deacetylase controlling axon regeneration. EMBO J 31(14):3063–3078. doi:10.1038/emboj.2012.160
39. Mathias R, Greco T, Oberstein A, Budayeva H, Chakrabarti R, Rowland E, Kang Y, Shenk T, Cristea I (2014) Sirtuin 4 is a lipoamidase regulating pyruvate dehydrogenase complex activity. Cell 159(7):10. doi:10.1016/j.cell.2014.11.046
40. Chini CC, Escande C, Nin V, Chini EN (2010) HDAC3 is negatively regulated by the nuclear protein DBC1. J Biol Chem 285(52):40830–40837. doi:10.1074/jbc.M110.153270
41. Cristea IM, Williams R, Chait BT, Rout MP (2005) Fluorescent proteins as proteomic probes. Mol Cell Proteomics 4(12):1933–1941. doi:10.1074/mcp.M500227-MCP200
42. Barrow KM, Perez-Campo FM, Ward CM (2006) Use of the cytomegalovirus promoter for transient and stable transgene expression in mouse embryonic stem cells. Methods Mol Biol 329:283–294. doi:10.1385/1-59745-037-5:283
43. Qin JY, Zhang L, Clift KL, Hulur I, Xiang AP, Ren BZ, Lahn BT (2010) Systematic comparison of constitutive promoters and the doxycycline-inducible promoter. PLoS One 5(5):e10611. doi:10.1371/journal.pone.0010611
44. Fakhrai H, Shawler DL, Van Beveren C, Lin H, Dorigo O, Solomon MJ, Gjerset RA, Smith L, Bartholomew RM, Boggiano CA, Gold DP, Sobol RE (1997) Construction and characterization of retroviral vectors for interleukin-2 gene therapy. J Immunother 20(6):437–448
45. Bebenek K, Abbotts J, Wilson SH, Kunkel TA (1993) Error-prone polymerization by HIV-1 reverse transcriptase. Contribution of template-primer misalignment, miscoding, and termination probability to mutational hot spots. J Biol Chem 268(14):10324–10334
46. Hsia SC, Shi YB (2002) Chromatin disruption and histone acetylation in regulation of the human immunodeficiency virus type 1 long terminal repeat by thyroid hormone receptor. Mol Cell Biol 22(12):4043–4052
47. Tsai YC, Greco TM, Boonmee A, Miteva Y, Cristea IM (2012) Functional proteomics establishes the interaction of SIRT7 with chromatin remodeling complexes and expands its role in regulation of RNA polymerase I transcription. Mol Cell Proteomics 11(2):M111 015156. doi:10.1074/mcp.M111.015156
48. Greco TM, Guise AJ, Cristea IM. (2016) Determining the Composition and Stability of Protein Complexes Using an Integrated Label-Free and Stable Isotope Labeling Strategy. Methods Mol Biol. 1410:39-63. doi:10.1007/978-1-4939-3524-6_3
49. Wisniewski JR, Zougman A, Nagaraj N, Mann M (2009) Universal sample preparation method for proteome analysis. Nat Methods 6(5):359–360. doi:10.1038/nmeth.1322
50. Wisniewski JR, Zougman A, Mann M (2009) Combination of FASP and StageTip-based fractionation allows in-depth analysis of the hippocampal membrane proteome. J Proteome Res 8(12):5674–5678. doi:10.1021/pr900748n
51. WHO/UNAIDS (2013) UNAIDS report on the Global AIDS Epidemic. UNAIDS. Global Report. WHO
52. Rappsilber J, Mann M, Ishihama Y (2007) Protocol for micro-purification, enrichment, pre-fractionation and storage of peptides for proteomics using StageTips. Nat Protoc 2(8):1896–1906. doi:10.1038/nprot.2007.261
53. Shaw PG, Chaerkady R, Zhang Z, Davidson NE, Pandey A (2011) Monoclonal antibody cocktail as an enrichment tool for acetylome analysis. Anal Chem 83(10):3623–3626. doi:10.1021/ac1026176
54. Svinkina T, Gu H, Silva JC, Mertins P, Qiao J, Fereshetian S, Jaffe JD, Kuhn E, Udeshi ND, Carr SA (2015) Deep, quantitative coverage of the lysine acetylome using novel anti-acetyl-lysine antibodies and an optimized proteomic workflow. Mol Cell Proteomics. doi:10.1074/mcp.O114.047555
55. van der Mijn JC, Labots M, Piersma SR, Pham TV, Knol JC, Broxterman HJ, Verheul HM, Jimenez CR (2015) Evaluation of different phospho-tyrosine antibodies for label-free phosphoproteomics. J Proteome. doi:10.1016/j.jprot.2015.04.006
56. Thingholm TE, Jorgensen TJ, Jensen ON, Larsen MR (2006) Highly selective enrichment of phosphorylated peptides using titanium dioxide. Nat Protoc 1(4):1929–1935. doi:10.1038/nprot.2006.185
57. Engholm-Keller K, Birck P, Storling J, Pociot F, Mandrup-Poulsen T, Larsen MR (2012) TiSH – a robust and sensitive global phos-

phoproteomics strategy employing a combination of TiO2, SIMAC, and HILIC. J Proteome 75(18):5749–5761. doi:10.1016/j.jprot.2012.08.007

58. Di Palma S, Zoumaro-Djayoon A, Peng M, Post H, Preisinger C, Munoz J, Heck AJ (2013) Finding the same needles in the haystack? A comparison of phosphotyrosine peptides enriched by immuno-affinity precipitation and metal-based affinity chromatography. J Proteome 91:331–337. doi:10.1016/j.jprot.2013.07.024
59. Melo-Braga MN, Ibanez-Vea M, Larsen MR, Kulej K (2015) Comprehensive protocol to simultaneously study protein phosphorylation, acetylation, and N-linked sialylated glycosylation. Methods Mol Biol 1295:275–292. doi:10.1007/978-1-4939-2550-6_21
60. Grozinger CM, Schreiber SL (2000) Regulation of histone deacetylase 4 and 5 and transcriptional activity by 14-3-3-dependent cellular localization. Proc Natl Acad Sci U S A 97(14):7835–7840
61. Vannini A, Volpari C, Filocamo G, Casavola EC, Brunetti M, Renzoni D, Chakravarty P, Paolini C, De Francesco R, Gallinari P, Steinkuhler C, Di Marco S (2004) Crystal structure of a eukaryotic zinc-dependent histone deacetylase, human HDAC8, complexed with a hydroxamic acid inhibitor. Proc Natl Acad Sci U S A 101(42):15064–15069. doi:10.1073/pnas.0404603101
62. Nielsen TK, Hildmann C, Dickmanns A, Schwienhorst A, Ficner R (2005) Crystal structure of a bacterial class 2 histone deacetylase homologue. J Mol Biol 354(1):107–120. doi:10.1016/j.jmb.2005.09.065
63. Bottomley MJ, Lo Surdo P, Di Giovine P, Cirillo A, Scarpelli R, Ferrigno F, Jones P, Neddermann P, De Francesco R, Steinkuhler C, Gallinari P, Carfi A (2008) Structural and functional analysis of the human HDAC4 catalytic domain reveals a regulatory structural zinc-binding domain. J Biol Chem 283(39):26694–26704
64. Zhang Y (2008) I-TASSER server for protein 3D structure prediction. BMC Bioinformatics 9:40. doi:10.1186/1471-2105-9-40
65. Roy A, Kucukural A, Zhang Y (2010) I-TASSER: a unified platform for automated protein structure and function prediction. Nat Protoc 5(4):725–738. doi:10.1038/nprot.2010.5
66. Yang J, Yan R, Roy A, Xu D, Poisson J, Zhang Y (2015) The I-TASSER Suite: protein structure and function prediction. Nat Methods 12(1):7–8. doi:10.1038/nmeth.3213
67. Huynh QK (2011) Evidence for the phosphorylation of serine259 of histone deacetylase 5 by protein kinase Cdelta. Arch Biochem Biophys 506(2):173–180. doi:10.1016/j.abb.2010.12.005
68. Mathias RA, Guise AJ, Cristea IM (2015) Post-translational modifications regulate Class IIa histone deacetylase (HDAC) function in health and disease. Mol Cell Proteomics 14(3):456–470. doi:10.1074/mcp.O114.046565
69. Diner BA, Li T, Greco TM, Crow MS, Fuesler JA, Wang J, Cristea IM (2015) The functional interactome of PYHIN immune regulators reveals IFIX is a sensor of viral DNA. Mol Syst Biol 11(1):787. doi:10.15252/msb.20145808

Chapter 6

Analysis of Expression and Functions of Histone Deacetylase 6 (HDAC6)

Miao Li, Yan Zhuang, and Bin Shan

Abstract

Histone deacetylase 6 (HDAC6) is a member of class IIb HDAC family. HDAC6 exists predominantly in the cytoplasm and deacetylates mainly non-histone proteins in the cytoplasm. Via its deacetylase and ubiquitin binding domains, HDAC6 regulates microtubules, cytoskeleton, intracellular trafficking, and cellular responses to stress. HDAC6 plays a central role in physiology and pathobiology in various organs and tissues. Herein we describe the methods for analysis of expression and function of HDAC6 in diverse physiological and pathological contexts.

Key words Histone deacetylase 6, Acetylation, Stress, Chaperone, Ubiquitin

1 Introduction

Histone deacetylase 6 (HDAC6) is a member of class IIb HDAC family and distinct from other HDACs because it contains two deacetylase domains and an ubiquitin-binding domain called ubiquitin carboxyl-terminal hydrolase-like zinc finger domain (ZnF-UBP) at its C terminus [1]. The human HDAC6 gene is located on the X chromosome (chrX:48,802,080-48,824,970) in the human genome assembly GRCh38/hg38 as illustrated on the UCSC genome browser [2]. The human HDAC6 principal transcript (RefSeq NM_006044) consists of 29 exons and is translated into the HDAC6 protein of 1215 amino acids and a molecular mass of 131 kDa. We identified one minor isoform of HDAC6 mRNA that is produced through alternative splicing and translated into the 1063 amino acid long HDAC6 protein with a molecular mass of 114 kDa [3]. HDAC6 possesses one nuclear localization signal motif as well as two nuclear export signal motifs and exists predominantly in the cytoplasm (Fig. 1). Distinct from its orthologs in C. elegans, drosophila, and mouse, human HDAC6 harbors a tetradecapeptide repeat domain (SE14) located between the C-terminal deacetylase domain and the C-terminal ZnF-UBP motif (Fig. 1)

Sibaji Sarkar (ed.), *Histone Deacetylases: Methods and Protocols*, Methods in Molecular Biology, vol. 1436, DOI 10.1007/978-1-4939-3667-0_6,

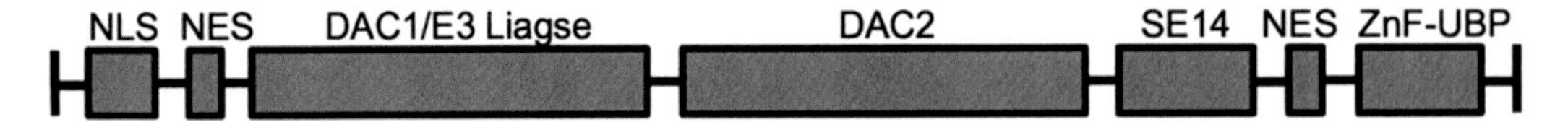

Fig. 1 Functional domains of the human HDAC6 protein. The functional and signaling domains of the human HDAC6 protein are illustrated as the indicated *boxes* that reflect the size and position of each domain. *NLS* nuclear localization signal, *NES* nuclear export signal, *DAC* deacetylase, *E3 ligase* E3 ubiquitin ligase, *SE14* tetradecapeptide repeat domain, *ZnF-UBP* ubiquitin carboxyl-terminal hydrolase-like zinc finger domain

[4]. SE14 repeat confers the human-specific leptomycin B-resistant cytoplasmic retention. HDAC6 functions through its two deacetylase domains, ubiquitin-binding Znf-UBP domain, and the newly discovered E3 ubiquitin ligase domain within its N-terminal deacetylase domain (Fig. 1) [5–7]. Consistent with its subcellular localization HDAC6 predominantly deacetylates non-histone proteins in the cytoplasm. Representative substrates deacetylated by HDAC6 are α-tubulin, heat-shock protein 90 (HSP90), and cortactin [8–10]. The proximal locations of two deacetylase domains and the linker region between them are essential to efficient deacetylation of HDAC6's substrates [11].

According to a transcriptome profile of various human tissues using Affymetrix HG133A array human HDAC6 is ubiquitously expressed and exhibits the highest expression in the pineal gland [12]. HDAC6 expression can be regulated transcriptionally under certain conditions. Estrogen up-regulates HDAC6 expression in estrogen receptor-positive breast cancer cells and up-regulation of HDAC6 mediates estrogen-induced increase in cell motility [13]. HDAC6 expression is also induced upon oncogenic Ras transformation and required for Ras-induced transformation [14]. On the other hand, HDAC6 expression is induced during senescence of colon cancer cells induced by the chemo prevention agent ursodeoxycholic acid [15].

HDAC6 regulates a myriad of molecular and cellular processes in the cell via its deacetylase and ubiquitin binding ZnF-UBP domains. Through its deacetylase domains HDAC6 promotes cell motility by deacetylating α-tubulin and cortactin that are master regulators of dynamic organization of microtubules and cytoskeleton during migration [8, 9]. Consistently the HDAC6-specific inhibitor tubacin increases acetylation of α-tubulin and decreases cell motility [7]. HDAC6-mediated deacetylation of α-tubulin is activated during TGF-β1 induced epithelial-mesenchymal transition in human lung epithelial cells and inhibition of HDAC6 attenuates the TGF-β1-induced epithelial-mesenchymal transition [16]. On the other hand HDAC6 mediates chemotactic motility of T lymphocytes independent of its deacetylase activity [16]. Besides microtubules, HDAC6 is phosphorylated and activated by Aurora A kinase and in turn deacetylates α-tubulin n the primary cilium to promote ciliary disassembly [17]. Moreover, HDAC6 is essential to autophagy-mediated cilia shortening in the mouse airway epithelial cells during exposure to cigarette smoke [18].

Besides cell motility HDAC6 regulates various cell processes via diverse substrates. One important substrate of HDAC6 is the chaperone protein heat-shock protein 90 (HSP90). HDAC6-mediated deacetylation of HSP90 is required for HSP90's chaperone of glucocorticoid and androgen receptors upon their engagement with their cognate ligands [10, 19]. Another fundamental cellular process regulated by HDAC6 is redox status in the cell. HDAC6 deacetylates peroxiredoxin I and II, and thereby decreases their reducing activity and resistance to superoxidation [20].

The ubiquitin-binding Znf-UBP domain at the C terminus of HDAC6 is essential to HDAC6-regulated cell responses to stress. In misfolded protein-induced stress HDAC6 binds polyubiquitinated misfolded proteins and recruits misfolded protein cargos to dynein motors for transport along microtubules to aggresomes for clearance [21]. In cell responses to environmental stress HDAC6 is recruited to stress granules in which a reversible translational suppression takes place upon stress [22]. HDAC6's ubiquitin-binding Znf-UBP domain is essential to the formation of stress granules. Moreover, HDAC6-dependent retrograde transport of cargos containing ubiquitinated proteins on microtubules is essential to the efficiency and selectivity of autophagic degradation [23]. In a Znf-UBP domain-dependent manner HDAC6 recruits a cortactin-dependent actin-remodeling machinery to mediate autophagosome–lysosome fusion and autophagic degradation, which is activated as a compensatory mechanism when the ubiquitin–proteasome system is impaired in the stressed cells [24, 25].

Although nuclear HDAC6 accounts for a minor portion of the total HDAC6 in the cell, nuclear HDAC6 regulates fundamental cellular processes in the nucleus. HDAC6 acts as a transcriptional co-repressor for its interacting transcription factors, such as RUNX2 and sumoylated p300 [9, 10]. HDAC6 regulates splicing in response to genotoxic stress as HDAC6 stabilizes serine/arginine-rich splicing factor 2 protein levels and activity [26]. HDAC6 can potentially regulate DNA damage repair because HDAC6-mediated deacetylation of Ku70 impairs recruitment of Ku70 to the chromatin fractions [27]. Moreover, HDAC6 sequentially deacetylates and ubiquitinates MutS protein homolog 2 and thereby promotes degradation of MutS protein homolog 2, which results in significantly reduced cellular sensitivity to DNA-damaging agents and DNA mismatch repair activities [5].

The critical role of HDAC6 in fundamental cellular, physiological, and pathological processes demands characterization of HDAC6 expression, localization, deacetylase activity, ubiquitin binding activity, etc. Herein we describe immunoprecipitation and immunoblots that are the most frequently used methods to characterize

HDAC6 expression and functions. To characterize HDAC6 expression and function in a biological context an investigator can carry out reciprocal immunoprecipitation and immunoblotting using the antibodies specific for HDAC6, its interacting partners, and modified moieties such as acetyl lysine and ubiquitin.

2 Materials

2.1 Cell Culture

1. Cell lines or primary cell culture of an investigator's interest.
2. Cell culture medium and apparatus appropriate for the cell type of an investigator's interest.
3. The plasmid, viral vectors, and siRNA oligos to overexpress or knockdown HDAC6. Wild type HDAC6 and the HDAC6 mutants with defective deacetylase and/or ubiquitin binding activity are available in either untagged or FLAG or EGFP tagged version from AddGene (addgene.org). The tagged HDAC6 expression vectors are useful tools for immunoprecipitation and purification of HDAC6 and its interacting proteins.

2.2 Cell Lysis

1. NP-40 lysis buffer: NaCl 150 mM, NP-40 1.0%, Tris–HCl 50 mM, pH 8.0, 5 mM dithiothreitol, protease inhibitor cocktail (a sample composition as following: Aprotinin at 80 μM, Bestatin at 4 mM, E-64 at 1.4 mM, Leupeptin at 2 mM, and Pepstatin A at 1.5 mM, add fresh), and PMSF (final concentration 50 μg/ml, add fresh) (*see* **Note 5**).
2. Phosphate-buffered saline (PBS): 137 mM NaCl, 2.7 mM, KCl, 4.3 mM Na_2HPO_4, and 1.5 mM KH_2PO_4.
3. Protein assay: A bicinchoninic acid (BCA) based protein assay kit is recommended because of its compatibility with the detergents used during lysis procedure.
4. Cell scrapers.

2.3 Immunoprecipitation

1. Protein A/G Agarose.
2. Bovine serum albumin.
3. Species and isotype-matched control antibodies.

2.4 Immunoblotting

1. A mini-cell system for immunoblotting. A mini-cell system consists of a vertical electrophoresis box that separates proteins based on size and a cassette that holds the transferring sandwich for transferring proteins from gels to membranes. The size of the gel box and the transferring cassette fits a mini gel in size of around 8.3 × 6.4 cm.
2. LDS Sample Buffer (4×): 106 mM Tris–HCl, 141 mM Tris–Base, 2% LDS, 10% glycerol, 0.51 mM EDTA, 0.22 mM SERVA Blue G250, 0.175 mM phenol red, pH 8.5.

3. Precast Gels with 4–12 % gradient bis-acrylamide, paired with MOPS SDS Running Buffer for mid-size proteins (14–200 kDa) (Life Technologies). Adjust types of gels and buffers for large or small size proteins.
4. MOPS SDS Running Buffer (20×): 50 mM MOPS, 50 mM Tris–Base, 0.1 % SDS, 1 mM EDTA pH 7.7.
5. Transfer Buffer (20×): 25 mM bicine, 25 mM Bis–Tris, 1 mM EDTA, pH 7.2.
6. PVDF membrane of 0.2 or 0.45 μm.
7. Prestained protein standard.
8. Washing buffer: TBST (50 mM Tris, 150 mM NaCl, 0.05 % v/v Tween 20, pH 7.6, 0.1 % Tween 200 add fresh).
9. Blocking buffer: TBST+ 5 % nonfat dry milk or BSA.

2.5 Primary and Secondary Antibodies for Immuno-precipitation and Immunoblot

1. HDAC6-specific antibodies: Clone H-300 is recommended for human HDAC6 (Cat# sc-11420, Santa Cruz Biotechnology). Clone D21B10 is recommended for mouse HDAC6 (Cat# 7612, Cell Signaling Technology) (*see* **Notes 1** and **2**).
2. Primary antibodies capable of immunoprecipitating and immunoblotting proteins of the HDAC6-targeting or interacting proteins: acetylated α-tubulin (clone 6-11B-1, Cat# 6793, Sigma-Aldrich), pan-acetyl-lysine (Cat# ICP0380-100, ImmuneChem), ubiquitinylated protein (clone FK1, Catalog# 04-262, Millipore), α-tubulin (Cell Signaling Technology, Cat # 2144). The antibodies from the above vendors are recommended based on previous success of using these antibodies in the literature.
3. Secondary antibodies for immunoblot: IRDye 800CW, 680RD, or 680LT secondary antibodies that can be detected based on infrared fluorescence on an Imaging System. (Alternative: HRP-conjugated secondary antibodies, enhanced chemiluminescence kits, and an imaging system able to capture chemiluminescence).
4. Imaging System for detection of IRDye-conjugated secondary antibodies.

3 Methods

3.1 Cell Lysis

1. Manipulate HDAC6 expression and/or treat cells as desired. Transiently or stably overexpress wild type and mutant HDAC6 using plasmid or viral vectors. Knockdown HDAC6 expression using siRNAs or shRNAs targeting HDAC6 (alternative: using primary cell culture or tissue materials obtained from HDAC6 wild type and null mice).

2. Lyse cell culture using 1× LDS lysis buffer for straight immunoblotting or NP-40 buffer for immunoprecipitation. Scrape the cells off the culture plate using the lysis buffer of choice at 200 μl/well of per well of 6-well culture plate for a ~80% confluent culture (adjust the volume of the lysis buffer based on cell number and cell culture surface area). Collect the lysates into a microfuge tube. For LDS lysates, proceed to (see 3.1.3). For NP-40 lysates, proceed to (see 3.1.4) (*see* **Note 3**).
3. Heat LDS lysates at 70 °C for 10 min and proceed to immunoblotting immediately or store samples at −20 °C for later use (see 3.3).
4. Sonicate NP-40 lysates. Determine the protein concentration of NP-40 lysates using Protein Assay Kit per the provider's instructions. Proceed to immunoprecipitation immediately or store samples at −80 °C for later use (see 3.2).

3.2 Immunoprecipitation

1. Gently swirl the Protein A/G Agarose slurry to resuspend the agarose beads. Aspirate the volume needed based on the formula of 20 μl slurry for every 2 μg of antibody. Centrifuge the slurry at 500–1000 ×*g* for 2 min at 4 °C. Remove the supernatant (half of the starting volume) without disturbing the settled slurry. Add PBS (half of the starting volume) to reconstitute the slurry.
2. Preclear 500 μg–1 mg NP-40 lysates for each immunoprecipitation. Mix the lysates with 20 μl Protein A/G Agarose. Incubate by rotation for 30 min at 4 °C. Centrifuge samples at 500–1000 ×*g* for 5 min at 4 °C, transfer the precleared NP-40 lysates to a fresh microfuge tube.
3. Mix the precleared NP-40 lysates with the desired primary antibodies (*see* Subheading 2.5) at 2–5 μg antibody per 1 mg lysates in a 500 μl to 1 ml volume and incubate by rotation at 4 °C overnight.
4. Add Protein A/G Agarose at 20 μl slurry for every 2 μg of the primary antibody to the lysates–antibody mixture and incubate by rotation at 4 °C for 1 h.
5. Centrifuge the agarose–lysates–antibody mixture at 500–1000 ×*g* for 5 min at 4 °C to capture the immunocomplexes.
6. Remove the supernatant without disturbing the settled immunocomplexes (Optional: save as flow-through to assess immunoprecipitation efficiency). Wash the settled immunocomplexes with NP-40 buffer three times by repeating the following steps:
   (a) Gently mix the settled immunocomplexes with 1 ml NP-40 buffer,
   (b) Incubate the mixture for 30 min by rotation,
   (c) Centrifuge the agarose–lysates–antibody mixture at 500–1000 ×*g* for 5 min at 4 °C,
   (d) Remove the supernatant.

7. Elute the immunocomplexes by adding equal volume of 2× NuPAGE LDS lysis buffer and heating the mixture at 70 °C for 10 min. Centrifuge the immunocomplexes at 500–1000 × *g* at room temperature and transfer the supernatant to a fresh microfuge tube for immunoblots immediately or store the supernatant at −80 °C for later use (*see* **Note 4**).

3.3 Immunoblotting

1. Assemble mini-cell gel box for electrophoresis.
2. Load equal amount of total protein or equal volume of the eluted immunocomplexes of each sample along with prestained standard onto the gel.
3. Electrophorese at 200 V for 1 h.
4. Prepare transfer buffer containing 10% v/v methanol (add fresh) and soak the pads and filter paper. Pre-wet the PVDF membrane in methanol.
5. Assemble the transfer sandwich consisting of pads, filter paper, the gel, and the PVDF membrane. Remove air bubbles from the sandwich completely.
6. Insert the transfer sandwich into a mini-cell transferring cassette and immerse the transferring cassette vertically in the transferring buffer in a mini-cell system. Transfer at 30 V for 1 h at room temperature.
7. Separate the transferred PVDF membrane from the transfer sandwich and incubate it in 10 ml blocking buffer with agitation on a platform shaker for 1 h at room temperature.
8. Wash the PVDF membrane with 50 ml of TBST three times, 5 min/wash.
9. Incubate the PVDF membrane in 10 ml blocking buffer containing the desired primary antibody at a desired dilution with agitation on a platform shaker at 4 °C overnight (alternative: 1 h at room temperature). Use H-300 (human HDAC6-specific at 1:500 dilution). Use 6-11B-1 (acetylated α-tubulin-specific at 1:4000 dilution, *see* Fig. 2).
10. Wash the PVDF membrane with 50 ml of TBST three times, 10 min/wash.
11. Incubate the PVDF membrane in 10 ml blocking buffer containing the species appropriate secondary antibodies with gentle agitation for 1 h at room temperature. Use light-proof containers from this step and beyond to protect IRDye conjugated secondary antibodies. IRDye-conjugated secondary antibodies at 1:15,000 dilution.
12. Wash the PVDF membrane with 50 ml of TBST three times, 10 min/wash.

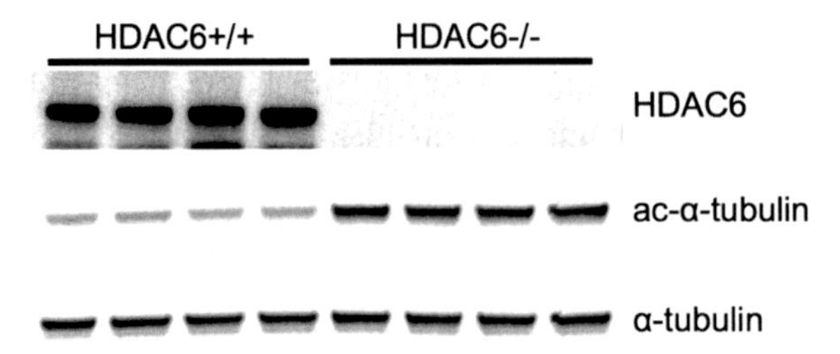

Fig. 2 Deacetylation of α-tubulin by HDAC6 in the mouse lung. Total protein lysates were extracted from the lung of wild type and HDAC6 null mice. Equal amount of protein was immunoblotted for HDAC6 (D21B10, 1:1000 dilution), acetylated α-tubulin (6-11B-1, 1:4000 dilution), and α-tubulin (1:1000 dilution)

13. Capture the secondary antibody-derived infrared fluorescence using an imaging system capable of capturing fluorescence. Adjust the imaging parameters, such as acquisition time, to achieve optimal signal over noise ratios. Alternatively an imaging system capable capturing chemiluminescence can be used if a horseradish peroxidase conjugated secondary antibody is used in **step 11** and detected using an enhanced chemiluminescence kit.

4 Notes

1. Human HDAC6 apparent molecular mass on a SDS gel is approximately 160 kDa instead of its molecular mass of 131 kDa. The exact cause of this aberrant migration pattern is unknown although the human specific tetradecapeptide repeat domain (SE14) is proposed as the culprit [4].
2. Clone D21B10 (Cat# 7612, Cell Signaling Technology) is highly recommended for detection of mouse HDAC6. This antibody yields the most consistent sensitivity in detection of mouse HDAC6 among the antibodies we tested (*see* Fig. 2).
3. Fractionation of cell lysates is recommended when examination of expression and function of HDAC6 in a particular subcellular compartment (membrane, cytoplasmic, nuclear, and cytoskeletal) is desired.
4. The immunocomplexes obtained using an HDAC6-specific antibody are ready for HDAC6 deacetylase assay using Histone Deacetylase Assay Kit, Fluorometric in an on Protein A/G agarose fashion. The immunocomplexes can also be eluted for discovery of the HDAC6-interacting proteins when coupled with Mass-Spec as exemplified in a recent report [9]. The immunocomplexes obtained using a pan-acetyl-lysine-specific antibody (Cat #ICP0380-100, ImmuneChem) from cells with variable expression of HDAC6 is recommended for discovery

of the protein that are deacetylated by HDAC6 when coupled with Mass-spec.

5. Phosphatase Inhibitor Cocktail can be added fresh to NP-40 buffer when phosphorylation status of HDAC6 or its interacting proteins is of an investigator's interest. This need is supported by ample evidence that HDAC6 catalytic activity is regulated by phosphorylation and HDAC6 regulates phosphorylation of its interacting proteins [17].

References

1. Bertos NR, Wang AH, Yang XJ (2001) Class II histone deacetylases: structure, function, and regulation. Biochem Cell Biol 79(3):243–252
2. Kent WJ, Sugnet CW, Furey TS, Roskin KM, Pringle TH, Zahler AM et al (2002) The human genome browser at UCSC. Genome Res 12(6): 996–1006. doi:10.1101/gr.229102, Article published online before print in May 2002
3. Zhuang Y, Nguyen HT, Lasky JA, Cao S, Li C, Hu J et al (2010) Requirement of a novel splicing variant of human histone deacetylase 6 for TGF-beta1-mediated gene activation. Biochem Biophys Res Commun 392(4):608–613. doi:10.1016/j.bbrc.2010.01.091
4. Bertos NR, Gilquin B, Chan GK, Yen TJ, Khochbin S, Yang XJ (2004) Role of the tetradecapeptide repeat domain of human histone deacetylase 6 in cytoplasmic retention. J Biol Chem 279(46):48246–48254. doi:10.1074/jbc.M408583200, M408583200 [pii]
5. Zhang M, Xiang S, Joo HY, Wang L, Williams KA, Liu W et al (2014) HDAC6 deacetylates and ubiquitinates MSH2 to maintain proper levels of MutSalpha. Mol Cell 55(1):31–46. doi:10.1016/j.molcel.2014.04.028
6. Grozinger CM, Hassig CA, Schreiber SL (1999) Three proteins define a class of human histone deacetylases related to yeast Hda1p. Proc Natl Acad Sci U S A 96(9):4868–4873
7. Haggarty SJ, Koeller KM, Wong JC, Grozinger CM, Schreiber SL (2003) Domain-selective small-molecule inhibitor of histone deacetylase 6 (HDAC6)-mediated tubulin deacetylation. Proc Natl Acad Sci U S A 100(8):4389–4394. doi:10.1073/pnas.0430973100
8. Hubbert C, Guardiola A, Shao R, Kawaguchi Y, Ito A, Nixon A et al (2002) HDAC6 is a microtubule-associated deacetylase. Nature 417(6887):455–458. doi:10.1038/417455a
9. Zhang X, Yuan Z, Zhang Y, Yong S, Salas-Burgos A, Koomen J et al (2007) HDAC6 modulates cell motility by altering the acetylation level of cortactin. Mol Cell 27(2):197–213. doi:10.1016/j.molcel.2007.05.033
10. Kovacs JJ, Murphy PJ, Gaillard S, Zhao X, Wu JT, Nicchitta CV et al (2005) HDAC6 regulates Hsp90 acetylation and chaperone-dependent activation of glucocorticoid receptor. Mol Cell 18(5):601–607. doi:10.1016/j.molcel.2005.04.021
11. Zhang Y, Gilquin B, Khochbin S, Matthias P (2006) Two catalytic domains are required for protein deacetylation. J Biol Chem 281(5):2401–2404. doi:10.1074/jbc.C500241200
12. Su AI, Wiltshire T, Batalov S, Lapp H, Ching KA, Block D et al (2004) A gene atlas of the mouse and human protein-encoding transcriptomes. Proc Natl Acad Sci U S A 101(16):6062–6067. doi:10.1073/pnas.0400782101
13. Saji S, Kawakami M, Hayashi S, Yoshida N, Hirose M, Horiguchi S et al (2005) Significance of HDAC6 regulation via estrogen signaling for cell motility and prognosis in estrogen receptor-positive breast cancer. Oncogene 24(28):4531–4539. doi:10.1038/sj.onc.1208646
14. Lee YS, Lim KH, Guo X, Kawaguchi Y, Gao Y, Barrientos T et al (2008) The cytoplasmic deacetylase HDAC6 is required for efficient oncogenic tumorigenesis. Cancer Res 68(18): 7561–7569. doi:10.1158/0008-5472.CAN-08-0188
15. Akare S, Jean-Louis S, Chen W, Wood DJ, Powell AA, Martinez JD (2006) Ursodeoxycholic acid modulates histone acetylation and induces differentiation and senescence. Int J Cancer 119(12):2958–2969. doi:10.1002/ijc.22231
16. Cabrero JR, Serrador JM, Barreiro O, Mittelbrunn M, Naranjo-Suarez S, Martin-Cofreces N et al (2006) Lymphocyte chemotaxis is regulated by histone deacetylase 6, independently of its deacetylase activity. Mol Biol Cell 17(8):3435–3445. doi:10.1091/mbc.E06-01-0008, E06-01-0008 [pii]
17. Pugacheva EN, Jablonski SA, Hartman TR, Henske EP, Golemis EA (2007) HEF1-dependent Aurora A activation induces disassembly of the primary cilium. Cell 129(7):1351–1363. doi:10.1016/j.cell.2007.04.035

18. Lam HC, Cloonan SM, Bhashyam AR, Haspel JA, Singh A, Sathirapongsasuti JF et al (2013) Histone deacetylase 6-mediated selective autophagy regulates COPD-associated cilia dysfunction. J Clin Invest 123(12):5212–5230. doi:10.1172/JCI69636
19. Ai J, Wang Y, Dar JA, Liu J, Liu L, Nelson JB et al (2009) HDAC6 regulates androgen receptor hypersensitivity and nuclear localization via modulating Hsp90 acetylation in castration-resistant prostate cancer. Mol Endocrinol 23(12):1963–1972. doi:10.1210/me.2009-0188
20. Parmigiani RB, Xu WS, Venta-Perez G, Erdjument-Bromage H, Yaneva M, Tempst P et al (2008) HDAC6 is a specific deacetylase of peroxiredoxins and is involved in redox regulation. Proc Natl Acad Sci U S A 105(28):9633–9638. doi:10.1073/pnas.0803749105
21. Kawaguchi Y, Kovacs JJ, McLaurin A, Vance JM, Ito A, Yao TP (2003) The deacetylase HDAC6 regulates aggresome formation and cell viability in response to misfolded protein stress. Cell 115(6):727–738
22. Kwon S, Zhang Y, Matthias P (2007) The deacetylase HDAC6 is a novel critical component of stress granules involved in the stress response. Genes Dev 21(24):3381–3394. doi:10.1101/gad.461107
23. Iwata A, Riley BE, Johnston JA, Kopito RR (2005) HDAC6 and microtubules are required for autophagic degradation of aggregated huntingtin. J Biol Chem 280(48):40282–40292. doi:10.1074/jbc.M508786200
24. Pandey UB, Nie Z, Batlevi Y, McCray BA, Ritson GP, Nedelsky NB et al (2007) HDAC6 rescues neurodegeneration and provides an essential link between autophagy and the UPS. Nature 447(7146):859–863. doi:10.1038/nature05853
25. Lee JY, Koga H, Kawaguchi Y, Tang W, Wong E, Gao YS et al (2010) HDAC6 controls autophagosome maturation essential for ubiquitin-selective quality-control autophagy. EMBO J 29(5):969–980. doi:10.1038/emboj.2009.405
26. Edmond V, Moysan E, Khochbin S, Matthias P, Brambilla C, Brambilla E et al (2011) Acetylation and phosphorylation of SRSF2 control cell fate decision in response to cisplatin. EMBO J 30(3):510–523. doi:10.1038/emboj.2010.333
27. Chaudhary N, Nakka KK, Chavali PL, Bhat J, Chatterjee S, Chattopadhyay S (2014) SMAR1 coordinates HDAC6-induced deacetylation of Ku70 and dictates cell fate upon irradiation. Cell Death Dis 5:e1447. doi:10.1038/cddis.2014.397

Chapter 7

Analysis of Histone Deacetylase 7 (HDAC7) Alternative Splicing and Its Role in Embryonic Stem Cell Differentiation Toward Smooth Muscle Lineage

Junyao Yang, Andriana Margariti, and Lingfang Zeng

Abstract

Histone deacetylases (HDACs) have a central role in the regulation of gene expression, which undergoes alternative splicing during embryonic stem cell (ES) cell differentiation. Alternative splicing gives rise to vast diversity over gene information, arousing public concerns in the last decade. In this chapter, we describe a strategy to detect HDAC7 alternative splicing and analyze its function on ES cell differentiation.

Key words HDAC7, Alternative splicing, Plasmid construction, Luciferase assay

1 Introduction

Alternative splicing is a specific feature in eukaryotic species, which has been classified as intron retention, exon skipping, alternative 5′ splice sites or alternative donor sites, and alternative 3′ splice sites or alternative acceptor sites [1]. Through alternative splicing, different mRNA species can be derived from a single gene, giving rise to different protein variants with different even opposite functions, which gives rise to vast diversity over gene information.

Histone deacetylases (HDACs), known as epigenetic regulators of gene transcription, are a family of enzymes that remove acetyl groups from lysine residues of histone proteins, which plays indispensable roles in nearly all biological processes [2]. Mammalian HDACs family, including 18 members, are grouped into four classes according to their homology with yeast histone deacetylases [3]. HDAC7, belonging to Class IIa. There are eight and four transcript variants in mouse and human HDAC7, respectively, which are derived from alternative transcription and splicing. Our previous study revealed that one of the mouse HDAC7 mRNA variants had two ATG codons separated by about 120 nucleotides which included three sequential stop codons. Our studies demonstrated

Sibaji Sarkar (ed.), *Histone Deacetylases: Methods and Protocols*, Methods in Molecular Biology, vol. 1436,
DOI 10.1007/978-1-4939-3667-0_7,

that this HDAC7 mRNA underwent further splicing during embryonic stem (ES) cell differentiation towards a smooth muscle cell (SMC) lineage which is essential in maintaining the cardiovascular system in embryonic [4]. Recent study also showed that stem/progenitor cell differentiation toward SMC contributed to neointima formation [5]. Therefore, targeting on HDAC7 splicing may be beneficial for vascular disease intervention.

In this chapter, we describe the strategy that we used to detect HDAC7 alternative splicing and analyze its function. The strategy includes the identification of the existence of different HDAC7 splicing isoforms, the cloning of the different HDAC7 cDNAs and transfection, luciferase reporter analysis, analysis of different HDAC7 isoforms cellular localization by immunofluorescence staining and western blot assay.

2 Materials

2.1 Cell Culture and ES Cell Differentiation

1. Mouse ES cells (ES-D3 cell line, CRL-1934; ATCC, Manassas, VA), HEK293 cells (ATCC).
2. Mouse collagen IV (5 μg/ml).
3. ES culture medium: Dulbecco's modified Eagle medium (DMEM) supplemented with 10% fetal bovine serum (FBS), 10 ng/ml recombinant human leukemia inhibitor factor (LIF), 0.1 mM 2-mercaptoethanol, 100 U/ml penicillin, and 100 μg/ml streptomycin.
4. Differentiation medium (DM): MEM alpha medium supplemented with 10%, 0.05 mM 2-mercaptoethanol, 100 U/ml penicillin, and 100 μg/ml streptomycin.
5. HEK293 culture medium: DMEM, supplemented with 10% FBS, 100 U/ml penicillin, and 100 μg/ml streptomycin.
6. Phosphate-buffered saline (PBS): 137 mM NaCl, 10 mM Phosphate, 2.7 mM KCl, no Calcium, no magnesium and a pH of 7.4.
7. 0.25% trypsin–EDTA.

2.2 PCR and Western Blot

1. RNeasy Mini Kit: For purification of total RNA from cells.
2. QIAshredder: For simple and rapid homogenization of cell lysates.
3. Improm-II RT kit: Reverse-transcribe RNA templates starting with total RNA into first-strand cDNA in preparation for PCR amplification.
4. RNase inhibitor: For inhibiting eukaryotic RNases.
5. Taq DNA polymerase: Synthesizes DNA from single-stranded templates in the presence of dNTPs and a primer.

6. Hypotonic Buffer: 10 mM Tris–HCl, pH 7.5; 10 mM KCl; 1 mM EDTA; protease inhibitor.
7. High Salt Buffer C: 10 mM Tris–HCl, pH 7.5; 420 mM KCl; 1 mM EDTA; 30% Glycerol; protease inhibitor.
8. Then ice-cold lysis buffer (50 mM Tris–HCl pH 7.5, 150 mM NaCl, 1 mM EDTA pH 8.0)
9. SYBR® Green PCR Master Mix: For PCR amplification.
10. Mouse anti-α-tubulin (Clone B-5-1-2, T 5168), rabbit-anti-HDAC7 (KG-17, H 2662), mouse anti-HA (Clone HA-7, H9658), mouse anti-FLAG (clone M2, F1804), rabbit anti-H4 (SAB4500311) and rabbit anti-MEF2C (SAB2103534) antibodies, and normal mouse IgG (I5381), normal donkey serum (D9663).
11. Protein G-agarose bead.

2.3 Plasmid Construction

1. pGL3-luc basic vector.
2. Renilla luciferase substrate.
3. 1× reporter lysis buffer.
4. Opti-MEM medium.
5. pShuttle2 and Adenoviral expression system 1.

2.4 Transient Transfection and Nucleofection

1. Fugene-6-Reagent: Multicomponent formulation for the transfection of eukaryotic cells.
2. HDAC activity assay kit (colorimetric): Detects HDAC activity.

3 Method

3.1 Detection and Verification of HDAC7 Alternative Splicing

3.1.1 ES Cell Differentiation

1. ES cells are normally maintained in gelatin-coated flask (0.04 % gelatine/PBS for 2h at room temperature) in ES medium, and subcultureed at ratio of 1:6 every other day.
2. For differentiation, ES cells are maintained in collagen IV-coated flasks or plates (5 μg/ml in PBS for 2 h at room temperature) in DM for time indicated. Medium is refreshed every other day.

3.1.2 RNA Extraction

HDAC7 alternative splicing is detected by routine RT-PCR, in which the total RNAs from differentiating cells at various time points are extracted using the RNeasy Mini Kit:

1. The cells are detached by scratching off using a cold plastic cell scraper in the presence of medium and pellet by centrifuging at 6400 × *g* at room temperature for 5 min. The pellet is resuspended in appropriate volume of Buffer RLT.

2. The cell lysates are applied to a QIAshredder column and spun down at 21130 × *g* at room temperature (RT) for 2 min. This step helps to shear chromosomal DNA.
3. One volume of 70 % ethanol is added to the sheared lysate, and mixed well by pipetting.
4. Up to 700 μl of the sample is transfered to an RNeasy spin column placed in a 2 ml collection tube.
5. The tube lid is closed gently, and centrifuged at 20000 × *g* at RT for 1 min. Discard the flow-through.
6. Remaining mixed lysates are applied, and repeat **steps 4** and **5**.
7. 700 μl Buffer RW is added to the RNeasy spin column. The tube lid is closed gently, and centrifuged at 20000 × *g* at RT for 1 min. The flow-through is discarded.
8. 500 μl Buffer RPE is added to the column. The tube lid is closed gently, and centrifuged at highest speed at RT for 1 min. The flow-through is discarded. Repeat once.
9. The tube is spun down at 20000 × *g* at RT for 2 min.
10. The RNeasy spin column is placed in a new 1.5 ml collection tube (supplied).
11. 30–50 μl RNase-free water is directly added to the spin column membrane. The tube lid is closed gently, and centrifuged at highest speed at RT for 1 min to elute the RNA.
12. The RNA concentration is measured using NanoDrop machine.

3.1.3 Reverse Transcriptase-Polymerase Chain

1. 2 μg RNA is reverse-transcribed into cDNA with random primer by MMLV reverse transcriptase (RT).
2. The RNA is heated at 70 °C for 5 min and then chilled on ice. A mixture of RT reaction is added to the denatured RNAs, giving to a 20 μl reaction containing 1× RT buffer, 1 mM dNTPs, 3 mM $MgCl_2$, 50 ng random primers, 1 U/μl RNasin plus, and 100 U MMLV-RT.
3. The RT reaction is conducted by a protocol of 25 °C for 5 min, 37 °C for 60 min, and 70 °C 5 min. PCR is performed according to standard procedures.
4. Briefly, a 25 μl volume contains 1× PCR buffer, 0.2 mM dNTPs, 2 mM MgCl2, 10 pmol of primers, 1 U Taq DNA polymerase, and 50 ng cDNA (relative to RNA amount).
5. PCR protocol: 95 °C 5 min to pre-denature cDNAs, 35 cycles of three steps of amplification at 94 °C 1 min, 55 °C 1 min, and 72 °C 1 min, and final extension at 72 °C 5 min.
6. PCR product is run on 2 % agarose gel containing 0.5 μg/ml ethidium bromide and image is taken under the BioSpectrum AC Imaging System.

With the primer set of HDAC7-s-1 versus HDAC7-s-2 (Table 1), we have demonstrated that HDAC7 mRNA underwent further splicing during ES cell differentiation toward SMC lineage in vitro (Fig. 1).

3.1.4 Expression Plasmid Construction

To verify the splicing events, we cloned the partially and fully spliced HDAC7 via PCR amplification into a modified pShuttle2-FLAG-HA vector, in which FLAG tag is inserted into downstream of the first ATG codon while HA tag is inserted into the upstream of stop codon in the second open reading frame. A short HDAC7-HA vector (the second open reading frame) was created by mutagenesis. An illustration of the 5′ terminal HDAC7 mRNA sequence is shown in Fig. 2a, in which primer positions are indicated. Figure 2b shows the illustration of the three vectors. If transfected into target cells, fully spliced HDAC7 plasmid will produce FLAG-HDAC7-HA double tagged protein, while partially spliced and short HDAC7

Table 1
Primers for HDAC7 cloning and splicing

Gene	Name of primer	Sequence	Length (bp)	Acc No.
HDAC7-c	HDAC7-cl	5′>cga tct ggt acc tgg atg cac agc c <3′ 5′>gtc agc tct aga ctg aca tca gag acg agg<3′	2603	AK036586
HDAC7-1	HDAC7-1-1 HDAC7-1-2	5′>gcc ggg gct gtg cat cca gg<3′ 5′>gcg ggc tgc cct gcc ctc cag<3′	2546	AK036586
HDAC7-2	HDAC7-2-1 HDAC7-2-2	5′>gcg ggc tgc cct gcc ctc cag<3′ 5′>gtc tcc cta tag tga gtc<3′	2531	AK036586
HDAC7-s	HDAC7-s-1 HDAC7-s-2	5′>cga tct ggt acc tgg atg cac agc c<3′ 5′>gct acg gca ctt cgc ttg ctc<3′	u-371 s-314	AK036586

HDAC7-c Primers for cloning. *HDAC7-1* Primers for Spliced HDAC7, *HDAC7-2* Primers for short HDAC7, *HDAC7-s* Primers to distinguish unspliced and spliced isoforms, *u* unspliced, *s* spliced.

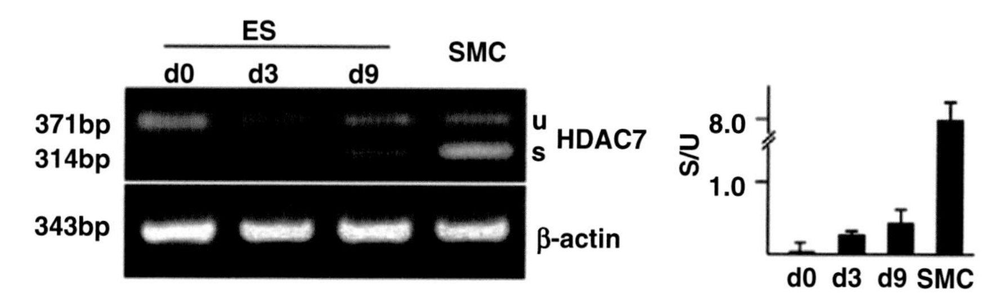

Fig. 1 HDAC7 undergoes further splicing during ES cell differentiation into SMCs. RT-PCR shows HDAC7 splicing in both undifferentiated and differentiated ES cells. The ratio of spliced to unspliced in the *right panel*. *U* unspliced isoform, *s* spliced isoform. The data presented are representative or means (± s.e.m.) of three independent experiments

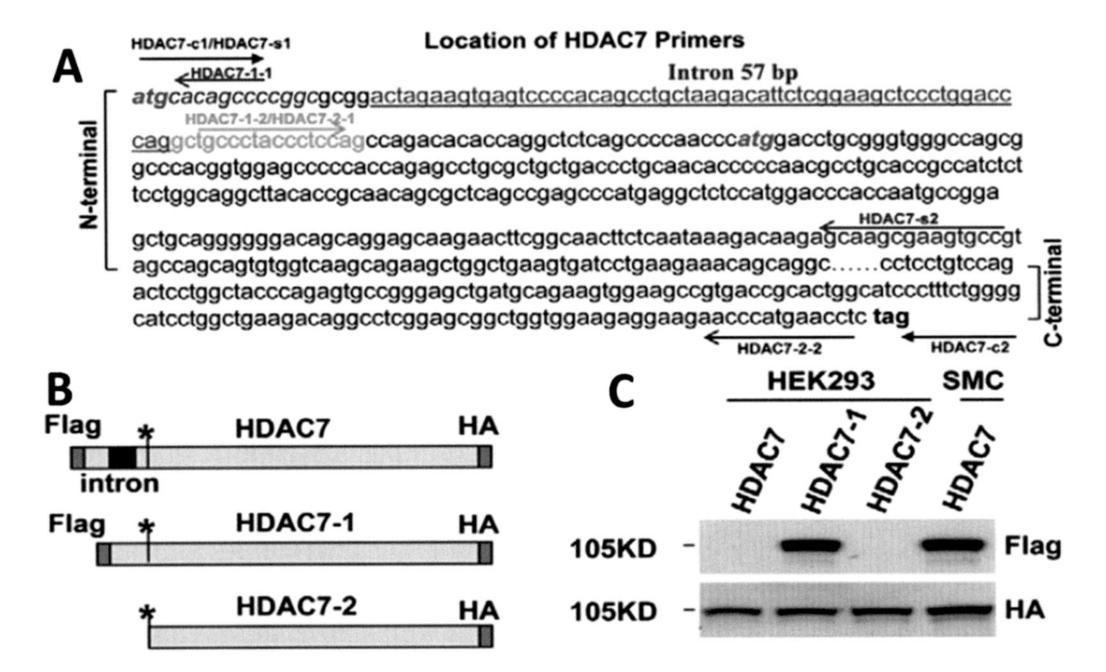

Fig. 2 Cloning of different HDAC7 isoforms. (**a**) A schematic illustration of the location of HDAC7 primers. (**b**) A schematic illustration of the cloned HDAC7 splicing isoforms. (**c**) Western blot confirmation of the HDAC7 isoform clones. HEK293 was transfected with pShuttle2-HDAC7, pShuttle2-HDAC7-1, and pShuttle2-HDAC7-2. SMCs are infected with ad-HDAC7 virus

plasmids only produce HDAC7-HA protein. If partially spliced HDAC7 undergoes further splicing, both HDAC7-HA and FLAG-HDAC7-HA will be detected by Western blot analysis.

HDAC7 Expression Plasmid Construction

1. Total RNAs of 3 day pre-differentiated ES cells are extracted using the RNeasy Mini Kit.
2. Full-length mouse HDAC7 cDNA fragment is obtained by RT-PCR amplification with the primer set HDAC7c-1 and HDAC7c-2 (Table 1).
3. The HDAC7 PCR product is digested with KpnI and XbaI and cloned into KpnI/XbaI sites of the modified pShuttle2-Flag-HA vector, designated pShuttle2-HDAC7. The fully spliced HDAC7, designated as pShuttle2-HDAC7-1 (Flag and HA tagged full length HDAC7) and the short HDAC7 (HA tagged), designated as pShuttle2-HDAC7-2 isoforms are created by PCR-based mutagenesis with specific primer sets (Table 1).
4. All the constructs are verified by DNA sequencing.

Adenoviral Vector Construction

To increase expression efficiency, HDAC7, HDAC7-1, and HDAc7-2 fragments are subcloned into adenoviral expression system 1 vector.

1. pShuttle2-HDAC7, pShuttle2-HDAC7-1, and pShuttle2-HDAC7-2 plasmid DNAs are digested with PI-Sce and I-Ceu.
2. The fragments are ligated to PI-Sce/I-Ceu digested and dephosphorylated adenoviral arm vector.
3. The resulting Ad-HDAC7, Ad-HDAC7-1, and Ad-HDAC7-2 viral DNA vectors are linearized with Pac I digestion and transfected into HEK293 cells for viral particle production, according to the protocol provided.

3.1.5 Transient Transfection and Adenoviral Infection

To verify HDAC7 isoforms, HEK is transfected with pShuttle2-HDAC7-1, and pShuttle2-HDAC7-2 vectors, while SMCs are infected with Ad-HDAC7 virus.

Transient Transfection

1. HEK293 cells are seeded at 1×10^5 cell/well in 6-well plate for 24 h and treated with serum-free DMEM for 1 h prior to transfection.
2. 1 μg HDAC7 isoform plasmids is diluted in 50 μl of Opti-MEM medium and mixed with 3 μl of Fugene 6 reagent in 50 μl of Opti-MEM medium, incubated at room temperature for 30 min and then added to one well of the 6-well plates in serum-free DMEM. Five hours later, one volume of DMEM containing 20 % FBS is added and incubated for 24 h, followed by refreshing medium and further incubation for 24 h.
3. After 48 h, the cells are subjected to Western blot assays.

Adenoviral Infection

SMCs are seeded at 1×10^5 cell/well in 6-well plate for 24 h prior to virus infection. Upon infection, the medium is removed, and 1 ml fresh growth medium containing 1×10^6 viral particle [at 10 multiplicity of infection (MOI)] is added and incubated for 6 h, and then cultured in normal growth medium for 24 h. The cells are then subjected to Western blot assays.

3.1.6 Western Blot Analysis

1. The cell culture plates, or flasks are placed on ice and washed with ice-cold PBS.
2. The cells are scratched off using a cold plastic cell scraper, and transferred into a precooled microcentrifuge tube.
3. The ice-cold lysis buffer 0.5 ml per 5×10^6 cells/60 mm dish/75 cm^2 flask is added, supplemented with protease inhibitors and 0.5 % Triton X-100, and lysed by sonication. Constant agitation is maintained for 30 min at 4 °C.
4. The cell lysates are spun at $16{,}000 \times g$ for 5 min in a 4 °C precooled centrifuge.
5. The tube is gently removed from the centrifuge and placed on ice. The supernatant is transfered to a fresh tube kept on ice, and the pellet is discarded.
6. The protein concentration is determined.

7. To reduce and denature: 5× SDS is added and each cell lysate in sample buffer is boiled at 100 °C for 5 min and aliquot. Lysates are stored at −20 °C.
8. Equal amounts of protein are loaded into the wells of the SDS-PAGE gel, along with molecular weight markers. 40 μg of total protein is loaded from cell lysate.
9. Samples are run on the SDS-PAGE gel for 1.5 h at 160 V.
10. The separated protein is transferred to nitrocellulose transfer membranes.
11. The membranes are blocked with TBS containing 5 % dry milk and 0.1 % Tween 20 on a shaker at room temperature for 1 h.
12. The membranes are incubated with appropriate dilutions of primary antibody in 5 or 2 % blocking solution overnight at 4 °C or for 2 h at room temperature.
13. The membranes are washed in three washes of TBST, 5 min each.
14. The membranes are incubated with the appropriate primary antibodies at 4 °C overnight.
15. The membranes are washed in three washes of TBST, 5 min each.
16. The membranes are incubated with the labeled secondary antibody (1:1000) in 5 % blocking buffer in TBST at room temperature on a shaker for 1 h.
17. The membranes are washed in three washes of TBST, 5 min each, then rinse in TBS.
18. The membranes are developed. Blots are stripped and re-probed with α-tubulin as loading control.

From the experiments described above, HDAC7-1 is detected by both anti-FLAG and anti-HA antibodies, while HDAC7 and HDAC7-2 are detected only by HA as expected (Fig. 2c). As shown in Fig. 1, SMCs shows higher HDAC7 splicing. Indeed, Ad-HDAC7 infection in SMCs produced both FLAG and HA bands, and the FLAG band is much stronger (Fig. 2c), confirming further splicing event in SMCs.

3.2 To Analyze the Potential Role of HDAC7 Splicing in SMC Differentiation

As described above, HDAC7 undergoes further splicing during ES cell differentiation toward to SMC lineage. We wondered whether this splicing was involved in this process. We used luciferase reporter system to assess the effect of overexpression of HDAC7 isoforms on SMC marker gene expression like SM22 and SMA.

3.2.1 Report Gene Plasmid Construction

The DNA fragments corresponding to mouse SM22 (*AH003214*) and SMA (*M57409*) gene promoter are amplified from mouse genomic DNA with primer set as follows:

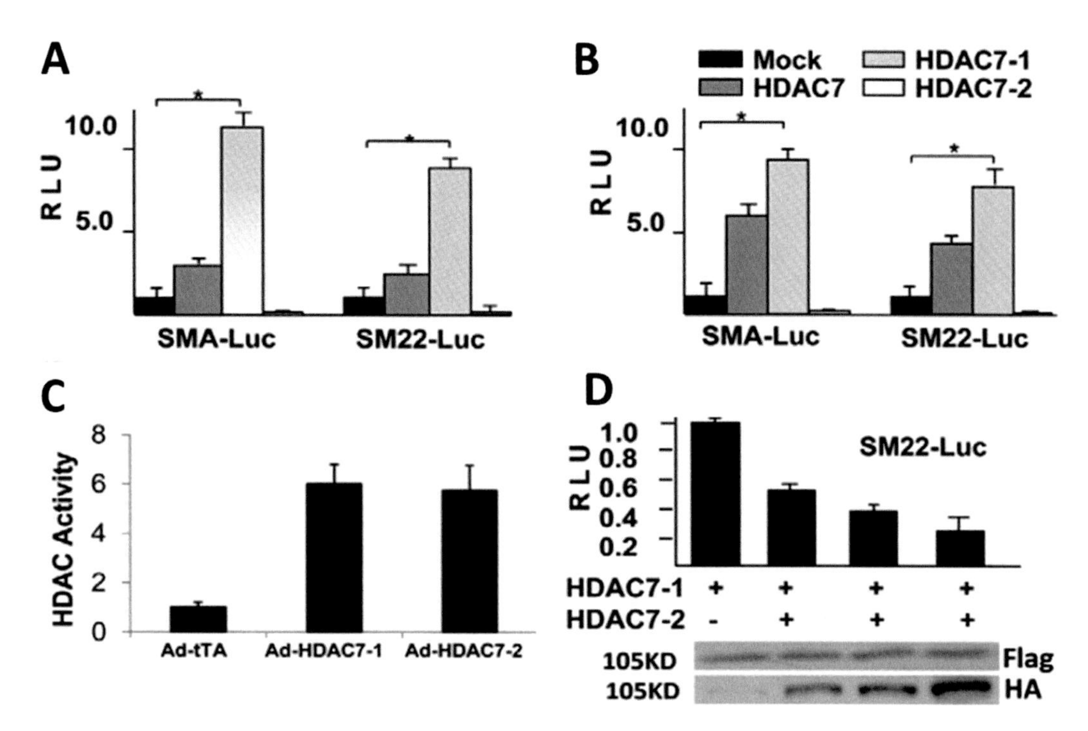

Fig. 3 Different HDAC7 isoforms exert different effects on SMC differentiation. (**a**, **b**) Luciferase reporter analysis shows the different effect of HDAC7 isoforms on SMA-Luc and SM22-Luc reporters in differentiated ES cells (**a**) and mature SMCs (**b**), respectively. *: $p < 0.05$. (**c**) Both HDAC7 isoforms possess similar deacetylase activity. Differentiated ES cells were infected with ad-HDAC7-1 and ad-HDAC7-2 viruses and incubated for 48 h, followed by HDAC activity assays. Ad-tTA was included as control. (**d**) HDAC7-2 suppresses HDAC7-1-induced SM22 reporter gene expression in a dose-dependent manner in ES cells. ES cells were cotransfected with 0.5 μg pShuttle2-HDAC7-1, and 0, 0.5, 1.0, 1.5 μg pShuttle2-HDAC7-2, pShuttle2 empty vector was included to compensate the total plasmid amount. HA and Flag represent the expression of the short and spliced forms, respectively. *RLU* relative luciferase unit. The data presented are representative or means (± s.e.m.) of nine independent experiments for (**a**), (**b**), (**d**) and three for (**c**)

SM22-P, 1350 bp, 5′-ttcaggacgtaatcagtg-3′ (nt 4-21) and 5′-agcttcggtgtctgggctg-3′ (nt 1371-1353); SMA-P, 1113 bp, 5′-tgcatgagccgtgggag-3′ (nt 16-32) and 5′-acttaccctgacagcgac-3′ (nt 1128-1111). The PCR product of SM22 and SMA gene promoter sequences are cloned into pGL3-luc basic vector, designated pGL3-SM22-Luc and pGL3-SMA-Luc, respectively. All the plasmids are verified by DNA sequencing. Reporter analysis has revealed that different HDAC7 isoforms had different effect on SM22 and SMA expression in ES or SMC cells (Fig. 3a, b, and d).

3.2.2 Transient Transfection

1. ES cells are seeded at 2×10^4cell/well in collagen IV-coated 24-well plates in DM 48 h and pretreated with serum free alpha-MEM (500 μl/well) for 1 h prior to transfection. SMCs are seeded at 2×10^4cell/well in gelatin-coated 24-well plates in DMEM 24 h and pretreated with serum free DMEM (500 μl/well) for 1 h prior to transfection.

2. One microgram pGL3-SM22-Luc or pGL3-SMA-Luc reporter plasmid, 1 μg HDAC7 isoform plasmids, plus 0.2 μg Renilla-Luc reporter plasmid are diluted in 50 μl of Opti-MEM medium and mixed with pre-diluted 7 μl of Fugene 6 in 50 μl of Opti-MEM medium, incubated at room temperature for 30 min and then added to three wells of the 24-well plates and incubated for 5 h. After that, the medium is removed and complete growth medium is added and incubated for 48 h, followed by luciferase activity assays.

3.2.3 Luciferase Reporter Assay

1. Cells are transfected with reporter plasmids and overexpression plasmids as described above in Subheading 3.3.
2. After 48 h transfection, the cells are washed once with PBS.
3. The cells are lysed with 200 μl/well 1× reporter lysis buffer.
4. All assay components (reagent and sample) are allowed to equilibrate to room temperature prior to assay.
5. 30 μl cell lysate is mixed with 100 μl firefly luciferase.
6. After at least 10 min, detect luciferase activity.
7. The relative luciferase activity unit (RLU) is defined as the ratio of firefly luciferase activity to that of Renilla luciferase activity with that of control group set as 1.0.

Data presented in Fig. 3a, b, and d revealed that both HDAC7 and HDAC7-1 plasmids increased, while HDAC7-2 decreased SM22 and SMA reporter gene expression in ES cells (Fig. 3a) and SMCs (Fig. 3b). HDAC7-2 exerted suppressive effect on HDAC7-1-induced SM22 reporter gene expression in ES cell in a dose-dependent manner (Fig. 3d).

3.3 To Explore the Underlying Mechanisms

From the data described above, we have found that fully spliced HDAC7 (HDAC7-1) promoted SMC differentiation while the short one (HDAC7-2) exerted suppressive effect. As HDAC7 is a class-II HDAC, its deacetylase activity is largely derived from associated class I HDACs [6]. As HDAC7-2 lacks the first 22 amino acids as compared to HDAC7-1, we wonder whether this shortage decreases its interaction with class I HDACs, therefore possessing less deacetylase activity. To test this, HDAC activity assays were performed.

3.3.1 Histone Deacetylase Activity Assay

1. ES cells are seeded on collagen-IV-coated dishes and cultured in DM for 3 days (70% confluence).
2. Ad-tTA (empty vector control), Ad-HDAC7-1, Ad-HDAC7-2 virus are added at 10 MOI in DM.
3. After 6 h, the virus solution is removed and fresh DM medium is added and incubated for 2 days.
4. The cells are lysed using the same procedure for Western blot **steps 1–6**.

5. 50 μg cell lysate is added to 96-well plate and subjected to HDAC activity assay.

Data presented in Fig. 3c shows that overexpression of HDAC7-1 and HDAC7-2 via adenoviral gene transfer significantly increases HDAC activity. There is no difference between these two isoforms concerning its HDAC activity. So the suppressive effect of HDAC7-2 seems not derived from HDAC activity.

HDAC7 is a shuttle protein, located in both cytoplasm and nucleus [7]. So the next step is to assess whether HDAC7-1 and HDAC7-2 differ in cellular location. To address this issue, immunofluorescence staining and cellular fraction assays were performed.

3.3.2 Indirect Immunofluorescence Staining

1. SMCs are infected Ad-HDAC7-1 or Ad-HDAC7-2 virus at 10 MOI for 6 h and incubated in complete growth medium for 24 h.
2. The cells are detached from the flasks by trypsin and seeded onto gelatin-coated slides in growth medium and incubated for 24 h.
3. The cells on slides are washed once with PBS quickly to remove medium and fixed with 4% paraformaldehyde solution at room temperature for 15 min.
4. After three times of PBS washing with 5 min each, the cells are permeabilized by using 1% Triton X-100/PBS at room temperature for 10 min.
5. After three times of PBS washing with 5 min each, the permeabilized cells are incubated with 5% donkey serum in PBS at room temperature for 30 min (blocking procedure to prevent nonspecific intracellular antibody binding).
6. Diluted mouse anti-HA antibody (1:100 in 5% donkey serum/PBS) is applied to the slides and incubated at 37 °C in humidified box for 1.5 h, followed by three washes with PBS, 5 min each.
7. The cells are washed with PBS three times.
8. Alexa Fluor 546 donkey anti-mouse IgG (Dako, 1:1000 in 5 % donkey serum/PBS) was applied and incubated at 37 °C in humidified box for 40 min.
9. After three washes with PBS, 5 min each, the slides are counterstained with DAPI for 3 min, followed by PBS washing three times.
10. Coverslip is applied together with fluorescent mounting medium and the staining is observed and images are taken under a confocal microscope.

As shown in Fig. 4a, HDAC7 1 was located in both cytoplasm and nucleus, while HDAC7-2 was mainly located in the cytoplasm.

3.3.3 Cellular Fraction

1. SMCs are infected with Ad-tTA or Ad-HDAC7-2 virus at 10 MOI for 48 h.

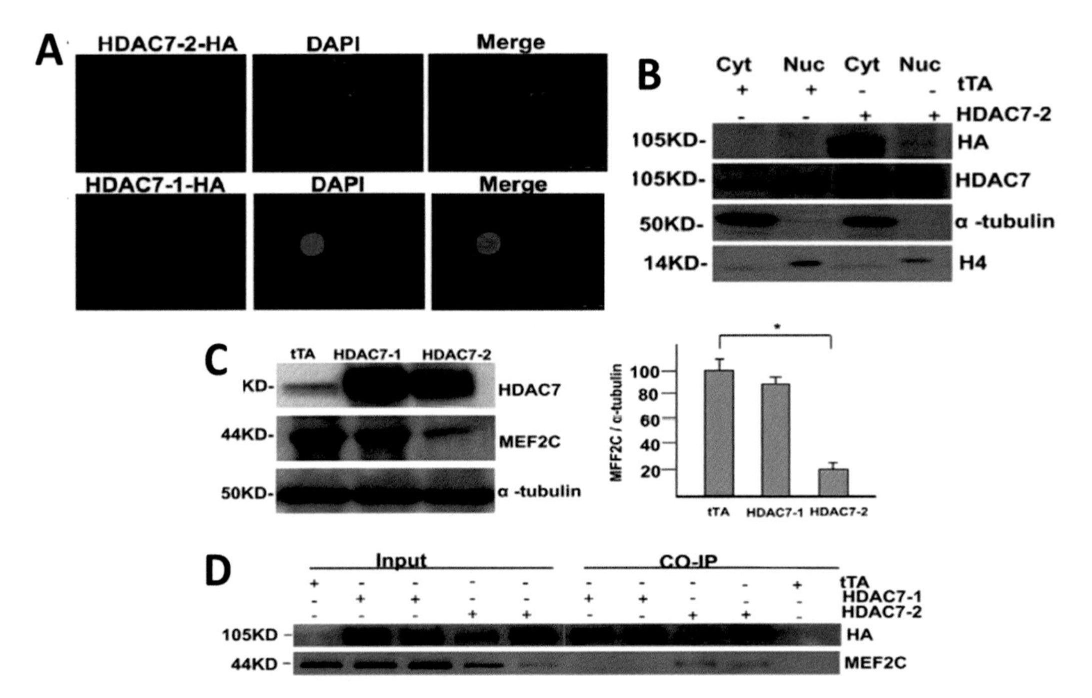

Fig. 4 Different HDAC7 isoforms show different cellular location and exert different effect on MEF2C. (**a**, **b**). HDAC7 isoforms shows different cellular location. SMCs were infected with ad-HDAC7-1, Ad-HDAC7-2 for 48 h, followed by immunofluorescence staining with anti-HA antibody (**a**) or cellular fraction and Western blot (**b**). Ad-tTA was included as control. Note that HDAC7-2 is mainly located in the cytoplasm. Scale bars: 50 μm. (**c**) HDAC7-2 decreased MEF2C protein level in SMCs as revealed by Western blot analysis. The *right panel* shows the relative amount of MEF2C against tubulin loading control. *: $p < 0/05$. € HDAC7-2 physically interacts with MEF2C as revealed by immunoprecipitation assays. Data presented are representative images or mean of three independent experiments

2. The cells are washed with ice-cold PBS containing 5% FBS, and then scratched off from the flask with rubber policemen.
3. The cells are pelleted by spinning at 6400 × g at 4 °C for 5 min.
4. Cell pellet is resuspended in 1 ml cold PBS, and transfered into 1.5 ml centrifuge tube.
5. Cells are spun down at 10,000 × g at 4 °C for 10 s, and the supernatant is thoroughly decanted.
6. The cell pellet is resuspended in 200 μl/75 ml-flask of hypotonic buffer, and incubated on ice for 15 min with vortexing briefly every 5 min.
7. 12.5 μl of 10% NP-40 is added and vortexed immediately at 20,000 × g for 10 s.
8. The mixture is spun down at 20,000 × g at 4 °C for 10 s.
9. The supernatant is transferred to a fresh tube; this is the cytoplasm fraction.

10. The nuclear pellet is washed once with ice-cold PBS and then resuspended in 70 μl of high salt buffer C and incubated on ice for 45 min with vortexing every 5 min.
11. After spinning down at 20,000 ×*g* at 4 °C for 5 min, and the supernatant is recovered. This is the nuclear extract.
12. The cytoplasm and nuclear fractions are subjected to Western blot analysis.

As shown in Fig. 4b, HDAC7-2 was mainly located in the cytoplasm, confirming the observation by immunofluorescence staining (Fig. 4a).

MEF2C is reported to play an essential role in SMC differentiation [8]. We wondered whether HDAC7-2 affected MEF2C and therefore suppressed SMC differentiation. First, we detected the MEF2C protein levels by Western blot assays following HDAC7-1 and HDAC7-2 overexpression via adenoviral gene transfer. As shown in Fig. 4c, HDAC7-2 significantly decreased MEF2C protein level. Thereafter, we tested whether there was a direct interaction between HDAC7 and MEF2C by co-immunoprecipitation.

3.3.4 Co-immunoprecipitation

1. ES cells are cultured in collagen IV-coated flasks in DM for 3 days, and then infected with Ad-tTA, Ad-HDAC7-1 or HDAC7-2 at 10 MOI and cultured for 48 h.
2. The cells are lysed as described in Western blot **steps 1–6**.
3. 1 mg whole lysates are pre-cleared with 2 μg normal mouse IgG and 10 μl of Protein G-agarose bead (Sigma) by incubation at 4 °C on a rotator for 1 h, followed by spinning at 10,000 ×*g* at 4 °C for 5 min to remove the beads.
4. The supernatant is incubated with 2 μg mouse anti-HA antibody at 4 °C on a rotator for 2 h, and 10 μl of Protein G-agarose beads is added and incubated for another 2 h.
5. The beads are pulled down by spinning at 400 ×*g* at 4 °C for 3 min, and washed five times with 1 ml PBS buffer, 5 min each.
6. The precipitate is resuspended in 25 μl of 1× SDS loading buffer, and incubated at 70 °C for 30 min, followed by spinning at 400 ×*g* at 4 °C for 3 min.
7. The supernatant is applied to Western blot probed with anti-MEF2C and HA. 50 μg cell lysate is included as input control.

As shown in Fig. 4d, HDAC7-2 but not HDAC7-1 binds to MEF2C, which might cause MEF2C degradation.

From the strategy and experiments described above, we have demonstrated that HDAC7 undergoes further splicing during ES differentiation toward SMC differentiation. The partially spliced HDAC7 mRNA produces a short HDAC7 isoform, which mainly locates in the cytoplasm and interacts with transcription factor MEF2C. The interaction may retain MEF2C in the cytoplasm and increase MEF2C degradation, therefore suppressing SMC marker gene expression. Further splicing will alleviate the production of short HDAC7 isoform, therefore abolishing the suppressive effect and leading to SMC marker expression.

Acknowledgement

This study was funded by British Heart Foundation project grant PG13-63-30419.

References

1. Sun X et al (2015) SplicingTypesAnno: annotating and quantifying alternative splicing events for RNA-Seq data. Comput Methods Programs Biomed 119(1):53–62, PubMed: 25720307
2. Lee JH, Hart SR, Skalnik DG (2004) Histone deacetylase activity is required for embryonic stem cell differentiation. Genesis 38(1):32–38, PubMed:14755802
3. Grozinger CM, Hassig CA, Schreiber SL (1999) Three proteins define a class of human histone deacetylases related to yeast Hda1p. Proc Natl Acad Sci U S A 96(9):4868–4873, PubMed: 10220385
4. Margariti A et al (2009) Splicing of HDAC7 modulates the SRF-myocardin complex during stem-cell differentiation towards smooth muscle cells. J Cell Sci 122(Pt 4):460–470, PubMed: 19174469
5. Tang Z et al (2012) Differentiation of multipotent vascular stem cells contributes to vascular diseases. Nat Commun 3:875, PubMed: 22673902
6. Haberland M et al (2009) The many roles of histone deacetylases in development and physiology: implications for disease and therapy. Nat Rev Genet 10(1):32–42, PubMed: 19065135
7. Kato HI et al (2004) Histone deacetylase 7 associates with hypoxia-inducible factor 1alpha and increases transcriptional activity. J Biol Chem 279(40):41966–41974, PubMed: 15280364
8. High FA et al (2007) An essential role for Notch in neural crest during cardiovascular development and smooth muscle differentiation. J Clin Invest 117(2):353–363, PubMed: 17273555

Chapter 8

Large-Scale Overproduction and Purification of Recombinant Histone Deacetylase 8 (HDAC8) from the Human-Pathogenic Flatworm *Schistosoma mansoni*

Martin Marek, Tajith B. Shaik, Sylvie Duclaud, Raymond J. Pierce, and Christophe Romier

Abstract

Epigenetic mechanisms underlie the morphological transformations and shifts in virulence of eukaryotic pathogens. The targeting of epigenetics-driven cellular programs thus represents an Achilles' heel of human parasites. Today, zinc-dependent histone deacetylases (HDACs) belong to the most explored epigenetic drug targets in eukaryotic parasites. Here, we describe an optimized protocol for the large-scale overproduction and purification of recombinant smHDAC8, an emerging epigenetic drug target in the multicellular human-pathogenic flatworm *Schistosoma mansoni*. The strategy employs the robustness of recombinant expression in *Escherichia coli* together with initial purification through a poly-histidine affinity tag that can be removed by the thrombin protease. This protocol is divided into two steps: (1) large-scale production of smHDAC8 in *E. coli*, and (2) purification of the target smHDAC8 protein through multiple purification steps.

Key words Histone deacetylase, Enzyme, Recombinant expression, Purification, *Schistosoma*

1 Introduction

Schistosomiasis, or bilharzia, is a parasitic disease caused by trematode flatworms of the genus *Schistosoma* (*S. mansoni*, *S. japonicum*, and *S. haematobium* are the main species of medical relevance) [1, 2]. According to the World Health Organization (WHO) statistics, schistosomes infect around 230 million people worldwide and cause at least 300,000 deaths yearly, with about 800 million people further at risk of infection [3]. The control of schistosomiasis is dependent on mass treatment with a single drug, praziquantel [4], and the consequent risk of the appearance of resistant strains raises the spectrum of widespread drug resistance. Ultimately, praziquantel-resistant schistosome strains have already been reported [5, 6], and these findings rendered the development of new anti-schistosomal drugs a strategic priority.

Sibaji Sarkar (ed.), *Histone Deacetylases: Methods and Protocols*, Methods in Molecular Biology, vol. 1436,
DOI 10.1007/978-1-4939-3667-0_8,

Schistosomes, like many eukaryotic pathogens, typically display various morphologically distinct stages during their complex life cycles. Epigenetic mechanisms fundamentally underlie the pathogens' morphological transformations, and the targeting of epigenetics-driven cellular programs therefore represents an Achilles' heel of human parasites. Today, zinc-dependent histone deacetylases (HDACs) belong to the most explored epigenetic targets, notably for anticancer therapies [7–9]. This fact significantly speeds up the search for new antiparasitic agents since drugs validated against cancers can be effectively tailored into antiparasitic therapeutics. Nevertheless, one of the key bottlenecks in antiparasitic drug discovery is recombinant production in large quantities of parasites' epigenetic targets for structure-based and pharmacological studies.

In this chapter, we describe an optimized protocol for the large-scale overproduction and purification of *Schistosoma mansoni* HDAC8 (smHDAC8), an emerging epigenetic drug target in this pathogenic organism [10–12]. Our protocol employs the robustness of the *Escherichia coli* expression system that enables cost-effective production, as well as scale up to industrial-scale fermentation production. Specifically, our protocol details the various parameters required during growth, induction and purification that enable the production of soluble smHDAC8 in milligram quantities, since standard parameters only lead to the production of this enzyme in insoluble inclusion bodies.

2 Materials

Prepare all solutions and media using ultrapure deionized water and analytical grade chemicals.

2.1 Cell Transformation

1. Expression plasmid vector pnEA/tH-smHDAC8 [10] (Fig. 1), where the full-length *smHDAC8* gene is inserted between the *Nde*I and *Bam*HI restriction sites of the pnEA-tH expression vector [13] and is in frame with a sequence encoding a C-terminal thrombin cleavage site followed by a poly-histidine affinity purification tag (*see* **Note 1**).
2. Chemically competent cells of *Escherichia coli* BL21(DE3) strain.
3. Ice bucket.
4. 42 °C water bath.
5. 2 × Luria broth (2 × LB) medium: for 1 L, weigh 20 g tryptone, 10 g yeast extract, and 20 g NaCl. Add distilled water to reach 1 L. Sterilize by autoclaving.
6. 37 °C shaking incubator.
7. LB agar.

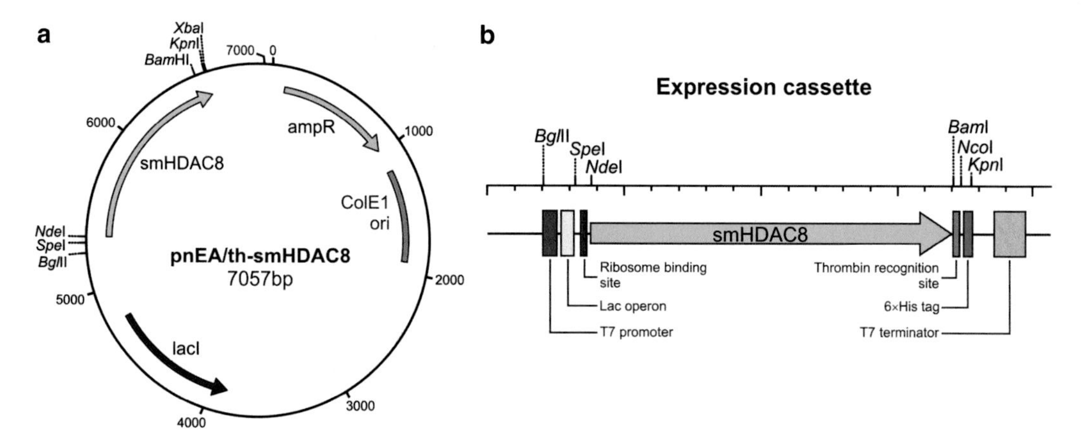

Fig. 1 Schematic representation of the plasmid used for the expression of smHDAC8 in *E. coli*. (**a**) Map of the pnEA/tH-smHDAC8 plasmid. The *smHDAC8* gene is inserted between *Nde*I and *Bam*HI restriction sites. Selection in *E. coli* is performed by the beta-lactamase ampicillin resistance gene (ampR). Origin of replication sequence (ColE1) is available for maintenance in *E. coli* cells, and the lacI gene is present for expression of the Lac repressor protein. (**b**) Details of the expression cassette. The *smHDAC8* gene is controlled by the T7 promoter and T7 terminator. The *smHDAC8* gene is cloned in frame with a sequence coding for a C-terminal thrombin cleavage site followed by a poly-histidine tag

8. Ampicillin 1000× stock solution; 100 mg/mL (in H_2O).
9. Standard petri dishes (diameter of 9 cm) and large petri dishes (diameter of 15 cm).
10. 37 °C incubator.

2.2 Cell Cultures

1. Salt medium (0.17 M KH_2PO_4, 0.72 M K_2HPO_4): for 100 mL, dissolve 2.31 g of KH_2PO_4 and 12.54 g of K_2HPO_4 in 90 mL of distilled water. Stir till the salts have dissolved, then adjust the volume of the solution to 100 mL with distilled water and sterilize by filtering.
2. Terrific-Broth (TB) rich medium: for 1 L, add 12 g tryptone, 24 g yeast extract and 4 mL glycerol to 900 mL distilled water and sterilize by autoclaving. Prior to use, add 100 mL of salt medium.
3. Ampicillin 1000× stock solution; 100 mg/mL (in H_2O).
4. 5-L flasks.
5. Thermostatic shaker Certomat BS-1.
6. Single-beam spectrophotometer model BioPhotometer Plus.
7. Isopropyl β-d-1-thiogalactopyranoside (IPTG; 1 M stock solution).
8. Zinc chloride (100 mM stock solution).
9. 1-L centrifugal bottles for Type JS4.2 rotor.
10. Centrifuge (J6MI floor model centrifuge equipped with Type JS4.2 rotor).

11. Resuspension buffer (10 mM Tris–HCl pH = 8.0; 50 mM KCl). For 1 L, dissolve 1.21 g of Trizma base and 3.73 g of KCl in 900 mL distilled water. After the buffer and salt have dissolved, adjust pH with HCl to reach a pH value of 8.0. Finally, adjust volume of the solution to 1 L with distilled water and sterilize by filtering.

2.3 Purification Steps

1. Lysis buffer (10 mM Tris–HCl pH = 8.0; 50 mM KCl). Same as the resuspension buffer used to resuspend the cell pellets at the end of the production step.
2. High-pressure homogenizer Microfluidizer Processor M-110EH.
3. Ultracentrifuge Beckman Coulter Optima L90K equipped with Type Ti-45 fixed-angle rotor.
4. Thick-wall ultracentrifuge tubes for Type Ti-45 fixed-angle rotor.
5. Talon Metal affinity resin.
6. 10-mL glass column with adaptor for use with a peristaltic pump.
7. Peristaltic pump model EP-1 Econo Pump.
8. Thrombin stock solution (1 U/μl in 25 mM Tris pH = 8.0 and 50% glycerol; kept at −20 °C).
9. Rolling mixer model RM-5.
10. Ion-exchange chromatography buffers. Low-salt buffer (10 mM Tris–HCl at pH = 8.0; 50 mM KCl) and high-salt buffer (10 mM Tris–HCl at pH = 8.0 and 1 M KCl). The low-salt buffer corresponds to the lysis buffer. The high-salt buffer is prepared as described for the lysis buffer, with the exception that 74.55 g of KCl are used to reach 1 M final concentration.
11. Polypropylene gravity-flow Econo-Pac column.
12. Bio-Rad Protein Assay reagent.
13. FPLC protein purification system.
14. 1-mL HiTrap Q FF column.
15. Gel filtration buffer: 10 mM Tris–HCl pH = 8.0; 50 mM KCl; 2 mM DTT. The buffer is prepared as the lysis buffer but is supplemented with 0.31 g of DTT prior to dissolution of the chemicals in water.
16. Column for gel filtration (16/60 Superdex 200).
17. Amicon Ultra centrifugal filter units with 30-kDa cutoff.
18. Apparatus for SDS-PAGE model Mini-PROTEAN Tetra System.
19. Single-beam spectrophotometer.

3 Methods

3.1 Cell Transformation

1. Mix 50 μl chemically competent BL21(DE3) *E. coli* cells with 50–100 ng of pnEA/tH-smHDAC8 plasmid expression vector and incubate the mix on ice for 20 min. Heat-shock the mix at 42 °C for 45 s and then incubate on ice for 2 min. Add 300 μL of 2×LB medium and incubate in a 37 °C shaker for 1 h. After incubation, spread 300 μl of bacterial suspension on a regular agar plate (9 cm) containing ampicillin (100 μg/mL). Incubate the plate overnight at 37 °C.
2. Inoculate large agar plates (15 cm) containing ampicillin (100 μg/mL) with several ampicillin-resistant colonies from the transformation in **step 1** (*see* **Note 2**). To do so, collect several colonies from the small agar plate with a kinked plastic tip and spread this inoculum homogeneously on the large agar plates. Incubate the plates overnight at 37 °C.
3. The next morning, resuspend the film-forming *E. coli* cells from the large agar plates. For resuspension, add approximately 10 mL of fresh sterile 2×LB medium per large agar plate (15 cm). Scratch the surface of the plate with a kinked Pasteur glass pipette to release the cells into the 2×LB medium. Once resuspension is done, transfer the liquid fraction into a 50-mL Falcon tube. Measure the OD_{600} value of this harvested inoculum.

3.2 Large-Scale Cultures

1. Transfer 1 L of Terrific-Broth (TB) medium in a 5-L flat-bottomed Erlenmeyer flask. Add ampicillin to reach a final concentration of 100 μg/mL. Use the harvested inoculum to inoculate the culture to start with an OD_{600} of approximately 0.2.
2. Grow the cells at 37 °C with shaking to high-density culture (OD_{600} approximately 4.0 – 6.0).
3. Add IPTG (0.5 mM final concentration) and zinc chloride (100 μM final concentration) and continue to incubate the cultures at temperature 37 °C.
4. After 1 h, recover the cells by centrifuging the cultures for 25 min at 4000×*g* at 4 °C (*see* **Note 3**) using 1 L centrifugation jars.
5. Resuspend the cell pellets in an ice-cold resuspension buffer (10 mM Tris–HCl pH = 8.0; 50 mM KCl). Use approximately 15 mL of this buffer to resuspended the cell pellet from 1 L culture. The cell suspension can be directly used for purification procedure or stored at –80 °C until further use (*see* **Note 4**).

3.3 Purification Procedure

1. The following protocol is provided considering the use of cell pellets from 3 L of cultures. First, if required, thaw the resuspended cell pellets. Adjust the volume of the cell resuspension to 40 mL

per liter of culture (i.e., final volume of 120 mL for 3 L of culture) using the lysis buffer (identical to the resuspension buffer).

2. Lyse the cell suspension using a Microfluidizer Processor at high pressure (18,000 psi) using a single round of lysis (*see* **Note 5**). After the lysis, centrifuge the disrupted cell suspension at 210,000 × *g* for 1 h and collect the supernatant in a ice-cold bottle.
3. Apply the supernatant to a column with 2 mL of Talon Metal affinity resin pre-equilibrated in lysis buffer. Briefly, connect the column with pre-equilibrated Talon resin to a peristaltic pump and pump the supernatant from **step 1** through the column with Talon resin at a flow rate of 4.0–5.0 mL/min. Every 30 min, disconnect the column from the peristaltic pump and mix the resin to release the excess pressure. After the loading, wash the column extensively with approximately 100 mL of the lysis buffer to remove nonspecifically bound proteins (*see* **Note 6**).
4. Release the smHDAC8 enzyme from the Talon resin by thrombin treatment. Briefly, resuspend the Talon resin with bound smHDAC8-His fusion protein with the lysis buffer and transfer it to a new sterile 15-mL Falcon tube. The volume of the resin suspension should be approximately 5 mL. Add 60 μl of thrombin (1U/μl) and place the tube on a rolling mixer overnight at 4 °C (*see* **Note 7**).
5. Next morning, separate the released smHDAC8 protein from the Talon resin particles by applying the resin suspension onto an Econo-Pac column and collect the unbound flow-through fraction into a fresh sterile 15-mL Falcon tube. Wash the resin with additional 3 mL of the lysis buffer to harvest all thrombin-released smHDAC8 enzyme. Check the presence and concentration of smHDAC8 in the flow-through fraction by the Bio-Rad Protein Assay.
6. Load the flow-through containing smHDAC8 enzyme from the **step 4** onto a 1-mL HiTrap Q FF column pre-equilibrated with low-salt ion-exchange chromatography buffer (10 mM Tris–HCl pH = 8.0; 50 mM KCl). Elute the bound protein with a gradient of KCl (50 mM to 1 M KCl): *see* Fig. 2a for a typical ion-exchange purification of smHDAC8. Identify fractions containing the smHDAC8 protein by SDS-PAGE.
7. Pool the peak fractions from the ion-exchange chromatography from **step 5** and load this sample onto a gel filtration column (16/60 Superdex 200) equilibrated with gel filtration buffer (*see* **Note 8**). Identify fractions containing the target protein by SDS-PAGE. *See* Fig. 2b for a typical gel filtration purification of smHDAC8.
8. Pool the peak fractions from gel filtration from **step 6**, and concentrate the smHDAC8 protein with an Amicon Ultra

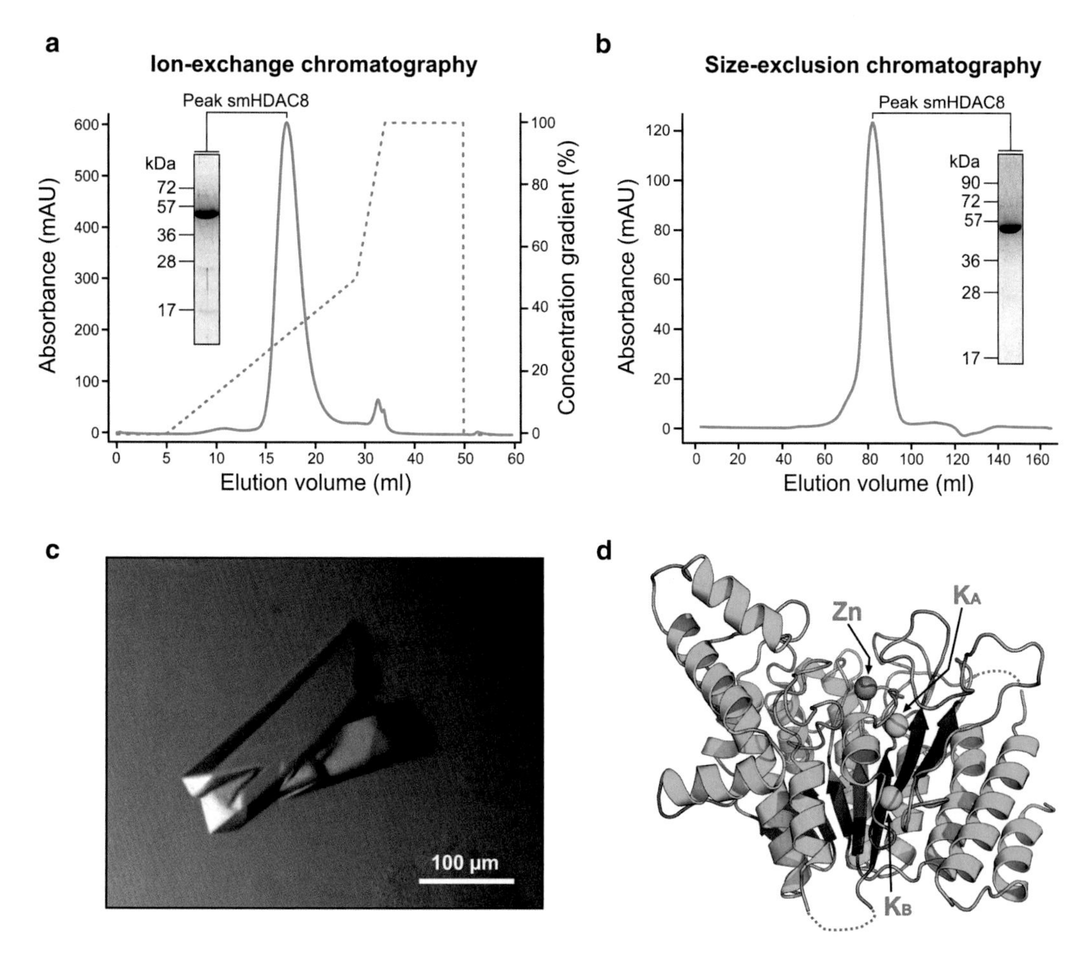

Fig. 2 Purification and crystallization of smHDAC8. (**a**) Chromatogram of ion-exchange purification of smH-DAC8. The gradient used for this purification step is displayed. (**b**) Chromatogram of gel filtration purification of smHDAC8. (**c**) Picture of an smHDAC8 crystal. (**d**). Atomic structure of smHDAC8 represented as ribbons. *Orange sphere* catalytic zinc ion, *blue spheres* potassium ions (K_A and K_B)

centrifugal filter unit to reach a final concentration of 2.5 mg/mL (*see* **Note 9**). Check purity of the purified smHDAC8 enzyme by SDS-PAGE and determine protein concentration by the Bio-Rad Protein Assay reagent (*see* **Note 10**).

9. Flash-freeze the final product with liquid nitrogen and store at –80 °C (*see* **Note 11**).

4 Notes

1. Initial affinity purification of smHDAC8 is suboptimal in presence of an N-terminal poly-histidine tag. Therefore, a C-terminal poly-histidine tag is used. The presence of a *Bam*HI cloning site and a sequence encoding a thrombin protease cleavage site

before the sequence encoding the poly-histidine tag leaves several residues at the C-terminus of smHDAC8 after thrombin cleavage of the poly-histidine tag: GSLVPR. These residues do not affect the enzyme's activity [10].

2. The expression yields for a protein are often better when using as starter for large cultures colonies that have grown on petri dishes rather than a liquid preculture. This is particularly true when using ampicillin resistance. Since smHDAC8 is not produced in large quantities in *E. coli*, the use of colonies grown on petri dishes as starters is preferred. To have sufficient colonies for inoculating the large cultures, streaking of the initial transformed colonies on larger petri dishes is carried out. In general, one large petri dish is sufficient to inoculate one liter of large liquid culture.
3. Do not exceed 1 h incubation in the presence of IPTG: when too much smHDAC8 enzyme is produced in a cell, it goes into insoluble inclusion bodies. This is essential since during our pilot experiments longer incubation times led to an almost complete loss of soluble protein. Reducing the expression time decreased the quantity of smHDAC8 proteins produced per cell and the formation of inclusion bodies. Use of Terrific Broth medium with induction at high density compensates for the small quantity of smHDAC8 present in each cell.
4. The cell paste is immediately resuspended in a buffer that corresponds to the lysis buffer. This avoids resuspension of the frozen cell paste in the lysis buffer at the beginning of the purification procedure. Indeed, this resuspension appears more deleterious to the solubility of smHDAC8 than when the cells are resuspended in the lysis buffer prior to freezing. A possible explanation for this behavior is that thawing the already resuspended cells is less stringent. Importantly, we and others [14] noted that zinc-dependent HDACs lose their enzymatic activities very fast when inappropriately handled. We therefore recommend the resuspension of the cell pellet immediately after centrifugation, and storing the cell suspension at −80 °C promptly to minimize loss of enzyme activity. Note also that long storage of the cell pellets at −80 °C negatively impacts the quality of the smHDAC8 protein: old cell pellets lead to purified protein that appears well-behaved by SDS-PAGE analysis, but that has reduced enzymatic activity and poorer behavior during crystallization.
5. The use of a microfludizer for cell lysis is essential for keeping smHDAC8 soluble during lysis. Use of sonication leads to the loss of half of the protein yield compared to the use of the microfludizer.
6. The binding buffer (10 mM Tris–HCl pH = 8.0; 50 mM KCl) may be enriched with 5 mM imidazole to increase the stringency of washing and to eliminate protein contaminants. This

procedure causes however some loss of smHDAC8 due to partial elution from the affinity resin.

7. In this step, thrombin specifically cuts the fusion protein at the thrombin recognition site located between smHDAC8 and the poly-histidine tag. This cleavage results in the release of smHDAC8 enzyme from the affinity resin, whilst the poly-histidine tag and the remaining nonspecifically bound contaminants stay bound to the Talon resin. Using this procedure a good level of purity of the smHDAC8 enzyme is already obtained at the first purification step.
8. We noted that the presence of DTT in the gel filtration buffer may cause a problem (decreased enzymatic activity) in some high-throughput deacetylase screening assays. To avoid this problem, it is possible to replace DTT (2 mM) with TCEP (1 mM) in the gel filtration buffer.
9. We usually concentrate the smHDAC8 protein to 2.5 mg/mL with no observable protein precipitation/aggregation. This purification procedure yields an smHDAC8 enzyme exhibiting deacetylase activity similar, if not identical, to its human counterpart, human HDAC8 [10]. In addition, crystallization of the smHDAC8 at this concentration yielded diffraction-quality crystals that enabled structure determination of this enzyme (Fig. 2c, d) [10].
10. Our protocol typically yields 1.0–1.5 mg of highly pure smHDAC8 per 1 L of bacterial culture.
11. The purified smHDAC8 enzyme shows rather quickly a decrease of its activity and of its propensity to crystallize when kept for 3–4 days at 4 °C. Storage at −80 °C slows down but does not stop this process. Therefore, it is advisable to use the enzyme relatively rapidly even when stored at −80 °C.

Acknowledgements

This work and the authors of this manuscript have been supported by funding from the European Union's Seventh Framework Programme for research, technological development and demonstration under grant agreements nos. 241865 (SEtTReND) and 602080 (A-ParaDDisE). The authors are supported by institutional funds from the Centre National de la Recherche Scientifique (CNRS), the Institut National de la Santé et de la Recherche Médicale (INSERM), the Université de Strasbourg and the Université de Lille 2, the French Infrastructure for Integrated Structural Biology (FRISBI; ANR-10-INSB-05-01), and by Instruct as part of the European Strategy Forum on Research Infrastructures (ESFRI).

References

1. Brown M (2011) Schistosomiasis. Clin Med 11(5):479–482
2. Ross A, Bartley P, Sleigh A et al (2002) Schistosomiasis. N Engl J Med 346(16): 1212–1220
3. Gray D, Ross A, Li Y, Mcmanus D (2011) Diagnosis and management of schistosomiasis. BMJ 342:2651
4. Dömling A, Khoury K (2010) Praziquantel and schistosomiasis. ChemMedChem 5(9): 1420–1434
5. Doenhoff M, Cioli D, Utzinger J (2008) Praziquantel: mechanisms of action, resistance and new derivatives for schistosomiasis. Curr Opin Infect Dis 21:659–667
6. Doenhoff M, Kusel J, Coles G, Cioli D (2002) Resistance of *Schistosoma mansoni* to praziquantel: is there a problem? Trans R Soc Trop Med Hyg 96(5):465–469
7. Li Z, Zhu W (2014) Targeting histone deacetylases for cancer therapy: from molecular mechanisms to clinical implications. Int J Biol Sci 10(7):757–770
8. Campbell R, Tummino P (2014) Cancer epigenetics drug discovery and development: the challenge of hitting the mark. J Clin Investig 124(1):64–69
9. West A, Johnstone R (2014) New and emerging HDAC inhibitors for cancer treatment. J Clin Investig 124(1):30–39
10. Marek M, Kannan S, Hauser A et al (2013) Structural basis for the inhibition of histone deacetylase 8 (HDAC8), a key epigenetic player in the blood fluke *Schistosoma mansoni*. PLoS Pathog 9(9), e100364
11. Stolfa D, Marek M, Lancelot J et al (2014) Molecular basis for the antiparasitic activity of a mercaptoacetamide derivative that inhibits histone deacetylase 8 (HDAC8) from the human pathogen *Schistosoma mansoni*. J Mol Biol 426(20):3442–3453
12. Kannan S, Melesina J, Hauser A et al (2014) Discovery of inhibitors of *Schistosoma mansoni* HDAC8 by combining homology modeling, virtual screening, and in vitro validation. J Chem Inf Model 54(10):3005–3019
13. Diebold M-L, Fribourg S, Koch M, Metzger T, Romier C (2011) Deciphering correct strategies for multiprotein complex assembly by co-expression: application to complexes as large as the histone octamer. J Struct Biol 175(2):178–188
14. Olson D, Udeshi N, Wolfson N et al (2014) An unbiased approach to identify endogenous substrates of "histone" deacetylase 8. ACS Chem Biol 9(10):2210–2216

Chapter 9

Visualization of HDAC9 Spatiotemporal Subcellular Localization in Primary Neuron Cultures

Noriyuki Sugo and Nobuhiko Yamamoto

Abstract

Histone deacetylase (HDAC) 9 is one of class IIa HDACs which are expressed in developing cortical neurons. The translocation of HDAC9 from the nucleus to the cytoplasm is induced by neuronal activity during postnatal development, and is involved in regulation of various gene expressions. Visualization of HDAC9 subcellular localization is a powerful tool for studying activity-dependent gene expression. Here, we describe a time-lapse imaging method using fluorescent protein-tagged HDAC9 in dissociated cortical neurons. This method reveals dynamic HDAC9-mediated gene expression in response to various signals.

Key words HDAC9, cDNA cloning, Primary cortical neuron cultures, Time-lapse imaging, Transfection, Fluorescent imaging

1 Introduction

HDAC9 is involved in class IIa HDACs in mammalian cells [1], and has multiple alternative spliced isoforms [2]. A well-characterized splice variant known as HDAC9ΔCD/MEF2-interacting transcriptional repressor (hereafter referred to as HDAC9) does not have HDAC catalytic domain, but can mediate gene expression with HDAC1 and HDAC3 [3–5]. HDAC9 is expressed not only in the adult brain, skeletal and cardiac muscle [2, 6] but also in the developing brain [7].

HDAC9 translocates between the nucleus and cytoplasm in response to neuronal activity, similarly to other class IIa HDACs [6–8]. As HDAC9 inhibits gene expression by compacting chromatin, it is thought to act as an activity-dependent repressor. Indeed, activity-dependent translocation of HDAC9 increases expression of c-fos, an immediate early gene, and promotes dendrite growth [7]. In this chapter, we show a cloning method of mouse *Hdac9* cDNA by PCR and real-time imaging with EGFP-tagged HDAC9 in living cortical neuron cultures. Live cell imaging with fluorescent proteins is an excellent method to monitor

Sibaji Sarkar (ed.), *Histone Deacetylases: Methods and Protocols*, Methods in Molecular Biology, vol. 1436,
DOI 10.1007/978-1-4939-3667-0_9, © Springer Science+Business Media New York 2016

spatiotemporal behavior of the molecule. This approach enables to visualize subcellular localization of HDAC9 underlying chromatin regulation of gene expression.

2 Materials

2.1 PCR

1. KOD DNA polymerase (KOD -Plus-, TOYOBO).
2. PCR buffer (10×) as supplied by the manufacturer of the DNA polymerase (TOYOBO).
3. 25 mM $MgSO_4$ solution.
4. 2 mM dNTPs mixture.
5. Template DNA: First-strand cDNA synthesized by reverse transcriptase with total RNA from mouse brain at embryonic day (E) 12.5.
6. Forward and reverse primers (100 μM) in H_2O.
7. 0.2 ml thin-wall tube.
8. DNA size standard: Lambda DNA-*Hind*III Digest (e.g., New England Biolabs).
9. Gel-loading buffer (6×).
10. Thermal cycler.
11. Microwave oven.
12. TAE buffer (50×): Dissolve Tris base 242 g in 700 ml H_2O. Add 57.1 ml of Glacial acetic acid and 100 ml of 0.5 M EDTA (pH 8.0). Adjust the final volume to 1 L with H_2O.
13. Agarose gel (0.8 % agarose–TAE): Add 0.8 g agarose in 200 ml TAE buffer in an Erlenmeyer flask. Heat the slurry in a microwave oven until the agarose dissolves. Pour the warm agarose solution into the mold.
14. Equipment for agarose gel electrophoresis.
15. Power supply device.
16. Ethidium bromide solution: 0.5 μg/ml in TAE buffer.
17. Image capturing system (e.g., UV illuminator and CCD camera).
18. Gel cleanup kit (e.g., Wizard SV Gel and PCR Clean-UP System, Promega).
19. TA cloning kit (e.g., pGEM-T Easy Vector System, Promega).

2.2 EGFP Fusion Protein Construction

The coding region of mouse *Hdac9* cDNA was fused to *egfp* cDNA in frame at the 3′-terminus to generate EGFP tagged HDAC9 expression vector [7]. The *Xho*I-*Sma*I fragment of *Hdac9* cDNA digested from the cloning vector was cloned to the multi-cloning site of CAG promoter [9] driven EGFP C1 plasmid (Clontech) (Fig. 1, *see* **Note 1**).

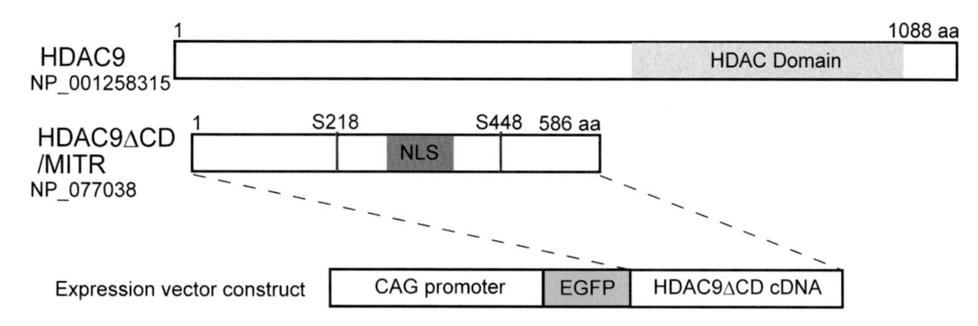

Fig. 1 Schematic depictions of HDAC9 isoforms and expression vector construct. Each isoform contains a nuclear localization signal (NLS) and phosphorylation sites of Ser-218 and Ser-418 associated with the subcellular localization. HDAC9ΔCD/MEF2-interacting transcriptional repressor (MITR) lacks a carboxyl-terminal catalytic domain (HDAC Domain)

2.3 Cortical Neuron Dissociation Culture

1. Cell culture medium: DMEM/F12 (11320-033, GIBCO, Life Technologies) supplemented with 10 % fetal bovine serum and B27 supplement (17504-044, GIBCO, Life Technologies).
2. Hanks' balanced salt solution (HBSS).
3. Ca^{2+}, Mg^{2+} free phosphate buffered saline (PBS).
4. 0.125 % trypsin and 0.02 % EDTA in PBS.
5. 4-well culture dish (176740, Nunc, Thermo Scientific).
6. 35 mm petri dish.
7. 100 mm petri dish.
8. Fine forceps.
9. Surgical scissors.
10. 0.1 mg/ml poly-L-ornithine solution: Dissolve poly-L-ornithine hydrobromide (Sigma-Aldrich, P3655) in 0.15 M borate buffer (pH8.5) to make 1.0 mg/ml solution. Then, dilute ten times with H_2O and sterilize by 0.22 μm pore size membrane filter. Store at 4 °C.
11. 15 ml conical centrifuge tube.
12. Glass Pasteur pipette.
13. Water bath at 37 °C.
14. Automatic cell counter (e.g., Logos Biosystems).
15. Binocular microscope.
16. Centrifuge.

2.4 Transfection

1. Lipofectamine 2000 transfection regent (11668-019, Invitrogen, Life Technologies).
2. Opti-MEM (31985-062, GIBCO, Life Technologies).
3. Transfection-grade plasmid DNA purification kit (e.g., PureLink HiPure Plasmid Filter Maxiprep kit, K2100-16, Invitrogen, Life Technologies).

2.5 Pharmacological Treatment

1. KCl depolarization solution: 170 mM KCl, 1.3 mM $MgCl_2$, and 0.9 mM $CaCl_2$, and 10 mM Hepes (pH 7.4). Store at 4 °C.
2. NaCl solution: 170 mM NaCl, 1.3 mM $MgCl_2$, and 0.9 mM $CaCl_2$, and 10 mM Hepes (pH 7.4). Store at 4 °C.
3. Tetrodotoxin (TTX; Wako Pure Chemical Industries, Osaka, Japan): 0.5 mg TTX is dissolved in 313 μl H_2O to make 5 mM solution and sterilized by 0.22 μm pore size membrane filter. Store at −20 °C.

2.6 Imaging Systems

1. Inverted fluorescence microscope (e.g., IX71 with 10×/0.3, 20×/0.45, or 40×/0.6 objective lens, Olympus) attached with a heating box maintained at 37 °C and a CCD camera (e.g., DP70, Olympus).
2. Software for adjusting brightness and contrast of the obtained images (e.g., Adobe Photoshop or ImageJ).

3 Methods

3.1 Hdac9 cDNA Cloning

1. To obtain a coding region of mouse *Hdac9* cDNA (GenBank accession number AF324492, 1.8 kb), a pair of primers was designed as follows.

   Forward primer carrying an *Xho*I site (bold): 5′-**CTCGAG**GAATGCACAGTATGATCAGC.

   Reverse primer carrying a *Sma*I site (bold): 5′-**CCCGGG**TGCCACCTATCTTATACTC.
2. Set up the following mixture in a 0.2-ml thin-walled amplification tube (Table 1, *see* **Note 2**):
3. Carry out PCR amplification in a thermal cycler by the following program (Table 2).

Table 1
Composition of reaction mixture for PCR

Component	Amount
Amplification buffer (×10), supplied by the manufacturer of the DNA polymerase	5 μl
2 mM dNTPs	5 μl
25 mM $MgSO_4$	2 μl
Forward primer (10 μM)	1.5 μl
Reverse primer (10 μM)	1.5 μl
KOD DNA polymerase	1 μl
Template DNA	1 μl
H_2O	To a final volume of 50 μl

Table 2
PCR condition

Cycle number	Denaturation	Annealing	Polymerization
1	2 min at 94 °C		
30 cycles	15 s at 94 °C	30 s at 60 °C	2 min at 68 °C
Last cycle			7 min at 68 °C

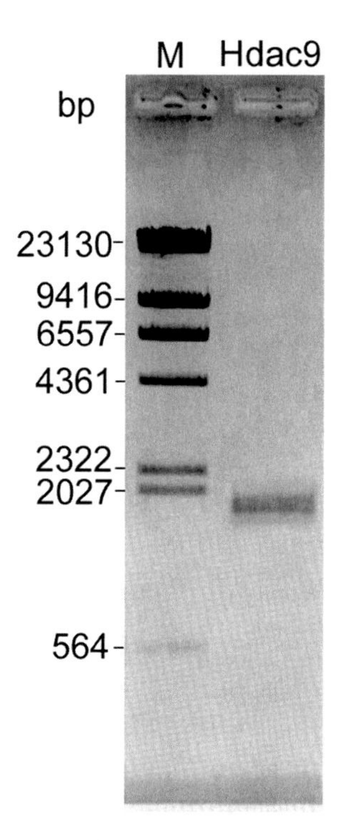

Fig. 2 PCR amplified *Hdac9* cDNA fragments in the gel. *Lane 1* (M) shows λ/*Hind*III DNA marker. *Lane 2* (Hdac9) shows the fragment of the cDNA

4. Mix 20 μl of PCR product with 4 μl the loading dye. Load the solution into the slots of the submerged 0.8 % agarose gel in a chamber. Load λ/*Hind*III DNA size marker into a slot on the same gel.
5. Apply a voltage of 100 V till the dye front has reached the bottom of the gel.
6. Stain the gel in the ethidium bromide solution at RT for 30 min.
7. Capture the image of the gel under UV transillumination. Dissect an 1.8 kb DNA fragment corresponded to *Hdac9* cDNA in a minimal volume of agarose gel using a razor blade (Fig. 2).

8. Recover DNA from the cut gel using a gel cleanup kit.
9. Clone the PCR products into a plasmid vector by using e.g., TA cloning Kit (*see* **Note 3**).
10. Confirm the cDNA sequence integrity.

3.2 Primary Cortical Neuron Cultures

1. Coat 4-well culture plates with the poly-L-ornithine solution at RT for 1 h.
2. Rinse the dishes with 0.5 ml of sterilized water three times, and then dry them completely. Store the coated dishes in a dark place at RT until use.
3. Under a binocular microscope, dissect cortical lobes from 8 to 15 mouse fetuses at E16 in a 100 mm petri dish with 15 ml of ice-cold HBSS.
4. Transfer the dissected cortical lobes to a 35 mm petri dish with 0.5 ml of ice-cold PBS and mince them with fine scissors.
5. Transfer the minced tissues to a 15 ml conical centrifuge tube with 3.5 ml 0.125% trypsin and 0.02% EDTA in PBS, and incubate them in a water bath for 5 min at 37 °C.
6. Add 4 ml of the cell culture medium to inhibit the action of trypsin, and dissociate the tissues thoroughly by pipetting 30–50 times with a fire-polished glass Pasture pipette. Remove the remaining tissue at the bottom of the tube.
7. Pellet the cells by centrifugation (200 × *g*, 5 min) and remove the supernatant. Resuspend the pellet gently in 4 ml of the cell culture medium. Repeat this step again to remove trypsin/EDTA completely.
8. Count the number of cells using a cell counter.

3.3 Transfection

1. Purify the plasmid by using transfection-grade plasmid DNA purification kits, and finally dissolve in sterile PBS (approx. conc. 5–10 μg/μl). Store at −20 °C until use.
2. Dilute 1.6 μg of the plasmid DNA with 50 μl Opti-MEM I medium.
3. Dilute 2 μl of Lipofectamine 2000 with 50 μl of Opti-MEM I medium. After incubation at RT for 5 min, combine the diluted DNA with the diluted Lipofectamine 2000. Mix gently and incubate at RT for 20 min.
4. Transfer the cell suspension containing $1–5 \times 10^5$ cells with the cell culture medium to a sterile 1.5 ml centrifuge tube. Centrifuge the cells (200 × *g*, 5 min).
5. Remove the supernatant and resuspend the pellet gently with 100 μl of Opti-MEM I medium containing DNA–Lipofectamine 2000 complexes, and placed the mixture on the center of the poly-L-ornithine-coated wells (*see* **Note 4**).

6. Incubate them at 37 °C for 1 h in an environment of 5 % CO_2 and humidified 95 % air (*see* **Note 5**).
7. Replace the transfection medium with 0.5 ml of the cell culture medium after the incubation and place in a CO_2 incubator until observation (*see* **Note 6**).

3.4 Live Cell Imaging

1. Place the 4-well culture plate on the inverted fluorescence microscope with an Olympus x40 objective lens, a filter set optimized for EGFP fluorescence and a heating box maintained at 37 °C.
2. Observe the cell morphology by phase-contrast images.
3. Optimize exposure time for image acquisition with a CCD camera. Capture all images of EGFP-HDAC9 transfected cells as quickly as possible to reduce photodamage, and then put back these cultures into a CO_2 incubator (*see* **Note 7**).
4. The subcellular localizations of EGFP-HDAC9 in the transfected cortical neurons are categorized into three classes showing mainly nuclear (N>C), both nuclear and cytoplasmic (roughly equal, N=C) and mainly cytoplasmic (N<C) localization, based on the differences in fluorescence intensities between the nucleus and the cytoplasm (*see* **Note 8**).
5. Perform these observations 3–6 times for 5–24 h after TTX or KCl treatment (*see* **Note 9**).

3.5 Pharmacological Treatment

1. Add pre-warmed 205 μl (0.41 volumes) of KCl (final conc. 50 mM) depolarization solution at 37 °C to 0.5 ml of the culture medium in the 4-well culture plate. Treat the control cultures with the pre-warmed same volumes of NaCl solution (*see* **Note 10**).
2. Add the sodium channel blocker TTX solution into the culture medium (final conc. 100 nM), to block action potentials in the cultured neurons (*see* **Note 11**).

4 Notes

1. Other fluorescent protein expression vectors are also available for tagging HDAC9.
2. High-fidelity DNA polymerases are also available from various manufacturers.
3. As KOD DNA polymerase has 3′–5′ exonuclease activity for proof reading, it is necessary to add a deoxyadenosine to the 3′ end of PCR products with Target Clone™ (TOYOBO), TA cloning kit.
4. Make a droplet on the culture dish to increase cell density for survival. This modified transfection method with Lipofectamine

2000 improves transfection efficiency for primary cortical neurons, compared to the manufacturer's instruction. If you need a larger scale of transfection, the transfection regents should be increased according to the number of cells.

5. The transfected cells will adhere to the dish for the incubation period.
6. One day after the transfection, EGFP-HDAC9 expression must be detected in the transfected cells.
7. To reduce photodynamic damage, it would be better to adjust light intensity with ND filters.
8. EGFP-HDAC9 is localized in the nucleus in most transfected cells until 5 days in vitro (DIV) (Fig. 3a). After 10 DIV when axonal processes and dendrites have formed, the tagged protein is detected in both nucleus and cytoplasm, although the remaining population still exhibits nuclear localization. After 14 DIV, when cortical cells have more elaborate dendrites with spines, EGFP-HDAC9 is present mostly in the cytoplasm (Fig. 3b).
9. Interval time for time-lapse imaging should be more than 2 h to reduce damage to the transfected cells. Use a microscope stage top incubator (e.g., Tokai Hit), if you need more frequent observations with a shorter interval.
10. Treat the cultures with the high KCl solution within 7 DIV when EGFP-HDAC9 predominantly localizes in the nucleus,

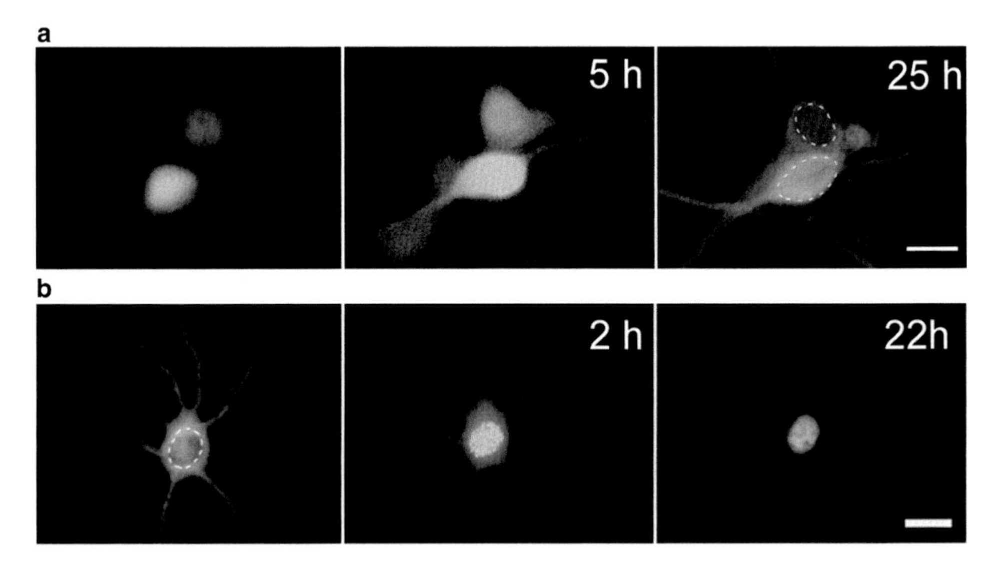

Fig. 3 Time-lapse imaging in EGFP-HDAC9 transfected cortical neurons after pharmacological treatments. (**a**) EGFP-HDAC9-expressing cortical neurons (5 DIV) were observed at 5 and 25 h after the addition of 50 mM KCl treatment. Scale bar: 10 μm. (**b**) Time-lapse imaging of EGFP-HDAC9-expressing cortical neurons (14 DIV) was performed at 2 and 22 h after TTX application (100 nM). A *dotted circle* indicates the nucleus. Scale bar: 20 μm

because spontaneous neuronal activity is much lower until 10 DIV in the present cultures. After addition of the KCl solution into the culture medium, EGFP-HDAC9 would gradually translocate from the nucleus to the cytoplasm in 24 h. The symptom of cytoplasmic localization should be detected within 5 h (Fig. 3a).

11. Spontaneous neuronal activity is increased after 10 DIV, when EGFP-HDAC9 predominantly localizes in the cytoplasm. After adding TTX to the culture medium, EGFP-HDAC9 would gradually translocate from the cytoplasm to the nucleus in 3 h. The symptom of nuclear localization would be found within 2 h (Fig. 3b).

Acknowledgment

This work was supported by JSPS KAKENHI Grant Nos. 16700286 and 18300105 (N.S. & N.Y.) and by The Nakajima Foundation (N.S.).

References

1. Yang XJ, Grégoire S (2005) Class II histone deacetylases: from sequence to function, regulation, and clinical implication. Mol Cell Biol 25(8):2873–2884
2. Zhou X, Marks PA, Rifkind RA, Richon VM (2001) Cloning and characterization of a histone deacetylase, HDAC9. Proc Natl Acad Sci U S A 98(19):10572–10577
3. Sparrow DB, Miska EA, Langley E, Reynaud-Deonauth S, Kotecha S, Towers N et al (1999) MEF-2 function is modified by a novel co-repressor, MITR. EMBO J 18(18):5085–5098
4. Zhou X, Richon VM, Rifkind RA, Marks PA (2000) Identification of a transcriptional repressor related to the noncatalytic domain of histone deacetylases 4 and 5. Proc Natl Acad Sci U S A 97(3):1056–1061
5. Zhang CL, McKinsey TA, Olson EN (2001) The transcriptional corepressor MITR is a signal-responsive inhibitor of myogenesis. Proc Natl Acad Sci U S A 98(13):7354–7359
6. Méjat A, Ramond F, Bassel-Duby R, Khochbin S, Olson EN, Schaeffer L (2005) Histone deacetylase 9 couples neuronal activity to muscle chromatin acetylation and gene expression. Nat Neurosci 8(3):313–321
7. Sugo N, Oshiro H, Takemura M, Kobayashi T, Kohno Y, Uesaka N et al (2010) Nucleocytoplasmic translocation of HDAC9 regulates gene expression and dendritic growth in developing cortical neurons. Eur J Neurosci 31(9):1521–1532
8. Chawla S, Vanhoutte P, Arnold FJ, Huang CL, Bading H (2003) Neuronal activity-dependent nucleocytoplasmic shuttling of HDAC4 and HDAC5. J Neurochem 85(1):151–159
9. Niwa H, Yamamura K, Miyazaki J (1991) Efficient selection for high-expression transfectants with a novel eukaryotic vector. Gene 108(2):193–199

Chapter 10

Expression and Function of Histone Deacetylase 10 (HDAC10) in B Cell Malignancies

John Powers, Maritza Lienlaf, Patricio Perez-Villarroel, Susan Deng, Tessa Knox, Alejandro Villagra, and Eva Sahakian

Abstract

Histone deacetylase 10 (HDAC10) belongs to the class IIb HDAC family and its biological role remains mostly unidentified. A decreased HDAC10 expression has been reported in patients with aggressive solid tumors (Osada et al. Int J Cancer 112: 26–32, 2004; Jin et al. Int J Clin Exp Pathol 7: 5872–5879, 2014), suggesting that loss of HDAC10 expression might confer a survival advantage to malignant cells. Consequently, results from our lab suggests that overexpression of HDAC10 in aggressive mantle cell lymphoma (MCL) and chronic lymphocytic leukemia (CLL) Z138c and MEC1 cells, respectively, resulted in a rapid induction of cell death in vitro with only 5 % of cells being alive at 48 h, cell cycle arrest, and up-regulation of co-stimulatory molecules. Here we present several standard methods to study the function of HDAC10 in B cell malignancies.

Key words Histone deacetylase 10, qRT-PCR, PCR, Adenoviral overexpression, Genotyping, Western blotting, High-throughput flow screening

1 Introduction

Initially, the biological role of histone deacetylases (HDACs) was known to be limited to their effect on histones; however, in the past several years many groups have shown that their function encompasses more complex regulatory roles. These functions appear to be tissue specific, dependent on the cellular compartment and stage of cellular differentiation [1–3]. Among the HDACs, HDAC10 is a novel class IIb histone. This subclass of HDACs (HDAC6 and 10) are categorized due to their structural similarly where they both contain two deacetylase domains. However, what set these two HDACs apart is that the second deacetylase domain in HDAC10 is incomplete and nonfunctional [4]. HDAC10 is ubiquitously expressed and is found in both cytoplasm and the nucleus; additionally it derives as two spliced transcription variants which encode residues 658 and 669 [5–7].

Sibaji Sarkar (ed.), *Histone Deacetylases: Methods and Protocols*, Methods in Molecular Biology, vol. 1436,
DOI 10.1007/978-1-4939-3667-0_10,

Recently, several reports have suggested that expression of HDAC10 play a critical role in tumor suppression. For example, overexpression of HDAC10 in cervical cancer cells robustly inhibits metastasis through inhibition of matrix metalloproteinase (MMP) 2 and 9 expression [8]. Others have also demonstrated that decreased expression of HDAC10 in lung cancer [9] and gastric cancer [10] predict adverse and poor prognosis for the patients. In this chapter, we demonstrate several techniques to detect and manipulate the expression of this HDAC in MCL and CLL [11].

2 Materials

2.1 Animal Models

HDAC10 constitutively knock out (HDAC10KO) mice are available for purchase though TACONIC (Cat. # TF1813 and nomenclature MGI:2158340). This mouse has a 129/SvEv-C57BL/6 background [7, 12]. Primers for genotyping are available through TACONIC Biosciences.

2.2 General Reagents and Buffers

1. 10× phosphate buffered saline (PBS) 7.4 pH: 900 mL ddH_2O, 80 g NaCl, 2 g KCl, 14.4 g Na_2HPO_4 and 2.4 g KH_2PO_4. pH to 7.4 with HCl and bring volume up to 1 L. Autoclave and filter.
2. Tissue culture treated plates; 6, 12, and 24 wells.
3. Round bottom 96-well plates.
4. Propidium iodide (PI): Dilute to (1 mg/mL) with 1× PBS.
5. Chicken erythrocyte nuclei singlets (CEN).
6. Fc Block: anti-CD32 and anti-CD16 functional grade with a final antibody concentration of 0.5 μg/mL each.
7. Flow cytometric compensation beads.
8. Flow cytometry calibration beads: Unstained, mid-range and high.
9. Fixable yellow dye (405 excitation/575 emission).
10. Heat-inactivated human AB sera.
11. Trypsin: trypsin 0.25 %, 2.21 mM EDTA in 1× HBSS without sodium bicarbonate, calcium, and magnesium.
12. 225 serologic flask filter cap.
13. Cell scrapers.
14. 10 mM Tris pH 8.0.
15. Western blot lysis buffer: 8 mL ddH_2O, 280 mM NaCl, 50 mM Tris–HCl pH 8.0, 0.5 % Igepal, 5 mM $MgCl_2$, 10 % glycerol, and one tablet protease inhibitor adjust to 10 mL.
16. Blocking buffer for western blot: 5 % powdered nonfat Milk, 0.1 % Tween diluted in PBS.

17. Bicinchoninic acid protein assay kit (BCA).
18. Protein ladder: range of 10–250 kDa.
19. Anti-human/mouse HDAC10 antibody (Clone EPR3576).
20. 10× Tris–glycine–SDS (TGS) buffer: 900 mL ddH2O, 30.3 g Tris base, 144.0 g glycine, 10.0 g sodium dodecyl sulfate, adjust volume to 1 L and 0.22 μM (filter-sterilize).
21. Commercially available preassembled transfer kit with PVDF nitrocellulose.
22. 685 nm excitation/715 nm emission goat anti-mouse antibody.
23. 785 nm excitation/815 nm emission goat anti-rabbit antibody.
24. 10% Precast Polyacrylamide gels.
25. 5.0 M NaCl: 29.22 g NaCl in 100 mL water. Autoclave.
26. 1.0 M $MgCl_2$: 20.33 g $MgCl_2$ in 100 mL water. Autoclave.
27. Molecular grade glycerol.
28. 1.0 M Tris–HCl (pH 8.0): 12.14 g Tris–HCl in 90 mL water, pH to 8.0, bring volume to 100 mL and autoclave.
29. 10× Hanks balanced salt solution (HBSS): 40 g NaCl, 2 g KCl, 5 g glucose, 300 mg KH_2PO_4 and 237.5 mg Na_2HPO_4 in 400 mL of ddH_2O, Bring volume to 500 mL and autoclave.
30. Complete DMEM media: DMEM supplemented with 10% heat-inactivated fetal bovine serum (FBS), 2.05 mM L-glutamine, 100 IU/L penicillin, and 100 IU/L streptomycin.
31. Commercially available complete serum free media.

2.3 Cell Culture

All cells were cultured and maintained in RPMI supplemented with 10% FBS, 2.05 mM L-glutamine, 100 IU/L penicillin, and 100 IU/L streptomycin. Culture conditions were set at 5% CO_2 and 37 °C in serologic plates and flasks.

2.4 General Equipment

1. Miniature vortexer.
2. Tube rocker.
3. Water-bath sonicator.
4. Refrigerated ultracentrifuge (10,000 × *g* max).
5. Refrigerated microcentrifuge (14,000 × *g* max).
6. Refrigerated benchtop centrifuge (3300 × *g* max).
7. Shaker rotisserie (tube shaker/rotator).
8. Flow cytometer: Must be equipped with a high-throughput sampler, at least four parameters including forward and side scatter and the ability to obtain height, width, and area per parameter.
9. Western blot power supply: high current (up to 250 V) and adjustable mA and W.

10. Western blot vertical mini gel electrophoresis chamber: PAGE/SDS-PAGE appropriate.
11. Adjustable heat block.
12. Infrared fluorescence membrane imager.
13. 37 °C 5 % CO_2 humidified incubator.

3 Methods

3.1 Cell Cycle Analysis

Flow cytometry analysis of cell cycle is a versatile tool allowing for the identification of discrete sensitive cellular subsets affected by drug treatment, while reporting cell ploidy, cell cycle stage, and to a lesser degree, viability. When used in conjunction with gene knockdown or overexpression studies an understanding of that gene function in the cell cycle stage can be elucidated. The following steps are performed in 5 mL flow cytometry compatible tubes: (Fig. 1)

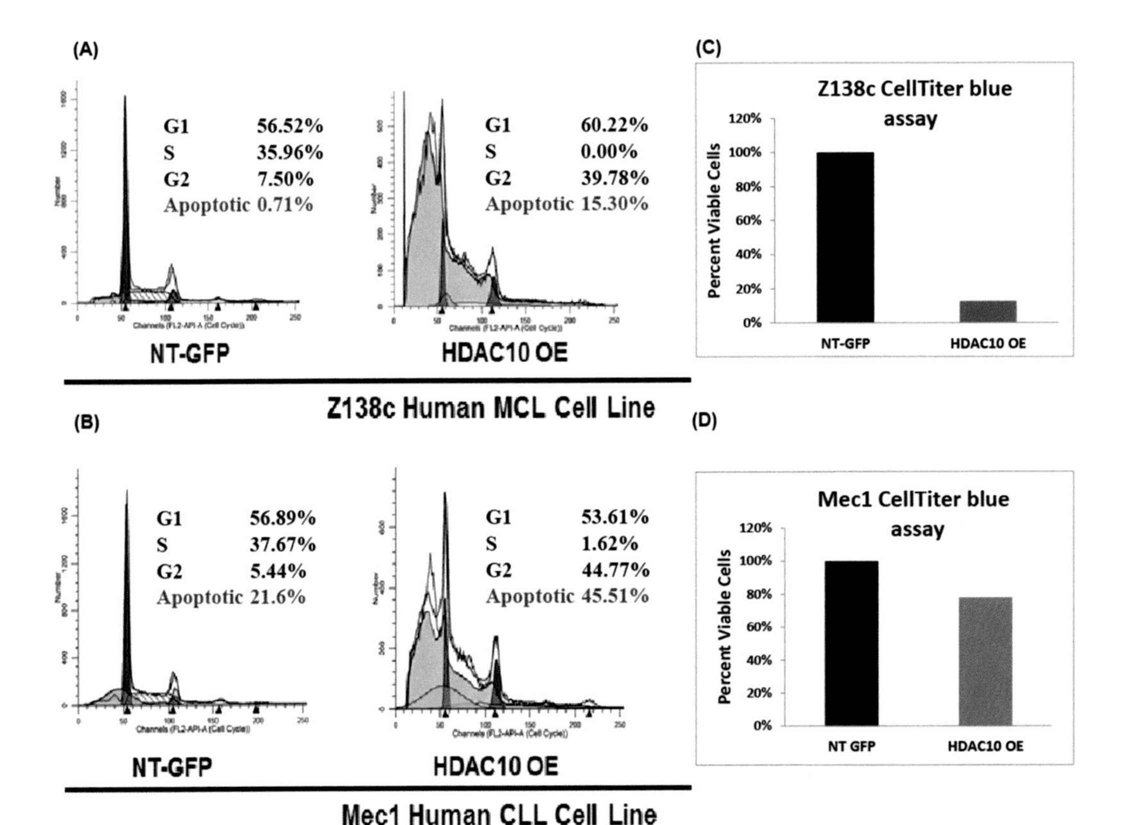

Fig. 1 Overexpression HDAC10 in B cell malignancies causes cell cycle arrest and consequently decreases viability. Z138c mantle cell lymphoma cells overexpressing HDAC10 show an increased G2 cell cycle arrest (**a**), and decrease cell (**c**), when compared to the non-target control counterpart. Mec1 chronic lymphocytic leukemia cells overexpressing HDAC10 show an increased G2 cell cycle arrest (**b**), and decrease cell (**d**), when compared to the non-target control counterpart

1. After the cells are transfected for the overexpression of HDAC10, 1×10^6 cells are collected (floating and adherent cells) *Note: This is of particular importance in adherent cell lines, since cell cycle arrest can cause non-adherence.*
2. Centrifuge samples at $500 \times g$ for 5 min at 4 °C.
3. Add 500 μL of PBS to each tube and thoroughly resuspend cells by flicking each tube.
4. Repeat **steps 2** and **3**.
5. Add 200 μL of PBS to each tube and thoroughly resuspend cells by flicking each tube.
6. Add one drop of chicken erythrocyte nuclei singlets (CEN) to each sample and to a tube by itself.
7. Add 2 mL of ice-cold 70% ethanol, 30 mL 1× PBS to 70 mL of ethanol (Stored at 20 °C for the experiment) to each tube and incubated at 20 °C for 2 h. Use caution and add the reagent drop-wise so undesired clumping of cells does not occur.
8. Samples are centrifuged at $500 \times g$ for 5 min at 4 °C.
9. Add 500 μL of PBS to each tube and resuspend cells by flicking each tube.
10. Repeat **steps** 7 and **8**.
11. The resulting cell pellet is flicked to resuspend in the residual volume in the tube.

    Add 400 μL of cell cycle staining mix, 9.7 mL 1× PBS, 180 μL propidium iodide stock (0.6 mM final), 20 μL RNase A Stock (0.2 mg/mL final), and 10 μL of Triton X-100 (0.1% v/v) (*see* **Note 1**).
12. Incubate tubes at 37 °C for 30 min.
13. Proceed to sample collection on Flow cytometer.
14. Run software in setup mode and display a dot plot forward scatter (FSC) versus side scatter (SSC), a dot plot for propidium iodide (FL2-H and FL2-A), a histogram for SSC and a histogram for propidium iodide (FL2-A).
15. Add an initial gate that covers the majority of cells in the FL2-H and FL2-A plot and display its contents in the histograms.
16. Adjust the voltage of the photomultiplier tube (PMT) SSC and FSC so the main cell population is in the dot plot and the peak of the SSC histogram of that population is at or around 200 Mean Fluorescent Intensity (MFI).
17. Adjust the voltage of the PMT for FL2 so the G/0 population is at or around 200 MFI. *Note: Two peaks and a short plateau in between them should be observed in a cell going thru normal cell cycle. The larger, less fluorescent peak is G/0.*

18. Adjust the voltage of FL4 so the resultant line of cells in the dot plot for FL2-H and FL2-A has a slope near to or equal to one.
19. Abort, de-select setup, and begin to acquire samples. Run all samples on a low flow rate. *Note: Increasing the flow rate causes wider peaks and less accurate measurements.*
20. Set the software to acquire 1×10^4 cells, of the gate described in **step 15**, for each sample and to acquire propidium iodide voltage height and area (FL2-H and FL2-A).
21. Normal cell cycle control samples should have three peaks with two peaks having a short plateau in between them; the larger, less fluorescent peak is G1/0, the short plateau is S-Phase and the smaller highly fluorescent peak is G2/M. When G1/0 is adjusted to 200 G2/M is typically found to be at 400. The very small peak around 50 or 75 is the chicken erythrocyte nuclei. This is used as an internal control for the assay in determining ploidy and identifying the correct peaks of G1/0 and G2/M.
22. Proceed to analysis. Several flow cytometric cell cycle analysis software are available and should be used to analyze samples.

3.2 High-Throughput Checkpoint Molecule Assay

Co-stimulatory and co-inhibitory molecules, or checkpoint molecules, have a varying pattern of expression depending on the tissue site, microenvironmental conditions, and cell type. It has come to light in recent years these molecules are influenced by epigenetic modifications one of which more recently described are HDACs. The following assay includes 30 of the most common checkpoint molecules and respective counterparts with an expandable area for inclusion of other molecules of interest.

1. Collect six million cells per treatment into an appropriately labeled 5 mL flow cytometer tube and bring volume to 5 mL with PBS.
2. Spin the cells at $500 \times g$ for 5 min at 25 °C.
3. Reconstitute in 900 μL of FACS buffer, 10 mL FBS in 490 mL 1× PBS.
4. For unstained control wells, remove 25 μL per sample, and to control plate wells (Table 1). An unstained control must be created for each cell line or cell type to be tested.
5. Spin down the cells at $500 \times g$ for 5 min at 25 °C.
6. Vortex residual volume and reconstitute in 1 mL of fixable yellow dye buffer, 10 μL fixable yellow reagent in 10 mL PBS.
7. Incubate for 30 min in the dark at 25 °C.
8. During the incubation time, prepare antibody mixes (as visualized in Table 2) in 500 μL amber centrifuge tubes or 200 μL eight tube strips.

Table 1
Control plate layout

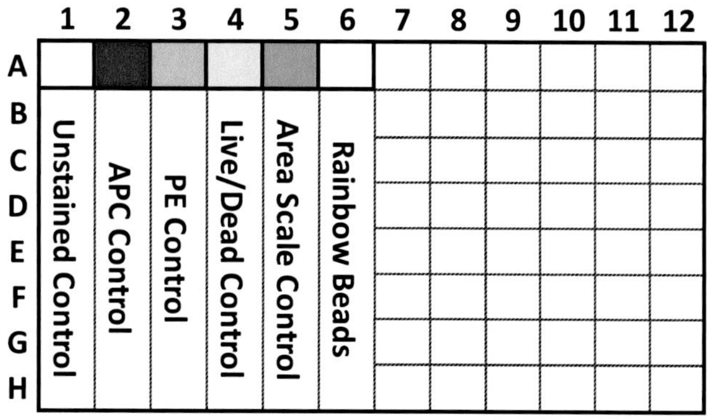

<table>
<tr><th></th><th>1</th><th>2</th><th>3</th><th>4</th><th>5</th><th>6</th><th>7</th><th>8</th><th>9</th><th>10</th><th>11</th><th>12</th></tr>
<tr><td>A</td><td></td><td></td><td></td><td></td><td></td><td></td><td></td><td></td><td></td><td></td><td></td><td></td></tr>
<tr><td>B</td><td rowspan="7">Unstained Control</td><td rowspan="7">APC Control</td><td rowspan="7">PE Control</td><td rowspan="7">Live/Dead Control</td><td rowspan="7">Area Scale Control</td><td rowspan="7">Rainbow Beads</td><td></td><td></td><td></td><td></td><td></td><td></td></tr>
<tr><td>C</td><td></td><td></td><td></td><td></td><td></td><td></td></tr>
<tr><td>D</td><td></td><td></td><td></td><td></td><td></td><td></td></tr>
<tr><td>E</td><td></td><td></td><td></td><td></td><td></td><td></td></tr>
<tr><td>F</td><td></td><td></td><td></td><td></td><td></td><td></td></tr>
<tr><td>G</td><td></td><td></td><td></td><td></td><td></td><td></td></tr>
<tr><td>H</td><td></td><td></td><td></td><td></td><td></td><td></td></tr>
</table>

9. Bring volume of cells, from **step** 7, to 5 mL with FACS buffer.
10. Spin cells at 500 × *g* for 5 min at 25 °C.
11. Reconstitute in 900 μL of FC block buffer, 200 μL human AB sera in 10 mL FACS Buffer.
12. Transfer 25 μL per well for each treatment into the experimental plate (Table 3). One plate accommodates three treatments or one control and two treatments. For example, an experiment consisting of seven treatments and a control would be three separate 96-well plates.
13. Transfer 25 μL of the remaining suspension of each treatment to the live/dead control and area scale control wells to the control plate.
14. Add one drop of compensation beads to the PE and APC control well along with 1 μL of one PE antibody and 1 μL of one APC antibody to the appropriate wells in the control plate.
15. Transfer 25 μL of the antibody mixes to the experimental plate and incubate both the control plate and experimental plate for 1 h at 25 °C.
16. Prepare one well labeled rainbow beads with one drop each of the flow cytometry calibration pre-stained beads.
17. Add 100 μL of FACS buffer and run the plates.
18. Carefully gate out debris/aggregates, Doublets and Dead cells and Collect 5×10^3 to 1×10^4 cells for analysis (*see* **Note 2**).
19. FCS files are exported and analyzed with a commercially available flow cytometry software and geometric median fluorescent intensity (gMFI) changes are reported and percent changes where appropriate. Also can be reported, recommended as overlay histograms (Fig. 2) (*see* **Note 3**).

Table 2
Antibody mix

Strip 1	Strip 2	Strip 3	Strip 4
PD-L1 (CD274)	Galectin-9	OX40L (CD252)	CD137
41BBL (CD137L)	TRAIL-R2 (CD262)	PD-1 (CD279)	TIM3
CD80	B7-H3 (CD276)	BTLA (CD272)	ICOS (CD278)
PD-L2 (CD273)	CD30L (CD153)	CD27	OX40 (CD134)
TRAIL-R1	ICOSL (CD275)	CD28	MHC II (HLA-DR, DQ, DP)
HVEM (CD270)	MHC I (HLA-A,B,C)	LAG3 (CD223)	FACS Buffer (Unstained)
CD70	B7-H4	CD40L (CD154)	
CD86	CD40	CTLA4 (CD152)	

# Treatments	# Plates	uL per antibody	FACs Buffer
1	1	1	49
2	1	1.5	73.5
3	1	2	98
4	2	2.5	122.5
5	2	3	147
6	2	3.5	171.5
7	3	4	196
8	3	4.5	220.5
9	3	5	245
10	4	5.5	269.5
11	4	6	294
12	4	6.5	318.5

Table 3
Plate layout per treatment

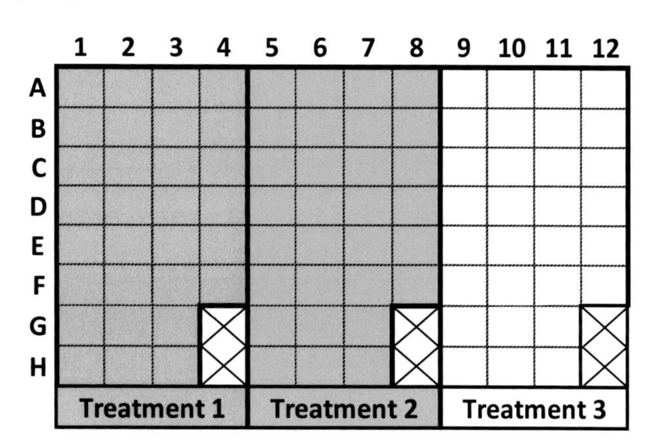

20. On the initial run of each cell line record the gMFI for each peak of the Rainbow beads and adjust the flow cytometer, detector voltages to match when the same cell line is run in any subsequent experiments.

3.3 Overexpression of HDAC10 by Recombinant Adenovirus

The overexpression of HDAC10, using a recombinant adenovirus, has been accomplished with a standard technique designed as an approach to transduce human cells. Adenoviral vectors may potentially be harmful, use caution and pay special attention to institutional security issues and follow all necessary biohazard precautions.

We received an aliquot of HDAC10 recombinant adenovirus generously provided by Dr. Seto's lab [13].

3.3.1 Testing and Large Scale Preparation of Adenovirus

1. Prepare a 6-well plate with confluent Ad293 cells in each well. Add the varying amount of adenovirus to each well, for example (Seen in the schematic representation below Fig. 3), 100, 20, 5, 1, 0.1, 0.01 μl. The ideal concentration of virus results in the detachment of 50% of the cells within 4 days. If the cells detach too quickly the virus was synthetized extremely fast and it may not be properly assembled. Slow production of the virus (5 days or more), will result in a poor virus titer. *Note the figure below*: (*In separate attachment*).
2. Using a 225 cm^2 tissue culture flask, expand the cells-line. Once the cells are 95–100% confluent infect them with the concentration of virus determined in **step 1** in a total volume of 50 mL complete DMEM media. Calculate the correct amount of virus using the following formula: **X μL of virus used to infect a 6-well plate × 225 cm^2/10 cm^2 = Amount of virus to be used/tissue culture flask (225 cm^2)**.

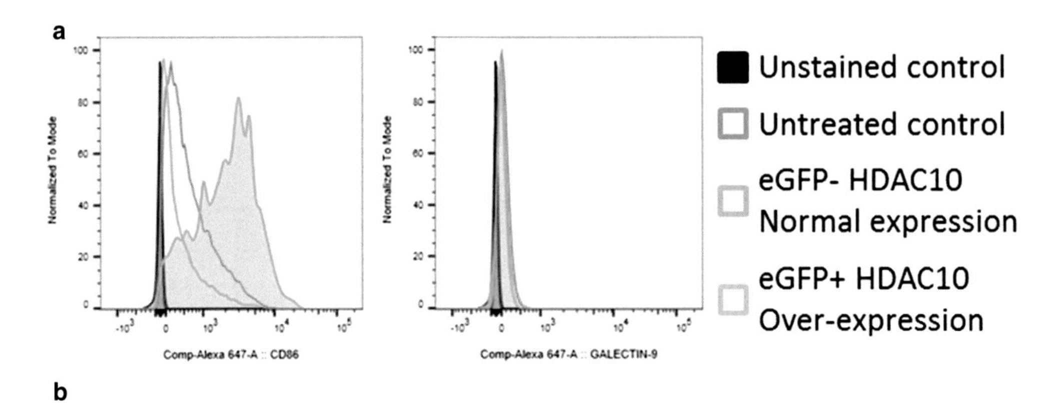

b

Cell line	OSU-CLL			
Target	HDAC10			
Condition	Basal	HDAC10 Normal Expression (eGFP-)	HDAC10 Over Expression (eGFP+)	RESULTS
PD-L1 (CD274)	141.8	192.6	213	No Change
41BBL (CD137L)	479.1	714.3	697.5	No Change
B7-1 (CD80)	2368.8	2341.6	5030	Increased
PD-L2 (CD273)	41.8	50.6	91	No Change
TRAIL-R1 (CD261)	231.1	292.3	290.5	No Change
HVEM (CD270)	589.8	613.6	832	Increased
CD70	1866.1	2874.3	5503.5	Increased
B7-2 (CD86)	410.8	623.6	1993	Increased
Galectin-9	105.8	149.6	142	No Change
TRAIL-R2 (CD262)	412.8	501.6	631	Increased
B7-H3 (CD276)	157.8	144.6	177	No Change
CD30L (CD153)	1187.1	1827.3	1682.5	No Change
ICOSL (CD275)	152.8	186.6	175	No Change
MHC I (HLA-A,B,C)	378.8	507.6	795	Increased
B7-H4 (VTCN1)	52.8	69.8	64.8	No Change
CD40	82.8	75.6	265	Increased
OX40L (CD252)	407.1	439.3	613.5	Increased
PD-1 (CD279)	102.8	117.6	108	No Change
BTLA (CD272)	1281.8	1398.6	1241	No Change
CD27	243.8	364.6	204	No Change
CD28	66.7	81.6	86	No Change
LAG3 (CD223)	54	57.3	96	No Change
CD40L (CD154)	5.2	7.7	7.2	No Change
CTLA4 (CD152)	12	16.5	15.6	No Change
41BB (CD137)	107.8	111.6	403	Increased
TIM3 (CD366)	42.4	58.9	49.6	No Change
ICOS (CD278)	176.8	193.6	179	No Change
OX40 (CD134)	757.8	933.6	888	No Change
MHC II (HLA-DR,DQ,DP)	3327.8	4822.6	4274	Increased

Fig. 2 HDAC10 overexpression in OSU-CLL increases expression of several Checkpoint molecules. (**a**) Data visualized as overlay histogram gives the advantage of observing the full spread of expression of a given marker. An example of two potential conditions in the high-throughput assay. (**b**) gMFI data visualized as a heat map show the highest gMFI expression in *green* cells, Mid-level expression in *yellow* and Lowest expression in *red* cells. Care was taken to highlight the name of the Checkpoint molecules that showed a distinct change from both basal expression and normal expression (Transfected control fraction)

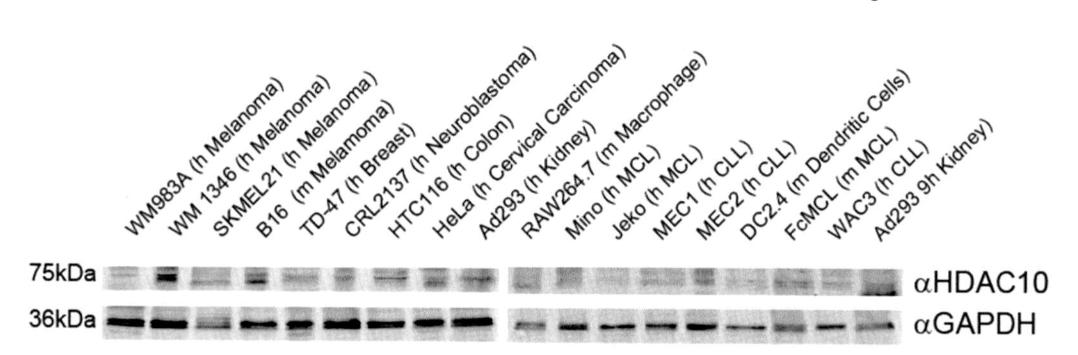

Fig. 3 Schematic representation of adenoviral particle distribution. Adenoviral distribution to determine the ideal concentration of virus for infection of Ad293 cells

3. After 4–5 days harvest the cells by pipetting up and down with a 10 mL sterile pipet. It is recommended to use this size of pipet because it gives more pressure to the surface in order to achieve complete detachment of the cells.
4. Add the harvested cell suspension to a 250 mL centrifuge tube and centrifuge in a Refrigerated centrifuge at $1800 \times g$ for 15 min at 4 °C.
5. Vacuum to remove the media and add 36 mL of 10 mM Tris pH 8.0.
6. Next, freeze at −20 °C, thaw at 37 °C in a water bath and vortex. Repeat this freeze/thaw cycle four times. It is extremely important that you achieve complete freeze and complete thaw and 30 s of vortex each time, this ensures the complete release of the virus from the internal cell compartments.
7. After harvesting, centrifuge at $1800 \times g$ and collect the supernatant.

3.3.2 Gradient Assembling

1. The assembling of the gradient will be prepared by using low density cesium chloride according to concentrations described in Table 4.
2. For each large scale amplification, prepare 70 mL of stock solution for both cesium chloride densities. The cesium chloride must be diluted in 10 mM Tris pH 8.0. For example, in order to prepare a stock solution of 70 mL measure 49.413 g of high density cesium chloride and add to 70 mL 10 mM Tris pH 8.0.
3. Using a 10 mL pipette, add 12 mL of low density cesium chloride into the bottom of the SW-28 tube.
4. Add 12 mL of high density cesium chloride into the bottom of the SW-28 tube adding it slowly to form the gradient. Use caution and try to not disturb the low density cesium chloride.

Table 4
Cesium chloride

	Density (p(g/mL))	Concentration of the salt
High density	1.45	0.7059 g
Low density	1.2	0.2857 g

5. Add 12 mL of the virus lysate over the other two layers low/high density cesium. Do this step slowly to avoid disturbing the already formed gradient.
6. Once the three gradients are formed, balance your tubes and centrifuge using the SW-28 rotor. Be careful because the rotor needs to be loaded without disturbing the gradients. If necessary, prepare a balance tube with the same components as the other tubes only instead of virus you would add 10 mM Tris at **step 5**. Another common procedure is to check each tube has the same weight as the one in the front in the centrifuge.
7. Centrifuge the tubes for 2.5 h at 50,000 × *g* at room temperature. It is very important to use the rubber seals on your buckets. This procedure will avoid evaporation of the sample due to the vacuum during centrifugation.
8. When the centrifuge is finished, carefully remove the rotor and the buckets, and then remove the tubes. Observe the cloudy phase in the top and one or two bands on the bottom. Remove the band at the bottom using a syringe and 18 G needle. Once the band has been extracted, dilute the sample in a 1:1 ratio with 10 mM Tris pH 8.0.
9. To assemble the second gradient in SW-28 tube, repeat **steps 3–5**; however, this time use 3.5 mL of light cesium, 3.5 mL of heavy cesium and 3 mL of your diluted sample extracted with the syringe from the previous steps.
10. Centrifuge your samples in a SW-41 rotor for 16-–8 h at 4 °C at 50,000 × *g*. For this centrifugation the rotor must be pre-cooled in a 4 °C room. Remember to put the rubber seals to avoid evaporation and to check the weight of each tube before place inside the rotor.
11. After centrifugation repeat **step 8**, removing the lower band and, this time, dilute the sample with the virus in a 1:1 ratio of virus to buffer (100 mM NaCl, 10 mM Tris pH 8.0, 5 mM KCl, 1 mM $MgCl_2$, 10% glycerol). At this point you can save your virus at −20 °C until you have your dialysis buffer ready.
12. The total volume of dialysis buffer needed is 6 L. The dialysis buffer's composition is the same as the virus dilution buffer.

Using Pierce dialysis cassettes (one cassette per virus), perform the dialysis procedure using 2 L of buffer (ice cold) and leaving the sample in the cold room stirring with a stir bar for 1 h. After 1 h, change the dialysis buffer and add new buffer. Repeat this three times. After the dialysis procedure, change the buffer and remove the cassette. Remove your sample and aliquot it into 1 mL microfuge tubes.

13. Once aliquoted, save your virus into a −80 °C freezer until use for infection.

3.3.3 Cell Transduction

1. Once you have generated an adenoviral stock with a suitable titer, you are ready to transduce the adenoviral construct into the mammalian cell line Z138c. We recommend using a range of MOIs (*MOI*: *Multiplicity of Infection). MOI is equal to the number of viral particles per cell, for example an MOI of 1 is equal to a single cell infected with a single viral genome* (e.g., 0, 0.5, 1, 2, 5, 10, 20, 50) to determine the MOI required to obtain optimal expression of your recombinant protein for your particular application. To decide the amount of virus needed for a certain MOI, use **Formula A**: # cells × desired MOI = total PFU (*Plaque Forming Units*) needed. Then use **Formula B**: (total PFU needed)/(PFU/mL) = total mL of virus required to reach your desired dose.
2. Plate your mammalian cells Z138c in complete RPMI media at approximately 50 % confluence. Incubate at 37 °C overnight in the incubator.
3. On the day of transduction (Day 1), thaw your adenoviral stock and dilute (if necessary) the appropriate amount of virus (at a suitable MOI) into fresh complete medium, for a 6-well plate use 350:1 final volume for infection/well. Do not vortex. Remember it is important to have 0 = MOI as a negative control to check your cell growth and death.

   Example: For an infection with a 200 MOI, and your adenovirus has a titer of 1.3×10^{11} PFU/mL and each well contains 1.8×10^{6} cells. Calculate as follows: **Formula A**: (1.8×10^{6} cells) × (200 MOI) = 3.6×10^{8} PFU, **Formula B**: (3.6×10^{8} PFU desired)/(1.3×10^{11} PFU/mL) = 0.0028 mL or 2.8:1. Add 348:1 media and the 2.8:1 of virus to the tube which will be used for the infection.
4. For transduction it is recommend to use the minimum concentration of FBS that the cells can withstand; usually this procedure is performed using RPMI media containing only 2 % FBS or equivalent serum free media.
5. Remove the culture medium from your cells. Mix the medium containing the adenovirus gently by pipetting. Add an additional 650 μL of low FBS medium to the cells. Gently swirl the plate to disperse the medium. Incubate at 37 °C overnight in the incubator.

6. The next day (Day 2), remove the medium containing virus and replace with 2 mL fresh complete RPMI medium.
7. Harvest the cells and measure the expression of recombinant HDAC10 protein by western blot. Check the expression of the protein after 48, 72, and 96 h post transduction.

3.4 Western Blot Protocol: Expression of Recombinant HDAC10 Protein

Western blot is a methodology which allows the observation of the expression of determined proteins from a complex extracted from a group of cells. This technique compares samples of interest with a standard of protein of know molecular weight. The technique is performed using an immobilized platform, the SDS-gel, in which the sample of interest is loaded and later these samples are transferred to a solid support which is known as the membrane [14]. Later, this membrane is incubated with antibodies of interest (Fig. 4).

1. (For adherent cells) Place the plate directly onto ice. Remove media and wash the plate with 1 mL PBS. Scrape the cells and transfer into 1.5 mL microcentrifuge tube.
2. Centrifuge at 500 × *g* for 5 min; remove the PBS and repeat once.
3. Add approximately 50 μL lysis buffer (whole cell extract) to each tube.
4. Sonicate the samples two times for 8 min each. Keep your samples in ice all the time.
5. After sonication, place tubes into ice.

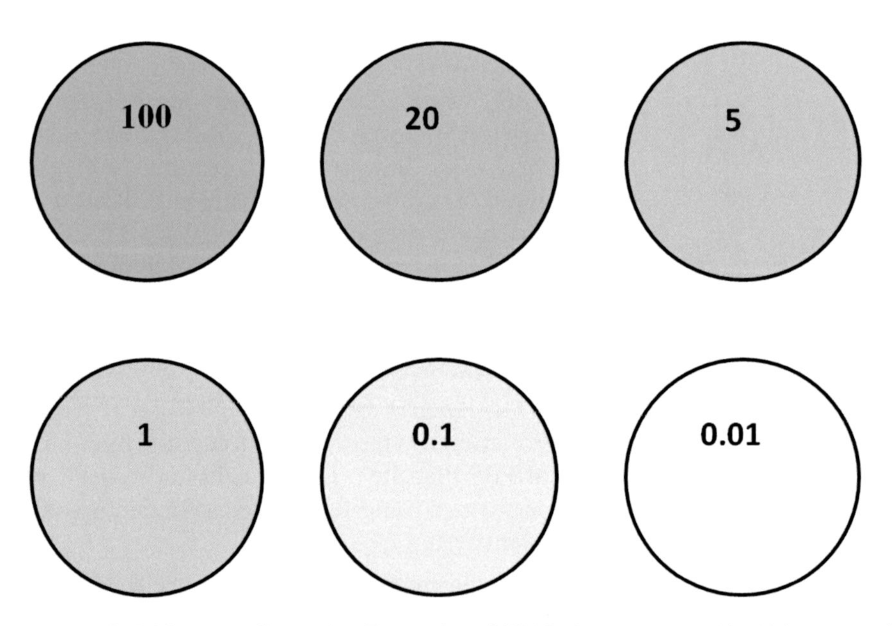

Fig. 4 Expression of HDAC10 in malignancies. Expression of HDAC10 was assessed in 13 human and murine malignant cells lines using immunoblotting analysis

6. Centrifuge the tubes for 10 min at full speed.
7. Collect the supernatant and measure the protein concentration using protocol outlined in the BCA kit.
8. After measuring the concentration, adjust your samples to 1 μg/μL of protein. Add 6× loading buffer at final concentration of 1× to each tube and complete the final volume with water.
9. Place the samples in the heat block at 100 °C for 5 min.
10. Centrifuge for one minute at max speed, 14,000 ×*g*, at room temperature.
11. Set up running gels and add 1× TGS Running Buffer (900 mL ddH_2O + 100 mL 10× Tris–Glycine/SDS) to the electrophoresis chamber. Only prepare enough for current experiment.
12. Add 5 μL of ladder in the first well and 10 μL of each sample into the subsequent wells.
13. Set the power supply at a constant 30 mAMP or 80 V. The gel will run for approximately 90 min on ice.
14. After **step 13** is finished, continue with the membrane transfer. Place the nitrocellulose membrane in Transfer Buffer until is ready to use. Empty the contents of soaking trays, rinse and dry. Use a wet transfer or fast transfer device. For fast transfer devices follow the manufacturer's instructions. If using a wet transfer, run the apparatus at 300 mAMP for 2 h.
15. Remove the nitrocellulose membrane from the cassette or from the fast transfer device and place into 5 mL of 5 % blocking solution for 30 min at room temperature on a rocker (5 % milk in 1× PBS-0.1 % TWEEN 20).
16. Remove the blocking solution and add the primary antibody at the appropriate dilution in the blocking solution, for 2 h at room temperature or incubate overnight on rocker at 4 °C.
17. The next day, wash three times for 10 min in 1× PBS–0.1 % Tween 20.
18. Apply secondary antibody for 1 h at room temperature on a rocker in 5 % blocking solution. The secondary antibody dilution and incubation depends on what method are you using to develop the membrane. For infrared emitting antibodies incubate for a minimum of 45 min, and the dilution is 1:20,000.
19. Wash three times for 10 min in 1× PBS–Tween.
20. Dry the membrane between filter paper, making sure to protect against the light with foil. To develop the membrane scan the membrane using an Infrared Fluorescence Membrane Imager following the manufacturer's protocol.

4 Notes

1. 4′,6′-diamidino-2-phenylindole (DAPI), a DNA specific dye, can be used as an alternative dye to PI. DAPI is free of any adverse RNA staining that can occur with PI, hence the required incubation with RNase A in the PI protocol. This incubation step can be avoided and RNase removed from the staining mix for a resultant DAPI staining mix of 1× PBS with 1 μg/mL DAPI and Triton X-100 (0.1 % v/v). All subsequent steps are the same. DAPI is excited with an alternative laser to PI and the appropriate detector must be utilized.
2. Utilizing the following gating strategy will ensure an aggregate-free, viable, single cell analyses: Dot plot one is Forward Scatter (Area) versus Side Scatter (Area). The population of interest will be a dense cellular cluster at around the 200–300 level. Gate tightly around this population. Dot plot two is Forward Scatter (Area) versus viability marker (Area). Gate tightly around the dense cellular cluster that is low expressing of the viability marker. Dot plot three is Forward Scatter (Height) versus Forward Scatter (Area). Tightly gate around the cellular population that has a slope of one. Cell clusters divergent from this slope are doublets, two cells stuck together. Dot plot four is Side Scatter (Height) versus Side Scatter (Area). Again, tightly gate around the population that has a slope of one, typically it is 99 % of the population; this is a secondary check point to ensure purity. The resultant population is subsequently divided into eGFP+ (HDAC10 overexpressing) and eGFP- (HDAC10 Native expression).
3. The majority of checkpoint molecules on cell lines result in only graduated shifts in gMFI intensity and not a bimodal distribution of an unexpressed and highly expressed cellular population. Care should be taken in identifying what type of distribution occurs with each checkpoint molecule in the cell of interest.

Acknowledgments

The authors gratefully acknowledge the flow cytometry core facilities at H. Lee Moffitt Cancer Center and their extended technical support for their project.

References

1. Villagra A, Sotomayor EM, Seto E (2010) Histone deacetylases and the immunological network: implications in cancer and inflammation. Oncogene 29(2):157–173
2. Sahakian EW, Karrune; Villagra, Alejandro; and Sotomayor, Eduardo (2013) Epigenetic approaches: emerging role of histone deacetylase inhibitors in cancer immunotherapy. *Cancer immunotherapy* (2 edn). Academic Press (Elsevier)
3. de Ruijter AJ, van Gennip AH, Caron HN, Kemp S, van Kuilenburg AB (2003) Histone deacetylases (HDACs): characterization of the classical HDAC family. Biochem J 370(Pt 3):737–749
4. Lee JH, Jeong EG, Choi MC et al (2010) Inhibition of histone deacetylase 10 induces thioredoxin-interacting protein and causes accumulation of reactive oxygen species in SNU-620 human gastric cancer cells. Mol Cells 30(2):107–112
5. Fischer DD, Cai R, Bhatia U et al (2002) Isolation and characterization of a novel class II histone deacetylase, HDAC10. J Biol Chem 277(8):6656–6666
6. Guardiola AR, Yao TP (2002) Molecular cloning and characterization of a novel histone deacetylase HDAC10. J Biol Chem 277(5): 3350–3356
7. Kao HY, Lee CH, Komarov A, Han CC, Evans RM (2002) Isolation and characterization of mammalian HDAC10, a novel histone deacetylase. J Biol Chem 277(1):187–193
8. Song C, Zhu S, Wu C, Kang J (2013) Histone deacetylase (HDAC) 10 suppresses cervical cancer metastasis through inhibition of matrix metalloproteinase (MMP) 2 and 9 expression. J Biol Chem 288(39):28021–28033
9. Osada H, Tatematsu Y, Saito H, Yatabe Y, Mitsudomi T, Takahashi T (2004) Reduced expression of class II histone deacetylase genes is associated with poor prognosis in lung cancer patients. Int J Cancer 112(1):26–32
10. Jin Z, Jiang W, Jiao F et al (2014) Decreased expression of histone deacetylase 10 predicts poor prognosis of gastric cancer patients. Int J Clin Exp Pathol 7(9):5872–5879
11. Sahakian E, Shah BD, Powers J et al (2011) The opposing role of Histone Deacetylase 10 (HDAC10) and HDAC11 in proliferation/survival of Mantle Cell Lymphoma (MCL) and Chronic Lymphocytic Leukemia (CLL). Blood 118(21):598
12. Carninci P, Kasukawa T, Katayama S et al (2005) The transcriptional landscape of the mammalian genome. Science 309(5740): 1559–1563
13. Villagra A, Cheng F, Wang HW et al (2009) The histone deacetylase HDAC11 regulates the expression of interleukin 10 and immune tolerance. Nat Immunol 10(1):92–100
14. Cheng F, Lienlaf M, Wang HW et al (2014) A novel role for histone deacetylase 6 in the regulation of the tolerogenic STAT3/IL-10 pathway in APCs. J Immunol 193(6): 2850–2862

Chapter 11

Functional Analysis of Histone Deacetylase 11 (HDAC11)

Jie Chen*, Eva Sahakian*, John Powers, Maritza Lienlaf, Patricio Perez-Villarroel, Tessa Knox, and Alejandro Villagra

Abstract

The physiological role of histone deacetylase 11 (HDAC11), the newest member of the HDAC family, remained largely unknown until the discovery of its regulatory function in immune cells. Among them, the regulation of cytokine production by antigen-presenting cells and the modulation of the suppressive ability of myeloid-derived suppressor cells (MDSCs) (Sahakian et al. Mol Immunol 63: 579–585, 2015; Wang et al. J Immunol 186: 3986–3996, 2011; Villagra et al. Nat Immunol 10: 92–100, 2009). Our earlier data has demonstrated that HDAC11, by interacting at the chromatin level with the IL-10 promoter, down-regulates *il-10* transcription in both murine and human APCs in vitro and ex vivo models (Villagra et al. Nat Immunol 10: 92–100, 2009). However the role of HDAC11 in other cell types still remains unknown. Here we present several methods that can potentially be used to identify the functional role of HDAC11, assigning special attention to the evaluation of immunological parameters.

Key words Histone deacetylase 11, Chromatin immunoprecipitation, qRT-PCR, PCR, Flow cytometry, Phenotyping, Genotyping

1 Introduction

In the last decade, epigenetic modifications have captured special attention due to their multifaceted functionality. Noteworthy among the epigenetic modifiers, histone acetyltransferases (HATs) and histone deacetylases (HDACs) have gained particular notice as it has been demonstrated by many that these epigenetic modifiers can be targeted by highly selective small-molecule inhibitors. This characteristic raised a renewed interest in these epigenetic modifiers as potential therapeutic targets, including cancer and immune-related diseases. Functionally, HDACs alter chromatin by acetylation/deacetylation of histone tails resulting in transcriptionally

The original version of this chapter was revised. The erratum to this chapter is available at: DOI 10.1007/978-1-4939-3667-0_22

*Authors contributed equally to this manuscript.

Sibaji Sarkar (ed.), *Histone Deacetylases: Methods and Protocols*, Methods in Molecular Biology, vol. 1436, DOI 10.1007/978-1-4939-3667-0_11, © Springer Science+Business Media New York 2016

active or inactive chromatin, respectively—ultimately leading to expression and/or suppression of numerous genes [1]. These proteins have been shown to play an important role in the regulation of genes involved in the inflammatory response [1–3] and cancer [4, 5], where they promote loss or gain of cellular functions, not naturally given to a particular cell type. Our approach has been to first identify the role of certain HDACs in the transcriptional regulation of specific components of immune-related pathways, which can be evaluated individually by the analysis of the anti/pro-inflammatory responses against singular stimuli.

HDAC11 was discovered in 2002 [6], and because of its shared characteristic with class I and class II HDACs it was classified in a new and independent group known as class IV HDACs, where it is the only member. Interestingly, most of the reported functions for this deacetylase are related to the immune function [1, 7, 8]. Therefore, in this manuscript, our focus is to present detailed methods for examining the function of HDAC11 in immune cells.

2 Materials

2.1 Animal Models

C57BL/6 (WT) mice were purchased from Jackson laboratories, Tg-HDAC11-eGFP [9] reporter mice were provided by Nathaniel Heintz through the Mutant Mouse Regional Centers, and HDAC11KO was kindly supplied by Dr. Edward Seto at H. Lee Moffitt Cancer Center. Mice were kept in pathogen-free conditions and handled in accordance with approved protocols by the Institutional Animal Care and Use Committee (IACUC) at the University of South Florida.

2.2 General Reagents and Buffers

1. 10× PBS 7.4 pH: In 900 mL ddH_2O, add 80 g NaCl, 2 g KCl, 14.4 g Na_2HPO_4, and 2.4 g KH_2PO_4. pH to 7.4 with HCl and bring volume up to 1 L (autoclave and filter).
2. Tissue culture-treated plates: 6, 12, and 24 wells.
3. Ethanol (99.98 %, ACS/USP/Kosher grade).
4. Mouse Biotin Selection/Mouse CD4+ T cell isolation Kit.
5. 5 M NaCl: Dissolve 58.44 g of NaCl in 80 mL ddH_2O, bring volume up to 100 mL (autoclave and filter).
6. 1 M Tris–HCl pH 8.0: In 80 mL ddH_2O, add 12.11 g of Tris base, use HCl to adjust pH value to 8.0, then bring volume up to 100 mL (autoclave and filter).
7. 0.5 M EDTA: In 80 mL ddH_2O, add 18.61 g of disodium EDTA·2H_2O, use NaOH to adjust pH value to 8.0, then bring volume up to 100 mL (autoclave and filter).
8. 10 % SDS: Dissolve 10 g of SDS in 80 mL of ddH_2O, and then bring volume up to 100 mL (autoclave and filter).

9. TE buffer: 496 mL ddH_2O, 5 mL 1 M Tris–HCl pH 8.0, 1 mL 0.5 M EDTA pH 8.0 (autoclave and filter).
10. Mouse tail lysis buffer: In water, consists of 10 mM Tris–HCl PH 8.0, 50 mM NaCL, 25 mM EDTA PH 8.0, 0.5 % SDS. Proteinase K has to be added to lysis buffer right before use.
11. Heat-activated Taq DNA polymerase.
12. 1× Ammonium-chloride-potassium (ACK) lysis buffer: In 900 mL of ddH2O, add 8.3 g NH_4Cl, 1 g $KHCO_3$, and 200 μL of 0.5 M EDTA. Adjust pH to 7.2–7.4 and bring the total volume to 1 L. Autoclave and filter the solution.
13. FACS buffer: In 980 mL of 1× PBS, add 20 mL heat-inactivated FBS.
14. 1× 4′,6-Diamidino-2-phenylindole (DAPI) viability dye: Dissolve 10 mg of DAPI with 360.6 μL for a 100 mM stock solution, aliquot into 20 μL fractions in 200 μL tubes. Store at −80 °C for long-term storage or −20 °C for short term. For a 10× stock for viability solution dilute to 10 μM in FACS buffer, add 1 μL of 100 mM DAPI stock solution per 10 mL of FACS buffer.
15. Fc block: Anti-CD32 and anti-CD16 functional grade with a final antibody concentration of 0.5 μg/mL each.
16. Flow cytometer compensation beads.
17. SYBR Green Supermix.
18. TRIzol reagent.
19. Chloroform (99 %).
20. Isopropyl alcohol (99.5 %, ACS grade plus).
21. cDNA synthesis using RNA strand.
22. Rabbit anti-HDAC11 antibody.
23. Biotin-conjugated normal rabbit IgG-B.
24. Protein A agarose beads.
25. 37 % Formaldehyde solution.
26. 1.25 M Glycine: 9.38 g Glycine in 100 mL water. Filter.
27. 1.0 M KCl: 7.46 g KCl in 100 mL water. Autoclave.
28. 1.0 M $MgCl_2$: 20.33 g $MgCl_2$ in 100 mL water. Autoclave.
29. 5.0 M NaCl: 29.22 g NaCl in 100 mL water. Autoclave.
30. 2.0 M LiCl: 8.48 g LiCl in 100 mL water. Autoclave.
31. 20 % SDS: Dissolve 20 g of SDS in 80 mL of ddH_2O, bring volume up to 100 mL.
32. Phenol/chloroform/isoamyl alcohol pH 6.7, phenol: chloroform:isoamyl alcohol = 25:24:1.
33. Commercially available PCR Purification Kit.

34. Cross-linking buffer: 2.7 mL 37% formaldehyde was added into 97.3 mL 1× PBS pH 7.2, to reach 1% formaldehyde final concentration in a 100 mL final volume.
35. Stop buffer: 10 mL 1.25 M Glycine was added into 90 mL 1× PBS pH 7.2 to reach a 0.125 M glycine final concentration in a 100 mL final volume. It is preferred to prepare cross-linking buffer and stop buffer freshly before each experiment. Add protease inhibitor cocktail tablets just before use.
36. Lysis buffer: In mQwater, consists of 50 mM HEPES pH 7.8, 3 mM $MgCl_2$, 20 mM KCl, 0.25% Triton X-100, 0.5% IGEPAL.
37. Wash buffer: In water, consists of 10 mM Tris–HCl pH 8.0, 1 mM EDTA pH 8.0, 200 mM NaCl.
38. Sonication buffer: In water, consists of 50 mM HEPES pH 7.8, 140 mM NaCl, 1 mM EDTA, 0.5% SDS, 1% Triton X-100.
39. Dilution buffer I: In water, consists of 50 mM HEPES pH 7.8, 140 mM NaCl, 1 mM EDTA pH 8.0, 1% Triton X-100.
40. Dilution buffer II: In water, consists of 50 mM HEPES pH 7.8, 500 mM NaCl, 1 mM EDTA pH 8.0, 1% Triton X-100.
41. LiCl buffer: In water, consists of 20 mM Tris–HCl pH 8.0, 250 mM LiCl, 1 mM EDTA pH 8.0, 0.5% Triton X-100.
42. TE buffer: In water, consists of 10 mM Tris–HCl pH 8.0, 1 mM EDTA pH 8.0.
43. Elution buffer: In water, consists of 50 mM Tris–HCl pH 8.0, 1 mM EDTA pH 8.0, 1% SDS.

All buffers were filtered, and kept cold. Protease inhibitor cocktail tablets were added before use (**items 34–43**).

2.3 Cell Culture

All cells were cultured and maintained in RPMI supplemented with 10% FBS, 2.05 mM L-glutamine, 100 IU/L penicillin, and 100 IU/L streptomycin. Culture conditions were set at 5% CO_2 and 37 °C in serologic plates and flasks.

2.4 Equipment

1. Mini vortex.
2. Tube-rocker (for 15 or 50 mL tubes).
3. Water bath tube-top sonicator (for chromatin shearing).
4. Ultracentrifuge (for 1.5 mL tubes).
5. Refrigerated centrifuge (for 15 or 50 mL tubes).
6. Tube shaker/rotator (for 1.5 mL tubes).
7. Quantitative real-time PCR thermal cycler system.
8. PCR thermal cycler system.
9. Dounce homogenizer, 7 mL.
10. Flow cytometer (with a minimum of 14-color parameter).

3 Methods

3.1 PCR Genotyping of HDAC11 Knockout (HDAC11KO) Tg-HDAC11 eGFP Reporter (Tg-HDAC11-eGFP) Mice

The Tg-HDAC11 eGFP reporter mouse model was developed by engineering a bacterial artificial chromosome (BACs) insertion containing a promoter region of HDAC11 and a terminal eGFP reporter (Tg-HDAC11-eGFP) [10]. This model allows us to follow the dynamic changes in HDAC11 transcriptional activation, by the expression of eGFP. HDAC11KO mice were established using Lox/CRE technology to remove a floxed exon 3, a portion of the histone deacetylase 11 with catalytic function.

3.1.1 Mouse Tail Genomic DNA Preparation

1. Cut 0.2 cm piece of HDAC11 knockout (HDAC11KO) mouse tail and put the tail snip in 1.5 mL Eppendorf tube.
2. Add 400 μL LYSIS BUFFER and 10 μL Proteinase K (0.5 mg/mL working concentration, 20 mg/mL stocking concentration) into each tube and incubate at 55–60 °C overnight.
3. Add 200 μL 5 M NaCl to each tube and vortex immediately for 10 s.
4. Centrifuge sample tubes at maximum speed for 20–30 min at room temperature.
5. Transfer the supernatant (about 500–600 μL) to a new 1.5 mL Eppendorf tube.
6. Add 1 mL room-temperature ethanol (100%) to each tube, and invert for several times until stringy DNA precipitate is visible.
7. Centrifuge sample tubes at 14,000 × *g* for 1 min, and discard the supernatant.
8. Wash the pellets once with 70% ethanol. Centrifuge sample tubes at 14,000 × *g* for 1 min, and discard the supernatant.
9. Dry the pellets for 10–15 min on the bench top.
10. Add 50–100 μL TE buffer to the pellet and dissolve at 4 °C overnight.

3.1.2 PCR

1. *PCR reaction*: For each 25 μL PCR reaction, use 50 ng of mouse genomic DNA, 1 μL of HDAC11 promoter mouse genotyping primer forward and 1 μL of eGFP mouse genotyping primer reverse (*see* Table 1). As shown in Table 2, each mouse colony has to be tested with its own specific set of primers.

   The HDAC11KO mouse genotyping primers were designed as 1 forward primer and 2 reverse primers by our laboratory as described. Primer 1 forward is designed to be located at upstream of HDAC11 exon 3 and primer 2 reverse is located at downstream of HDAC11 exon 3; primer 1 reverse is located within HDAC11 exon 3. During PCR, 3 primers

Table 1

Subpopulation	Surface markers
Test A: Common myeloid progenitors (CMP)	CD3-,CD19-,NK1.1-,CD11B-,Ly6G-,CD45+, CD117^{High}, CD127-
Test A: Myeloblasts	CD45^{int}, SSC^{low}, CD117+,Ly6G-
Test A: Promyelocytes	CD45^{int}, SSC^{high}, CD117+, Ly6G-
Test B: Triple-negative DCs	CD3-, CD19-,NK1.1-, CD11C+, CD11B-, CD8-, CD4-
Test B: Double-negative DCs	CD3-, CD19-, NK1.1-, CD11C+, CD11B+, CD8-, CD4-
Test B: CD4+ DCs	CD3-, CD19-, NK1.1-, CD11C+, CD11B+, CD8-, CD4+
Test B: CD8+ DCs	CD3-, CD19-, NK1.1-, CD11C+, CD11B+, CD8+, CD4-
Test C: Classical monocytes	CD3-, Nk1.1-, CD115+, CD45R-, MHCII-, Ly6C-, CD43-
Test C: Nonclassical monocytes	CD3-, Nk1.1-, CD115+, CD45R, MHCII-, Ly6C+, CD43+
Test D: Neutrophils	CD3-,CD19-,NK1.1-,CD11C-,CD11B+, Ly6G+
Test D: Eosinophils	CD3-,CD19-,NK1.1,CD11C-,CD11B+,Ly6G-,SSC^{high},Ly6C$^{low/-}$

Table 2
Primers for HDAC11 knockout mouse genotyping and HDAC11 exon qRT-PCR

Purpose	Primer name	Sequence
Genotyping	Primer 1 forward	CTGTGGAGGGAGAGTTGCTC
Genotyping	Primer 1 reverse	GTTGAGATAGCGCCTCGTGT
Genotyping	Primer 2 reverse	GTGGACAGGACAAGGGCTAA
Genotyping	H11 promoter forward	GAGGCGTGTGTTCTGCGTA
Genotyping	eGFP reverse	GTAGGTCAGGGTGGTCACGA
qRT-PCR	H11 exon 3 forward	AGAGAAGCTGCTGTCCGATG
qRT-PCR	H11 exon 3/4 reverse	AGGACCACTTCAGCTCGTTG
qRT-PCR	Mm_Hdac11_1_SG	(Qiagen) Exons 8–9

were added to each reaction. For a wild-type mouse, primer 1 forward and primer 1 reverse will produce a DNA product of 437 base pair indicating the existence of exon 3. Meanwhile, primer 1 forward and primer 2 reverse will produce another DNA product which is over 780 base pair. A homozygous HDAC11KO mouse with a missing exon 3 will produce a

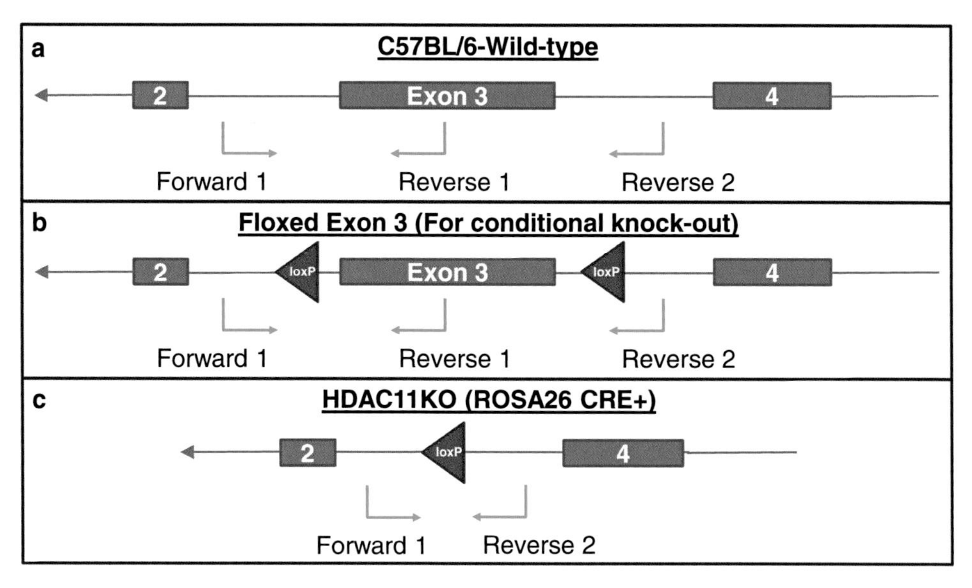

Fig. 1 Total and conditional HDAC11KO genotyping. Three genotyping primers were specifically designed to identify the loxP sites downstream and upstream of exon 3 and one primer found within the exon. Schematic representations of the three genotyping primer locations on a C57BL/6 wild-type HDAC11 allele (**a**), a conditional HDAC11KO allele (**b**), and a total HDAC11KO allele are displayed above. Genotyping PCR will result in a 437 bp fragment for a wild-type allele (**a**), a 471 bp fragment for a loxP allele (**b**), and a 278 bp fragment for a total HDAC11KO allele (**c**)

DNA product of 278 base pair made from primer 1 forward and primer 1 reverse. A heterozygous HDAC11 knockout mouse, both DNA fragments of 437 base pair and 278 base pair will be observed on a 1.5 % agarose gel (Fig. 1).

The Tg-HDAC11-eGFP mouse genotyping primers were designed by our laboratory. Briefly, HDAC11 promoter forward is designed to be located within the 3′ end of the HDAC11 promoter sequence and eGFP reverse is located within the first 200 bp of the 5′ end of the eGFP coding sequence. For a wild-type mouse, no DNA product is produced. For a homozygous or heterozygous Tg-HDAC11-eGFP mouse, a DNA product of approximately 400 base pair will be observed on a 1.5 % agarose gel.

2. *PCR program*: A thermocycler with gradient ability is required for this protocol. The step programming description can be found in Table 3.

   This technique has been used as a standard operating procedure by our lab and collaborators pertaining genotyping of these mouse colonies [7, 8].

3.2 Phenotyping of Tg-HDAC11-eGFP Reporter Mice

In this mouse model, expression of HDAC11 message is determined by flow cytometry analysis of eGFP reporter gene expression [10]. Only mice with a positive eGFP/HDAC11 PCR band are used as breeders and are subsequently checked for eGFP

Table 3
Universal touchdown genotyping PCR

Step	Temperature (°C)	Time		Actions per step
1	94	3	minutes	
2	94	30	seconds	
3	70	30	seconds	Decrease 0.5 °C per cycle
4	72	30	seconds	
5				Go to **step 2** 40 times
6	94	30	seconds	
7	50	30	seconds	
8	72	30	seconds	
9				Go to **step 6** 15 times
10	72	2	minutes	
11	4	Infinite		Hold

expression via flow cytometry. Breeders and pups, of age, have a submandibular bleed (SMB) performed to determine eGFP expression prior to breeding or experimental design.

3.2.1 Peripheral Blood Preparation

1. A submandibular bleed of an anesthetized mouse is performed to obtain, at minimum, 100 μL of blood in a heparinized tube and rocked.
2. The peripheral blood (PB) is subsequently transferred to a 5 mL cytometry tube.
3. The PB is lysed with 1 mL of ACK lysis buffer for 3 min.
4. The mix is then diluted with 3 mL of PBS and centrifuged for 3 min at 300 × *g*.
5. Remove the supernatant and wash the resultant pellet twice with PBS (*see* **Note 1**).
6. The resultant pellet is reconstituted in 500 μL of FACS buffer and run on a flow cytometry machine collecting for forward scatter (FSC), side scatter (SSC), eGFP, and a viability dye, typically DAPI or propidium iodide.
7. 10,000 events of the granulocyte population, as determined in the FSC and SSC plot, are collected and all events are saved (*see* **Note 2**).

3.3 Multiparameter Flow Cytometric Analysis to Dynamically Visualize the HDAC11 Expression Profile in Myelopoiesis

3.3.1 Extraction of Bone Marrow Cells

Bone marrow aspirate (BMA) was removed from Tg-HDAC11-eGFP using the following steps:

1. Collect mouse bone marrow cells by flushing the tibias and femurs with 10 mL ice-cold PBS.
2. Centrifuge the cells at 500 × *g* for 5 min at 4 °C.
3. Remove supernatant, reconstitute the cells in 3 mL ACK lysis buffer, and incubate at room temperature for 4 min.
4. Add ice-cold PBS and centrifuge the cells at 500 × *g* for 5 min at 4 °C.
5. Remove supernatant. Reconstitute the cells in 10 mL ice-cold PBS. Let cell suspension go through a 45 μm strainer to remove remaining large chunks.
6. Centrifuge single-cell suspension at 500 × *g* for 5 min at 4 °C.
7. Remove supernatant. Reconstitute cell pellet in FACS buffer for future use.

3.3.2 Flow Cytometry (Staining and Analysis)

Cells were labeled (as indicated in the chart below) with distinct cell surface markers that have been extensively described as specific for the cellular subpopulation under analysis (Fig. 2). Always create one test more than required for pipette error (i.e., seven samples would require a master mix for eight samples).

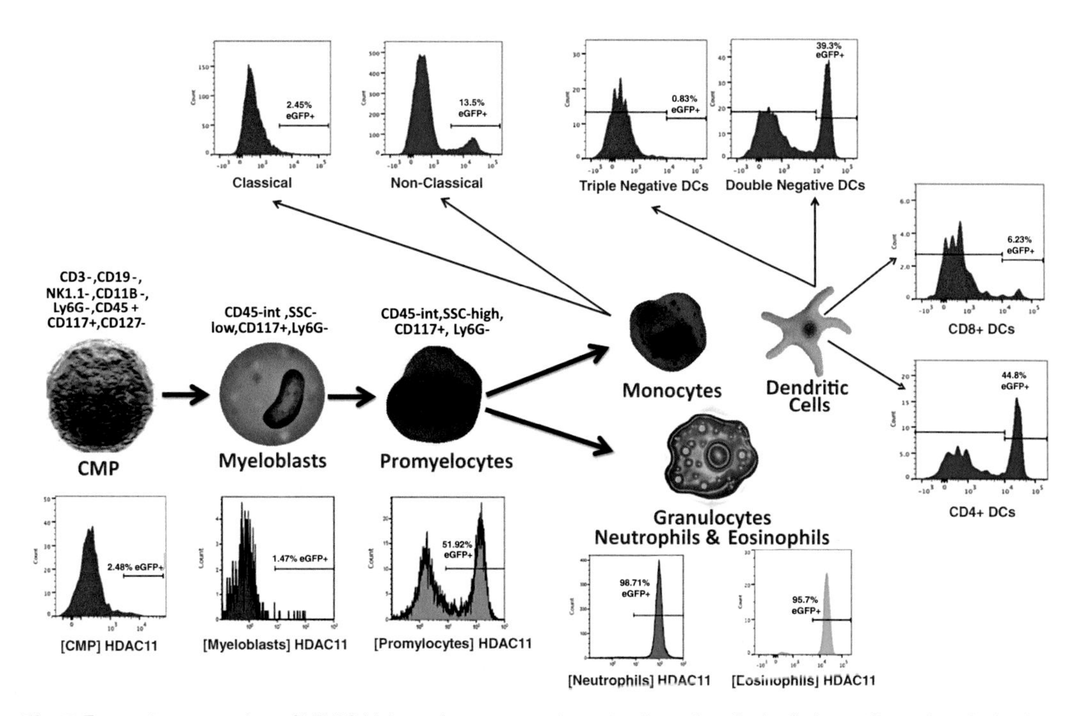

Fig. 2 Dynamic expression of HDAC11 in various compartments of myelopoiesis. Schematic and analytic visualization of HDAC11 message in myelopoiesis. Using HDAC11-eGFP transgenic mouse, in a multiparameter flow cytometric analysis (FACS-Aria (BD) device). In this figure, expression eGFP protein coincides with transcriptional activation of HDAC11

1. A single-cell suspension of BMA is created at 3×10^6 cells per mL in FACS buffer. Note: At minimum 1 mL is required for the following study.
2. 2 μL of BD FC block is added per mL of BMA suspension and incubated for 15 min at room temperature.
3. 100 μL of each sample is added to a fresh 5 mL cytometry tube labeled for each test below:
   (a) Common myeloid progenitors (CMPs).
   (b) Dendritic cells (DCs).
   (c) Granulocytes.
   (d) Monocytes.
   (e) Control tube (remaining mix).
4. 400 μL of fresh FACS buffer is added per tube and centrifuged at $300 \times g$ for 5 min (except for control tubes).
5. Create the following control tubes for each sample type tested for each test:
   (a) Fluorescence minus one (FMO) for each type of fluorescence tested (five tubes for CMPs, DCs, and monocytes and three tubes for granulocytes).
   (b) Unstained sample.
   (c) Viability stain only (DAPI).
6. Create master mixes of the antibodies for each test to be run and bring the master mix volume up to equal a volume appropriate for 50 μL per tube (plus extra tube) (*see* Table 4 for each test requirements).
7. Reconstitute each test tube with 50 μL of the appropriate test master mix.
8. FMO control tubes will be labeled "(Test Name) FMO—(Name of Fluoresence)" and all the antibodies EXCEPT the one listed will be added to the tube in the same concentration as listed in each test plus 50 μL of FACS buffer (i.e., the tube labeled CMP FMO—APC would contain all antibodies of the CMP test except APC).
9. Single-fluorescence stain controls are made using UltraComp beads, and apply the same volume of each antibody used to a separate tube labeled for each type of fluorescence, in each test. One drop of UltraComp beads is added to each of these tubes.
10. All tubes are incubated in the dark for 1 h at room temperature and subsequently washed with 450 μL FACS buffer and centrifuged for 5 min at $300 \times g$ and supernatant is removed. Note: Unnecessary to wash single-color controls.

Table 4
Flow cytometry master mixes

A	uL per test	CD Marker or Designation	Fluorescent molecule
Common Myeloid Progenitor	1	CD3	eFluor 450
	1	CD19	v450
	1	NK1.1	v450
	1	NKG2D	v450
	1	CD11b	v450
	1	Ly6g	v450
	1	CD45	PE-Cy7
	1	CD34	AlexaFlour700
	1	CD127	PerCP-Cy5.5
	1	CD117 (c-Kit)	APC
	10uL per well		

B	uL per test	CD Marker or Designation	Fluorescent molecule
Dendritic Cells	1	CD3	eFluor 450
	1	CD19	v450
	1	NK1.1	v450
	1	NKG2D	v450
	1	CD11b	PE
	1	CD11c	APC
	1	CD4	PerCP
	1	CD8	AlexaFlour700
	8uL per well		

C	uL per test	CD Marker or Designation	Fluorescent molecule
Monocytes (Classical and Non-Classical)	1	CD3	eFluor 450
	1	CD19	v450
	1	NK1.1	v450
	1	NKG2D	v450
	1	CD115	APC
	1	Ly6c	AlexaFlour700
	1	CD43	PE
	1	MHCII	PE-Cy7
	8uL per well		

D	uL per test	CD Marker or Designation	Fluorescent molecule
Granulocytes (Neutrophils/Eosinophils)	1	CD3	eFluor 450
	1	CD19	v450
	1	NK1.1	v450
	1	NKG2D	v450
	1	CD11b	APC
	1	Ly6g	PE
	6uL per well		

See **Note 7**

11. Reconstitute all the test tubes and the DAPI control tube in a 1× DAPI staining mix—do not do this to the control tubes. Run samples on flow cytometer.
12. When running samples, gate out dead cells, doublets, and aggregates and collect until you have 10,000 cells of the smallest population of interest; save all events.

    Previous work done in our lab has already confirmed the co-expression of HDAC11 message and eGFP message using qRT-PCR using HDAC11 and eGFP primers (data not shown).

Table 1. This protocol has been used in our laboratory to study the effect of HDAC11 in MDSC function and differentiation [11] as well as in other compartments of hematopoiesis. The flow cytometry technique described above has been the standard operating procedure for most of the flow data published in our recent manuscripts [12–15].

3.4 Quantitative Real-Time PCR of HDAC11KO Mouse Model

HDAC11KO mice were established by excision of the catalytic region of this gene (exon 3). Therefore in this model, if the qRT-PCR primers are not designed in the exon 3 region, the results are truncated HDAC11 amplicons. Hence, we designed a special HDAC11 primer set that was within the exon 3 region of this gene.

3.4.1 Total mRNA Isolation and cDNA Synthesis

1. Peritoneal neutrophils were harvested from C57BL/6 and HDAC11 knockout mice.
2. Collect cells in 15 mL conical and wash once with PBS. Pellets were transferred into a new 1.5 mL Eppendorf tube.
3. Add 1 mL TRIzol reagent to each sample, and lyse for 10 min at room temperature.
4. Add 0.2 mL chloroform to each 1 mL sample. Shake vigorously by hand for 15 s and rest at room temperature for 3 min.
5. Centrifuge the samples at 12,000 × *g* for 15 min at 4 °C.
6. Transfer supernatant to a new 1.5 mL Eppendorf tube. Add 0.5 mL of isopropyl alcohol to each sample. Incubate at room temperature for 10 min.
7. Centrifuge the samples at 12000 × *g* for 10 min at 4 °C.
8. Discard supernatant. Wash pellet once with cold 75 % ethanol. Gently invert the samples 4–5 times.
9. Centrifuge the samples at 7500 × *g* for 8 min at 4 °C.
10. Discard supernatant. Let the pellet dry at room temperature for 10 min.
11. Add 20–40 μL DEPC-treated water to each sample.
12. Iscript™ cDNA synthesis kit was used following the manufacturer's protocol. In 20 μL volume for each reaction, use 1 μg total mRNA.

3.4.2 Quantitative Real-Time PCR

1. For each qRT-PCR reaction, add 0.5 μL cDNA from each C57BL/6 or HDAC11 mouse sample, 0.5 μL of each HDAC11 exon 3 forward and reverse primer (*see* Table 2), 5 μL Bio-Rad SYRB green super mix, and 3.5 μL DEPC-treated water.

   For cells from HDAC11 knockout mice, no HDAC11 exon 3 can be detected by qRT-PCR.
2. For qRT-PCR program (*see* Table 5), 58–60 °C melting temperature was used based on the HDAC11 exon 3 primer gradient study (Fig. 3).

This protocol has been the standard operating procedure for studies performed in the HDAC11KO mice and this procedure has been instrumental in the preparation of our latest manuscript identifying the regulatory role on HDAC11 in neutrophil function [16].

Table 5
Quantitative RT-PCR

Step	Temperature (°C)	Time		Actions per step
1	95	3	minutes	
2	95	30	seconds	
3	60	30	seconds	Read
4				Go to step 2 40 times
5	95	1	minute	
6	58	10	seconds	**Melt Curve** (Start Temp.) Read 0.5°C/cycle to End Temp.
7	95			**Melt curve** (End Temp.)

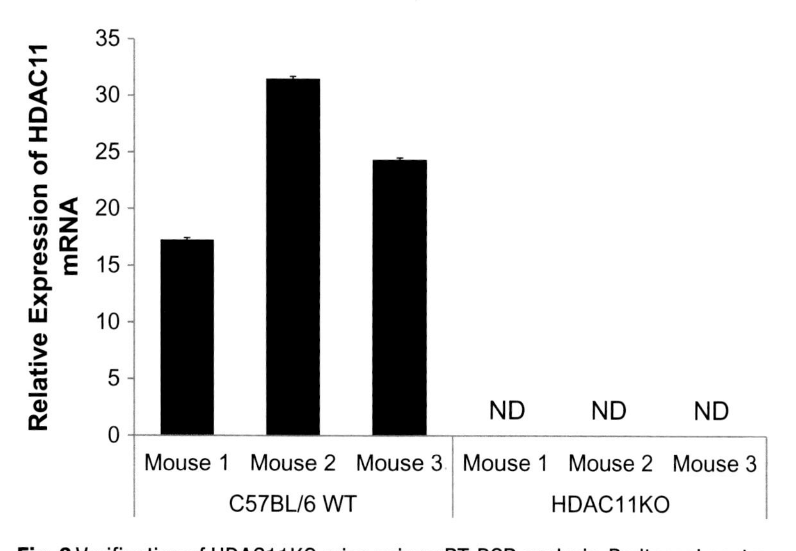

Fig. 3 Verification of HDAC11KO mice using qRT-PCR analysis. Peritoneal neutrophils (PNs) from 3 C57BL/6 (*left*) and HDAC11 KO (*right*) mice were harvested at 18 h post-thioglycolate injection at 5 %. Expression of HDAC11 mRNA was measured by qRT-PCR analysis

3.5 Chromatin Immunoprecipitation (ChIP) to Determine the Regulatory Role of HDAC11

3.5.1 Cells Cross-Linking

Sample Preparation

1. Approximately 20×10^6 cells from C57BL/6 and HDAC11 knockout mice per condition will be optimal.
2. Cells were collected in 50 mL conical tubes and washed twice with 10 mL 1× PBS.
3. Remove PBS by centrifuging at $1500 \times g$ for 5 min at room temperature.

4. Gently reconstitute cells in 20 mL of cross-linking buffer, set up the sample tubes on a tube-rocker or orbital rotator, and incubate for 10 min at room temperature (*see* **Note 3**).
5. Centrifuge samples at 1500 ×*g* for 5 min, at room temperature, and remove the supernatant.
6. Cells were gently reconstituted in 20 mL of stop buffer, and then incubated for 10 min on a tube-rocker at room temperature (**step 4**).
7. Centrifuge the samples at 1500 ×*g* for 5 min at 4 °C, and discard supernatant.
8. Reconstitute cells in 20 mL of PBS, gently invert tubes several times, spin at 1500 ×*g* for 5 min at 4 °C, and discard supernatant.
9. Repeat **step 8** if needed, and aliquot cells before centrifuge.
10. Proceed with chromatin preparation. Cell pellets from **step 9** can be stored at −80 °C until further use.

3.5.2 Chromatin Preparation

Sample Preparation

1. Remove sample tubes containing frozen cell pellets from −80 °C and allow to thaw on ice.
2. Prepare Dounce homogenizer (7 mL), set on ice.
3. Reconstitute each pellet of 20×10^6 cells in 2 mL of lysis buffer. Gently pipette up and down until completely homogenized. Incubate on ice for 10 min.
4. Transfer cells to the Dounce homogenizer and apply 25 strokes using the loose pestle (*see* **Note 4**). Transfer the nuclei to a new 15 mL conical and centrifuge at 1500 ×*g* for 5 min at 4 °C. Then discard supernatant.
5. Reconstitute nuclei pellet with 2 mL of wash buffer. Gently pipette up and down until completely homogenized. Incubate on ice for 10 min.
6. Pellet the nuclei by centrifuging at 1500 ×*g* for 5 min at 4 °C. Discard supernatant.
7. Add 600 μL sonication buffer to each tube, and gently pipette up and down until homogenized. Divide the sample by transferring 300 μL of nuclei sample to each 1.5 mL Eppendorf tube and proceed to the sonication section.
   Samples have to be kept on ice at all times.

Sample Sonication

1. For this specific protocol we used a Bioruptor™ XL from Diagenode. However, any bath sonicator with a timing programming will be suitable. Briefly, the water bath was precooled to 4 °C with crushed ice before each sonication cycle and a thin

layer of crushed ice was kept in the water bath for each cycle to avoid rapid temperature increase during sonication.

2. For each sonication cycle, use a program with eight pulses of 30 s followed by 30 s of resting time at 20 KHz frequency (300 W). To obtain 250 base pair to 500 base pair DNA fragments for the mouse peritoneal neutrophils, approximately eight cycles are required for each sample. However, this parameter has to be set for each cellular type (*see* **Note 5**). Centrifuge the samples at 16,000 ×*g* for 10 min at 4 °C. Transfer the supernatant to a new 1.5 mL Eppendorf tube (approximately 600 μL supernatant).
3. Take an aliquot (20 μL) from each sample and proceed to the *sample shear check* section. Either the rest of the samples can be stored at −80 °C or continue with immunoprecipitation.

Sample Shear Check

1. To each tube add 20 μL sample aliquot, 47.2 μL H_2O, and 2.8 μL of 5 M NaCl to reach a final concentration of 200 mM.
2. Incubate at 65 °C for 4 h or overnight.
3. Add RNase A to a final concentration of 20 μg/mL to each sample aliquot; incubate at 37 °C for 30 min.
4. Add Proteinase K to a final concentration of 100 μg/mL to each sample aliquot; incubate at 42 °C for 2 h.
5. Add 200 μL phenol/chloroform/isoamyl alcohol pH 6.7 to each sample aliquot, vortex for 15 s, and spin down at 16,000 ×*g* for 5 min.
6. Transfer 15 μL of the supernatant from the top layer; add 3 μL of 5× DNA loading dye with light xylene cyanol only.
7. Run samples on 1.5 % agarose gel at 40 V for 1.5 h.

3.5.3 Chromatin Immunoprecipitation

Sample Pre-clearance

1. Add 30 μL of protein A agarose beads to each sample, and incubate on a tube rotator for 1 h at 4 °C.
2. Centrifuge the samples at 6000 ×*g* for 5 min at 4 °C.
3. Transfer the supernatant to a new 1.5 mL Eppendorf tube. Take an aliquot of 50 μL from each sample as input. Input samples can be stored at −80 °C before reverse cross-linking.

Immunoprecipitation

1. Dilute 50 μL of each sample after preclearing with 450 μL dilution buffer I, to 500 μL final volume, one dilution per antibody, into a 1.5 mL Eppendorf tube.
2. During the first dilution tube of each sample, add a mixture of 5 μL (5 μg) rabbit-anti-HDAC11 antibody from Sigma and

20 μL (4 μg) rabbit-anti-HDAC11 antibody from BioVision. In the second dilution tube of each sample, add 9 μg normal rabbit IgG-B.

3. Set the sample tubes on a tube rotator overnight, at 4 °C, and cover the tubes with aluminum foil to avoid light exposure (*see* **Note 6**). Add 50 μL of protein A agarose beads to each sample, and incubate on a tube rotator for 4 h at 4 °C.

Sample Wash

1. After immunoprecipitation, centrifuge the samples at 6000 × *g* for 5 min at 4 °C, and discard supernatant.
2. Wash the pellet with 1 mL of each buffer in the following order:

   Twice with dilution buffer I.

   Twice with dilution buffer II.

   Twice with LiCl buffer.

   Twice with TE buffer.

   For each wash, centrifuge at 6000 × *g* for 5 min at 4 °C.

Elute and Reverse Cross-Linking

1. Discard supernatant from the last wash. Add 100 μL of elution buffer to each sample and incubate at 65 °C for 15 min. The immuno-complexes will be in the soluble fraction.
2. Centrifuge the samples at maximum speed for 20 s. Transfer supernatant to a new 1.5 mL Eppendorf tube.
3. Remove the input sample tubes from −80 °C and thaw on ice. Add 50 μL of elution buffer to each input tube to reach a final volume of 100 μL.
4. Add 4.16 μL 5 M NaCl (200 mM final concentration) to each IP and input sample tubes. Reverse cross-linking at 65 °C for 6 h or overnight.

DNA Purification

1. Add RNase A to a final concentration of 20 μg/mL to each sample; incubate at 37 °C for 30 min.
2. Add Proteinase K to a final concentration of 100 μg/mL to each sample; incubate at 42 °C for 2 h.
3. Use Qiagen™ DNA purification kit to purify each DNA sample.

   To obtain a higher amount of DNA, researchers can elute DNA twice from each column; use 50 μL elution buffer each time to obtain a higher amount of DNA.
4. Use 2 μL DNA for each qRT-PCR reaction. This is in a triplicate reaction, and the purified DNA must be kept in TE buffer pH 8.0.

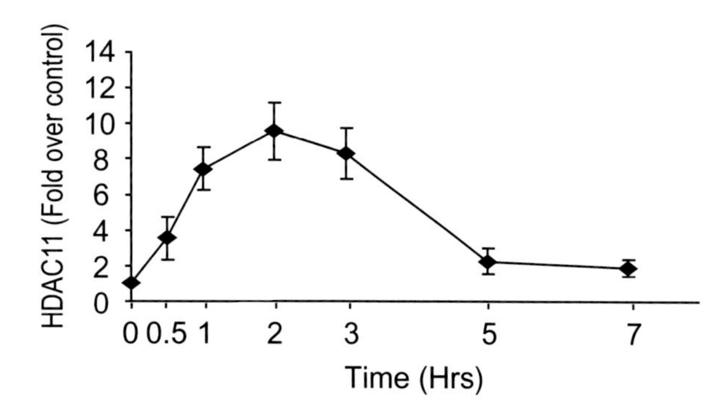

Fig. 4 ChIP analysis of HDAC11 in IL-10 promoter region. HDAC11 is recruited to the IL-10 gene promoter in macrophages under LPS stimulation. Macrophages were treated with LPS at 1.0 g/mL, and then harvested at 0 or 0, 0.5, 1, 2, 3, 5, and 7 h after treatment. Cells were then harvested and proceed to ChIP analysis using anti-HDAC11 cocktail and anti-rabbit IgG as described. Quantitative real-time PCR analysis was performed in the −87 base pair to −7 base pair region of the mouse gene IL-10 promoter. The recruitment of HDAC11 to IL-10 promoter was observed at 0.5 h and reached to a peak by 2 h after LPS stimulation [8] (representative figure from previously published studies)

It is recommended to determine the primer efficiency and melting temperature before running PCR on the ChIP samples.

This protocol has been used as a standard operating procedure by our lab and collaborators [8]. A prototypical experimental result is shown in Fig. 4. Briefly, we evaluated the recruitment of HDAC11 to the il10 promoter after stimulation of mouse macrophages with LPS (*see* **Note 7**).

4 Notes

1. If the resultant cell pellet is still robustly red in color, the ACK lysis treatment has to be repeated. Two lysis cycles are sufficient to remove enough red blood cells for analysis. This is directly related to the initial volume of red blood cells added to each tube.
2. 100 % of the granulocyte population will express eGFP and ~35–50 % (depending on the age of the mouse) of the lymphocyte population, also determined by FSC versus SSC, will express eGFP. We chose mice with eGFP expression greater than 10^3 mean fluorescent intensity unit with the aforementioned expression pattern as HDAC11-eGFP-positive mice.
3. Both chromatin shearing efficiency and immunoprecipitation efficiency can be affected by formaldehyde concentration and cross-linking time. In this protocol, we are using 1 % formaldehyde and 10-min incubation time for mouse peritoneal neu-

trophils. Cross-linking time and formaldehyde percentage should be optimized for other cell types, such as mouse T cells or cell lines [17, 18].

4. The Dounce homogenizer has to be set on ice at all times. It can be rinsed with cold PBS 3–5 times between different samples. The number of strokes could vary and should be optimized according to different cell types or cell lines. Briefly, take an aliquot of cells and stain with trypan blue and check under the microscope. Adjust stroke times until researchers can see that over 90% of the cells were stained but have kept a round nuclei. Decrease the number of strokes if broken nuclei were observed.
5. Sonication efficiency varies according to different cell types and equipment. It is recommended to optimize this step by a pilot experiment using a number of different sonication cycles on different cell types or cell lines. For some types of sonicator, it is recommended to check the water bath temperature and replace ice after each cycle, or the water bath could be kept on a heat-adjusted platform to maintain a 4 °C environment.
6. Using a mixture of two different HDAC11 antibodies has been confirmed with good immunoprecipitation efficiency on macrophages [8] and mouse peritoneal neutrophils (manuscript in submission). It is recommended to optimize IP antibody use for different cell types or cell lines.
7. HDAC11 has been observed to be differentially expressed in various compartments of myelopoiesis. This expression of HDAC11 appears to coincide with the granularity of cell subtype. HDAC11 also appears to have a role in the fate of cell differentiation in the process of myelopoiesis.

Acknowledgments

The authors gratefully acknowledge the flow cytometry core facilities at H. Lee Moffitt Cancer Center and their extended technical support for our project.

References

1. Villagra A, Sotomayor EM, Seto E (2010) Histone deacetylases and the immunological network: implications in cancer and inflammation. Oncogene 29(2):157–173
2. Foster SL, Hargreaves DC, Medzhitov R (2007) Gene-specific control of inflammation by TLR-induced chromatin modifications. Nature 447(7147):972–978
3. Woan KV, Sahakian E, Sotomayor EM, Seto E, Villagra A (2012) Modulation of antigen-presenting cells by HDAC inhibitors: implications in autoimmunity and cancer. Immunol Cell Biol 90(1):55–65
4. Miremadi A, Oestergaard MZ, Pharoah PDP, Caldas C (2007) Cancer genetics of epigenetic genes. Hum Mol Genet 16(R1):R28–R49

5. Schreiber RD, Old LJ, Smyth MJ (2011) Cancer immunoediting: integrating immunity's roles in cancer suppression and promotion. Science 331(6024):1565–1570
6. Gao L, Cueto MA, Asselbergs F, Atadja P (2002) Cloning and functional characterization of HDAC11, a novel member of the human histone deacetylase family. J Biol Chem 277(28):25748–25755
7. Sahakian E, Powers JJ, Chen J et al (2014) Histone deacetylase 11: a novel epigenetic regulator of myeloid derived suppressor cell expansion and function. Mol Immunol 63(2):579–585
8. Cheng F, Lienlaf M, Perez-Villarroel P et al (2014) Divergent roles of histone deacetylase 6 (HDAC6) and histone deacetylase 11 (HDAC11) on the transcriptional regulation of IL10 in antigen presenting cells. Mol Immunol 60(1):44–53
9. Gong S, Zheng C, Doughty ML et al (2003) A gene expression atlas of the central nervous system based on bacterial artificial chromosomes. Nature 425(6961):917–925
10. Heintz N (2001) BAC to the future: the use of bac transgenic mice for neuroscience research. Nat Rev Neurosci 2(12):861–870
11. Sahakian E, Powers JJ, Chen J et al (2015) Histone deacetylase 11: a novel epigenetic regulator of myeloid derived suppressor cell expansion and function. Mol Immunol 63(2):579–585
12. Wang H, Cheng F, Woan K et al (2011) Histone deacetylase inhibitor LAQ824 augments inflammatory responses in macrophages through transcriptional regulation of IL-10. J Immunol 186(7):3986–3996
13. Villagra A, Cheng F, Wang HW et al (2009) The histone deacetylase HDAC11 regulates the expression of interleukin 10 and immune tolerance. Nat Immunol 10(1):92–100
14. Cheng F, Lienlaf M, Wang HW et al (2014) A novel role for histone deacetylase 6 in the regulation of the tolerogenic STAT3/IL-10 pathway in APCs. J Immunol 193(6): 2850–2862
15. Cheng F, Wang H, Horna P et al (2012) Stat3 inhibition augments the immunogenicity of B-cell lymphoma cells, leading to effective antitumor immunity. Cancer Res 72(17): 4440–4448
16. Eva Sahakian JC, John J. Powers, Xianghong Chen, Kamira Maharaj, Susan L. Deng, Maritza Lienlaf, Hong Wei Wang, Andressa L. Sodré, Allison Distler, Limin Xing, Patricio Perez-Villarroel, Sheng Wei, Alejandro Villagra, Ed Seto, Eduardo M. Sotomayor, Pedro Horna and Javier Pinilla-Ibarz (2015) Essential regulatory role for histone deacetylase 11 (HDAC11) In neutrophil function. *J Leukoc Biol.* Under 2nd revision
17. Lee TI, Johnstone SE, Young RA (2006) Chromatin immunoprecipitation and microarray-based analysis of protein location. Nat Protoc 1(2):729–748
18. Nelson JD, Denisenko O, Bomsztyk K (2006) Protocol for the fast chromatin immunoprecipitation (ChIP) method. Nat Protoc 1(1): 179–185

Part II

Class III Deacetylases

Chapter 12

Sirtuin1 (SIRT1) in the Acetylation of Downstream Target Proteins

Ana R. Gomes, Jay Sze Yong, Khai Cheng Kiew, Ebru Aydin, Mattaka Khongkow, Sasiwan Laohasinnarong, and Eric W.-F. Lam

Abstract

Acetylation has been shown to be an important posttranslational modification (PTM) of both histone and nonhistone proteins with particular implications in cell signaling and transcriptional regulation of gene expression. Many studies have already demonstrated that SIRT1 is able to deacetylate histones and lead to gene silencing. It can also regulate the function of tumor suppressors including FOXO proteins and p53 by deacetylation. Here, we describe three experimental approaches for studying the modulation of the acetylation status of some of the known downstream targets of SIRT1.

Key words SIRT1, Acetylation, Immunoprecipitation, Western blotting, Site-directed mutagenesis

1 Introduction

Sirtuins or class III HDACs were first discovered in yeast and later found to be crucial for promoting longevity in higher eukaryotes [1, 2]. In mammals, this family has seven members and they function as NAD+-dependent deacetylases as well as ADP ribosyltransferases [3] targeting both histone and nonhistone proteins by modulating their acetylation levels (Table 1). SIRT1, the mammalian orthologue of the yeast silent information regulator 2 (Sir2), can mediate chromatin silencing and heterochromatin formation through the deacetylation of specific histones including H1, H3, and H4 [4] and modulates the activity of tumor suppressors such as p53 [5] and FOXO proteins [6, 7].

Although the role of SIRT1 in longevity is relatively well established, its mechanism of action and cellular substrates in cancer remain elusive. Overexpression of SIRT1 is found in various cancers including prostate [8], breast [9], and colon cancers [10] as well as acute myeloid leukaemia (AML) [11]. Functional studies have shown that SIRT1 interacts and deacetylates p53 and thereby

Sibaji Sarkar (ed.), *Histone Deacetylases: Methods and Protocols*, Methods in Molecular Biology, vol. 1436,
DOI 10.1007/978-1-4939-3667-0_12, © Springer Science+Business Media New York 2016

Table 1
Deacetylation of specific lysine residues by SIRT1 inhibits the activities of the above known downstream target genes

Target	Enzymatic activity	Residues	Function
Histone targets of SIRT1			
Histone H1	Deacetylation	K26 [4, 21]	Chromatin silencing, chromatin remodeling
Histone H3	Deacetylation	K9 [21], K14 [21], K56 [22]	Genomic stability, gene activity and heterochromatin silencing, histone incorporation into nucleosomal chromatin in DNA replication and repair, stem cell-specific transcriptional networks, chromatin responses to DNA damage, chromatin remodeling, DNA replication, stress resistance, cell cycle arrest
Histone H4	Deacetylation	K16 [21]	Transcription repression, chromatin silencing
Nonhistone targets of SIRT1			
Transcription factors and coregulators			
FOXO3a	Deacetylation	K242, K245, K262 [23]	Tumor suppressor, induction of cell cycle arrest, resistance to oxidative and genotoxic stress, inhibition of FOXO-induced apoptosis
FOXO1	Deacetylation	K245, K245, K262 [23]	Tumor suppressor, cell cycle arrest, apoptosis, adipogenesis
FOXO4	Deacetylation	K182, K185, K199, K211, K233L K403 [24]	Tumor suppressor, cell cycle arrest, apoptosis, neural differentiation
FOXO6	Deacetylation	K173, K176, K190, K202, K229 [24]	Tumor suppressor, cell cycle arrest, apoptosis, regulation of nervous system
p300/CBP	Deacetylation	CRD1 domain (K1020, K1024) [25]	Histone acetyltransferase, rate-limiting transcriptional cointegrator/cofactor of diverse transcription factors, metabolism, and cellular differentiation

PGC1α	Deacetylation	At least 13 residues	Transcriptional coactivator that regulates the genes that involved in energy metabolism, mitochondrial biogenesis
PPARγ	Deacetylation	K268, K293 [26] K154, K155 [27]	Regulates energy homeostasis by regulation of adipose tissues, promoting energy expenditure over energy storage, lipogenesis, differentiation, anti-inflammation, insulin sensitivity, cellular proliferation, apoptosis, and autophagy [27]
Nuclear factor (NF-κB)	Deacetylation	K218, K221, RelA/p65 at K310 [24, 28]	Adhesion, cell cycle, angiogenesis, apoptosis, regulation of innate and adaptive immune responses and carcinogenesis K218, K221: DNA binding activity K310: transcriptional activity of NF-kB
Autophagy proteins (ATG5, ATG7, ATG8)	Deacetylation	?	Autophagy proteins for degradation of proteins and organelles
Retinoblastoma protein (pRb)	Deacetylation	K873, K874 [29]	Cell cycle control via interactions with E2F transcription factors, role in regulating cellular differentiation, regulates p73 activity E2F: Cell cycle regulation, apoptosis by both p53-dependent or -independent mechanisms, activation of downstream pro-apoptotic target genes, including p53/p73, apoptosis in response to DNA damage, p53/p73 activation, nucleotide excision repair/DNA damage/DNA repair
c-Jun	Deacetylation	K271 [24]	Regulates AP-1 transcriptional activity, cell cycle progression, and apoptosis
c-Myc	Deacetylation	K382 [30]	Apoptosis
MyoD	Deacetylation	K99, K102, K104 [31]	Skeletal muscle differentiation
Myocyte-enhancing factor 2 (MEF2)	Deacetylation	K424 [32]	Muscle cell differentiation process, apoptosis

(continued)

Table 1
(continued)

Target	Enzymatic activity	Residues	Function
Tumor suppressors			
p53 p73 (high homology to p53) [30]	Deacetylation/ activated	K320, K373, K382 [23, 33]	Tumor suppressor, induce cell cycle arrest, apoptosis, senescence, DNA repair in response to DNA damage and oxidative stress
HIC1 (hypermethylated in cancer 1)	Deacetylation	K314 [34]	Transcriptional repressor which is involved in the regulation of growth control, cell survival and DNA damage response, tumor development and progression
Signaling proteins			
Ku-70	Deacetylation	K539, K542 [35]	Apoptosis, senescence, DNA damage, DNA repair protein
Xeroderma pigmentosum group A (XPA) Xeroderma pigmentosum group C (XPC)	Deacetylation	K63, K67 [36]	XPA is a core nucleotide excision repair (NER) factor essential for NER process in the UV-induced DNA damage repair pathway XPC is essential for initiating global genome NER by recognising the DNA lesion and recruiting downstream factors
Nijmegen breakage syndrome protein (NBS1)	Deacetylation	K233, K690 [37]	Involved in DNA damage repair pathway; phosphorylation of NBS1 by the ATM kinase is necessary for both activation of the S-phase checkpoint and for efficient DNA damage repair response
Heat-shock factor 1 (HSF1)	Deacetylation	K80 [38]	Protein homeostasis and the heat-shock response (HSR). Protect cells from protein-damaging stress associated with misfolded proteins and regulates the insulin-signaling pathway and aging
BCL6	Deacetylation	K379 [39]	Proto-oncogene, encodes transcriptional repressor, induction of apoptosis, neurogenesis

SMAD3	Deacetylation	K378 [40]	Crucial for intracellular signaling of transforming growth factor-B (TGF-B)
SMAD7	Deacetylation	K64, K70 [24]	Negative modulators of TGF-B family signaling: cell growth, differentiation, migration, apoptosis
MLH1 Pms2	Deacetylation	MLH1: K377 [41] Pms2: K431 [41]	Mismatch repair proteins involved in DNA repair to maintain genome integrity
Hypoxia-inducible factor (HIF) HIF-1α HIF-2α	Deacetylation	K647 [42]	HIF-1α: Control glycolysis in response to hypoxic conditions HIF-2α: Hypoxic stress responses
Phosphatase and tensin homologue (PTEN)	Deacetylation	K402 [43]	Tumor suppressor, lipid phosphatase, dephosphorylates phosphatidylinositols (PI3P/PIP,4P2/PI3,4,5P3), which are the product of PI3K, reverses the biological functions of PI3K, resulting in suppression of the mitogenic and antiapoptotic effects of PI3K
Androgen receptor (AR)	Deacetylation	K630, K632, K633 [44]	Nuclear receptor involved in endocrine dysfunction and prostate cancer PolyQ expansion of AR causes neurodegenerative disease spinal and bulbar muscular atrophy (SBMA)
β-Catenin	Deacetylation	K345 [24]	Proliferation through induction of Wnt signaling pathway involved in breast cancer
Phosphatidylinositol-4-phosphate 5-kinase (PIP5K)gamma	Deacetylation	K265/K268 [45]	Thyroid hormone secretion, regulation of exocytosis of TSH-containing granules
H2A.Z	Deacetylation	K115, K121 [46]	Promote cardiac hypertrophy

(continued)

Table 1
(continued)

Target	Enzymatic activity	Residues	Function
Others			
Liver X receptor (LXRα and LXRβ)	Deacetylation	LXRα: K432 [47] LXRβ: K433 [47]	Sterol sensors to regulate the cholesterol metabolism and hepatic lipid homeostasis
DNA methyl transferase (DNMT1)	Deacetylation	K1349, K1415 [48]	DNA methylation, chromatin organization, DNA repair, cell cycle regulation, apoptosis
Acetyl-CoA synthase (AceCS) 1	Deacetylation	K635 [49]	Mammalian metabolic enzyme involved in regulation of fatty acid synthesis through acetate incorporation
Endothelial nitric oxide (NO) synthase (eNOS)	Deacetylation	K496, K506 [50]	Links nutrient sensing with vascular physiology and its age-dependent derangement in cardiovascular diseases
Inhibitor of growth protein 3 (ING3)	Deacetylation	K181, K264 [41]	Tumor suppressor, regulating cell growth and tumorigenesis, cell survival, inflammation. Overexpression inhibits cell growth and promote apoptosis
Mini-chromosome maintenance complex (Mcm4 and Mcm6)	Deacetylation	Mcm4: K857, K818 [41] Mcm6: K173 [41]	Replicative helicases for DNA replication in the S phase of the cell cycle
Cyclin A2	Deacetylation	K67 [41]	Regulates proliferation by its involvement in cell cycle progression by interacting with CDK kinases

limits its tumor-suppressive functions. Accordingly, overexpression of SIRT1 decreases p53-dependent transcriptional activity in response to DNA damage [12–15]. Conversely, the tumor-suppressive functions of SIRT1 have also been proposed. *Sirt1*$^{-/-}$ mouse embryonic cells have been reported to display genetic instability due to defective and reduced DNA damage repair capacity [16]. Moreover, SIRT1-overexpressing transgenic mouse strains also show reductions in spontaneous carcinomas and enhanced mouse ageing, supporting that SIRT1 promotes genetic stability and suppresses tumor formation [17]. SIRT1 has also been demonstrated to be a downstream target of BRCA1, inhibiting the expression and function of the anti-apoptotic protein Survivin [18]. This apparent paradoxical dual role of SIRT1 in tumorigenesis can also be illustrated in the regulation of its cellular targets FOXO proteins. SIRT1 has been shown to bind and deacetylate FOXO3a resulting in a bypass of its classical transcriptional activation from pro-apoptotic genes (*Fas ligand* and *BIM*) to genes that regulate cell cycle arrest (*p27Kip1*) and DNA damage repair (*GADD45*) [19]. Evidence has also been provided to illustrate that targeting SIRTs and subsequently increasing the acetylation levels of p53 and FOXO proteins can be an effective strategy to enhance proliferative arrest and cell death [15] and to reverse drug resistance [20] in cancer. In this book chapter, some of the practical laboratory methodologies and techniques for studying the functions of SIRT1 will be described.

2 Materials

2.1 Chemicals, Reagents, and Equipment Used for Immunoprecipitation (IP)

1. IP lysis buffer [50 mM Tris–HCl pH 8.8, 150 mM NaCl, 5 mM EDTA, 1% IGEPAL CA-630], resolved in distilled water, 2 mM phenylmethanesulfonyl fluoride (PMSF), 1 mM NaF (sodium fluoride), 1 mM Na_3VO_4 (sodium orthovanadate), and protease inhibitor cocktail tablet.
2. Sample buffer 2× [(2% SDS, 25% glycerol, 62.5 mM Tris-Cl pH 6.8, 350 mM dithiothreitol (DTT), and bromophenol blue (~0.05 mg/ml)].
3. BCA Bradford Assay Solutions A and B (Pierce BCA Protein Assay Reagent A, and Pierce BCA Protein Assay Reagent B).
4. Tecan Sunrise Microplate Reader.
5. Antibodies: Acetylated-lysine antibody, FOXO3a antibody and FOXO3a antibody, SIRT1 antibody, negative control rabbit IgG1 and negative control rabbit immunoglobulin fraction.
6. Dynabeads Protein A/G, Protein A and Protein G.
7. DynaMag-2, Magnetic Particle Concentrator.

2.2 Chemicals, Reagents, and Equipment Used for Western Blotting

2.2.1 SDS Polyacrylamide Gel Electrophoresis

1. Resolving gel buffer: 1.5 M Tris pH 8.8.
2. Stacking gel buffer: 1.5 M Tris pH 6.8.
3. 10 % Sodium dodecyl sulfate (SDS) aqueous solution.
4. 25 % Ammonium persulfate (APS) aqueous solution.
5. Tetramethylethylenediamine (TEMED).
6. 30 % Acrylamide/0.8 % bisacrylamide.
7. Isopropanol.
8. Bio-Rad Mini-PROTEAN system—casting stand, appropriate casting frame, combs, and glass plates.
9. SDS-running buffer: 0.1 % (w/v) SDS, 25 mM Tris, 192 mM glycine.
10. Protein ladder: Novex Sharp Pre-stained Protein Standard.

2.2.2 Protein Transfer: Western Blot

1. Nitrocellulose membrane 0.45 μm.
2. Whatman 3 mm paper.
3. Ethanol.
4. Transfer buffer: 25 mM Tris, 190 mM glycine, and 20 % ethanol.
5. SDS-PAGE gel wet transfer apparatus.

2.2.3 Anti-Acetyl-Lysine Detection

1. Rabbit pan-specific acetylated-lysine antibody.
2. Rabbit FOXO3a antibody.
3. Anti-rabbit horseradish peroxidase conjugated secondary antibody.
4. 5 % (w/v) bovine serum albumin (BSA).
5. TBS-T: 20 mM Tris pH 7.6, 136 mM NaCl, 0.01 % (v/v) Tween.
6. Western Lighting® Enhanced Chemiluminescence (ECL) substrate kit.

2.3 Chemicals, Reagents, and Equipment Used for Site-Directed Mutagenesis

1. 1 μl 2.5 U/μl *PfuUltra* High Fidelity DNA polymerase.
2. 10× Reaction buffer.
3. Dpn I restriction enzyme.
4. Oligonucleotide control forward primer.
5. Oligonucleotide control reverse primer.
6. pWhitescript 4.5-kb control plasmid.
7. QuikSolution reagent.
8. dNTP mix.
9. XL10-Gold ultracompetent cells.

10. β-Mercaptoethanol mix (β-ME).
11. pUC19 control plasmid (0.1 ng/μl in TE buffer).
12. 14 ml BD Falcon polypropylene round-bottom tubes.
13. 5-Bromo-4-chloro-3-indolyl-β-d-galactopyranoside (X-gal).
14. Isopropyl-1-thio-β-d-galactopyranoside (IPTG).
15. Applied Biosystems GeneAmp PCR System 9700.
16. NaCl.
17. NZYM broth.
18. Tris–HCl.
19. EDTA.

3 Methods

3.1 Immunoprecipitation (IP)

Immunoprecipitation is a technique widely used to enrich the amount of a specific protein from cell lysates and to identify protein-protein interactions. The principle of IP relies on the binding of either mono- or polyclonal antibodies to its native target protein in a cell lysate. These complexes are immobilized onto magnetic beads (protein A or G). The complexes are later precipitated, eluted, and resolved by SDS-PAGE.

3.1.1 Lysate Preparation (*See* Notes 1–3)

1. When cells are ~80–90 % confluent (yields ~600–1000 μg total protein), harvest cells and wash the pellet in PBS before centrifuging at 1087 ×*g* for 5 min at 4 °C.
2. Discard the supernatant and centrifuge again (1087 ×*g*) briefly (30 s) at 4 °C to discard the excess of PBS. Pellets can be frozen at −80 °C until further use.
3. Lyse the pellets in IP lysis buffer on ice for 10 min and vortex 4–5 times throughout the incubation period.
4. The lysates are centrifuged at full speed (53248 ×*g*) for 10 min at 4 °C.
5. Discard the pellets and transfer the supernatants to new 1.5 ml microcentrifuge tubes.
6. Calculate protein concentrations by using BCA Bradford Assay; the reagents A and B are used in 1:50 ratio. Briefly, the sample is added to the plate containing the mix of reagents A + B followed by a 30-min incubation at 37 °C; the resultant purple color is measured at 562 nm in a TECAN microplate absorbance reader.
7. Pipette ~200–500 μg cell lysates per IP sample and take 1:10 of that amount to reserve for your input control.

3.1.2 Pre-clearing (See Note 4)

1. Incubate lysates with the corresponding 20 μl of protein A/G Dynabeads, previously equilibrated, at 4 °C using a rotating device between 4 and 6 h to minimize any nonspecific binding of beads.
2. Once the incubation period of beads with the IP lysates is completed, the lysates are transferred to a new set of 1.5 ml microcentrifuge tubes and the beads discarded by using the magnetic stand.

3.1.3 Antibody–Antigen incubation

1. Prepare fresh 20 μl of protein A/G Dynabeads washed as described previously.
2. Add the new A/G Dynabeads into the new set of 1.5 ml microcentrifuge tubes containing the lysates.
3. Lysates should be refilled up to 500 μl with PBS and incubated with 0.5–2 μg of antibody overnight at 4 °C on a rotator.

3.1.4 Immunoprecipitation

1. On the next day, the complex Dynabead-antibody-protein should have been achieved. Remove samples from the rotator and using the magnetic stand discard the supernatant and keep the Dynabeads.
2. Using the magnetic stand, gently wash the Dynabead with 500 μl PBS three times; after the final wash, transfer the IP samples into new 1.5 ml microcentrifuge tubes and remove all the residual PBS by centrifugation at 53248 × *g* for 1 min at 4 °C.
3. Thaw the input control on ice.
4. Add 20 μl of sample buffer 2× to the beads and to the input control; vortex and quickly centrifuge the samples for 1 min at full speed (53248 × *g*).
5. Denature proteins by boiling for 5 min at 95–100 °C in the heat block. To collect the samples, centrifuge for 1 min at full speed (53248 × *g*).
6. Samples are loaded on a SDS-PAGE gel and analyzed by Western blotting (Fig. 1).

3.2 Western Blotting

There are a variety of approaches available that can be used for identification and detection of acetylated proteins. For in vitro analysis of deacetylation of SIRT1 downstream targets, immunoprecipitation of SIRT1 from whole-cell lysates, followed by sodium dodecyl sulfate-polyacrylamide gel electrophoresis (SDS) and Western blot analysis with pan-specific acetylated lysine immunodetection is required.

Acetylation is a posttranslational modification crucial for both chromatin remodeling and gene expression activity of certain proteins. Western blotting analysis is the most convenient and

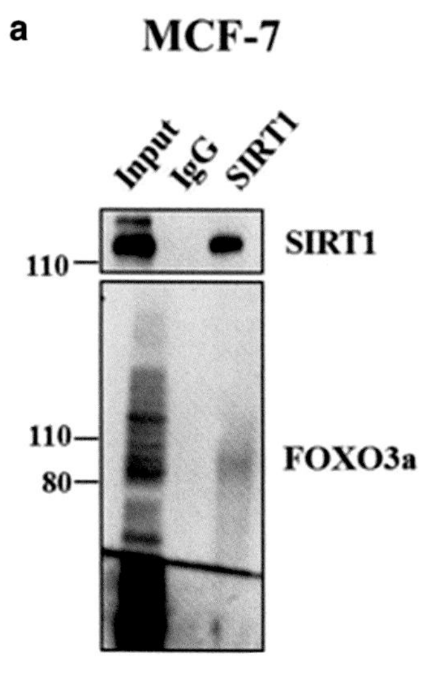

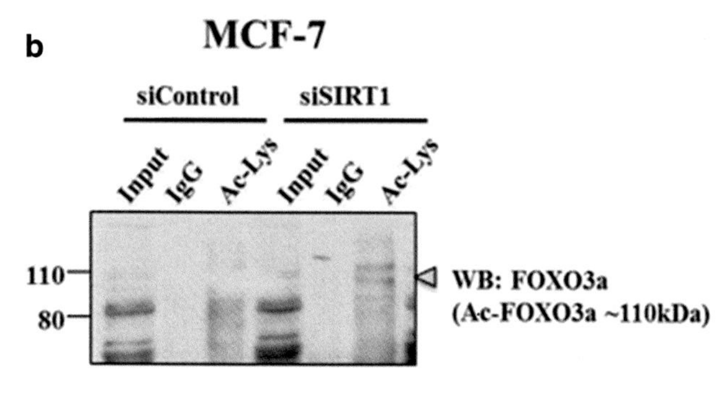

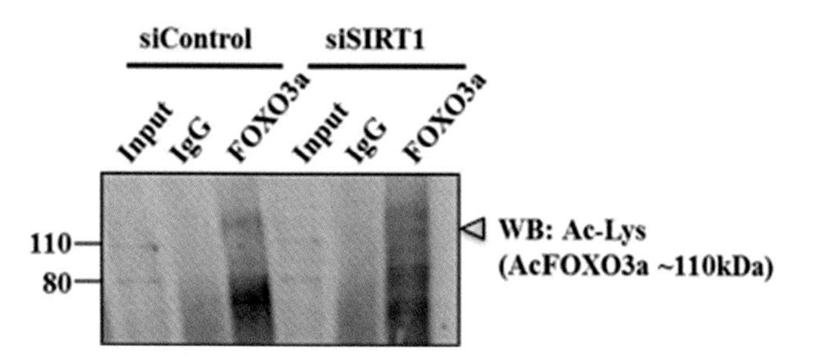

Fig. 1 SIRT1 modulates the acetylation status of FOXO3a in MCF-7 breast cancer cells. (**a**) MCF-7 cells and (**b**) MCF-7 transfected with control (siControl) or a SIRT1-siRNA-specific probe (siSIRT1) were subjected to immunoprecipitation with specific antibodies against acetylated lysine (Ac-Lys), FOXO3a or normal rabbit immunoglobulin G as a control. The immunoprecipitates were analyzed for the presence of the indicated proteins by Western blot

Table 2
Components and recipe of 7 % SDS-PAGE gel

	Resolving gel (7 %)	Stacking gel (5 %)
dH_20 (ml)	5.02	3.67
1.5 M Tris pH 8.8 (ml)	2.5	
1.5 M Tris pH 6.8 (ml)		0.42
30 % Acrylamide (ml)	2.33	0.83
10 % SDS (μl)	100	50
25 % APS (μl)	40	20
TEMED (μl)	10	10

cost-effective method to assess the acetylation status of a protein. This can be done by using either a site-specific antibody recognizing a specific lysine acetylation residue or a pan-specific acetylated antibody.

3.2.1 SDS-Polyacrylamide Gel Electrophoresis Preparation

1. Cast a 7% acrylamide (v/v) resolving gel (Table 2). The formation of polyacrylamide is based on a radical polymerization process. Radical formation from APS in combination with the catalytic properties of TEMED means that adding both last is highly recommended. Immediately pour the resolving mixture into the pre-assembled gel plate up to the first line of the casting frame. Add enough isopropanol to completely cover the top of the resolving gel to eliminate air pockets and allow consistent polymerization.
2. Depending on the ambient room temperature, polymerization can take up between 10 and 30 min. After polymerization of resolving gel, carefully rinse off all isopropanol using distilled water. Using filter paper, remove any residual water without touching the gel. Cast a 5 % stacking gel (Table 2) and immediately insert the 1.5 mm combs due to faster polymerization reaction. After approximately 10 min, stacking gel should be polymerized.
3. The complete gel is ready for electrophoresis immediately; however gels can also be wrapped in wet paper and stored overnight at 4 °C for next-day use.
4. Place the gel in an electrophoresis chamber and ensure that the combs are facing inwards. Fill the chamber with SDS-running buffer up to the top.
5. Remove combs and use a syringe and needle to carefully remove any floating residual polyacrylamide.

6. Load 4 μL of Novex Sharp Pre-stained Protein Standard ladder and 20 μL of sample into each corresponding well.
7. Connect the chamber after closing into a compatible power pack (Bio-Rad) and run the 7% gel for approximately 2 h at 90 V. Use bromophenol blue line as an indicator of electrophoresis progress.

3.2.2 Protein Transfer: Western Blot

1. Place a nitrocellulose membrane (usually 6 mm × 9 mm) in transfer buffer.
2. Soak two sponges and two Whatman papers (usually 7 mm × 10 mm) in transfer buffer and assemble the sandwich cassette configuration in a tray filled with transfer buffer.
3. Carefully open the glass plate and cut off the stacking gel and place gel over the top of the first Whatman paper.
4. Now place a labeled membrane on top of the gel and ensure that the membrane covers the whole gel.
5. Place the second piece of Whatman paper on top of the membrane followed by second sponge.
6. Use a roller to carefully remove any bubbles in between layers, as this will lead to inconsistent protein transfer.
7. Prepare wet transfer apparatus filled with transfer buffer and place sandwich caste into the transfer apparatus with ice. The correct orientation should be gel on the negative cathode and the membrane to be towards the anode.
8. Close lid and plug in the transfer chamber. Run the transfer for 90 min at constant 90 V.
9. Disable the apparatus and transfer the membrane into a tray with Ponceau temporary stain for ladder labeling.
10. Place the membrane into TBS-T and wash on a rocking platform until membrane has no more Ponceau and it is completely white.

3.2.3 Anti-Acetyl-Lysine Detection

1. Block the membrane in 5% (w/v) bovine serum albumin (BSA) for 30 min to inhibit nonspecific binding sites.
2. Incubate the membrane with rabbit pan-specific acetylated-lysine (Ac-Lys) antibody or the specific rabbit FOXO3a antibody diluted at 1:1000 in BSA; incubate overnight at 4 °C on a shaker.
3. The next day, wash the membrane three times for 5 min each wash with TBS-T in order to remove unbound antibodies.
4. Incubate the membrane with anti-rabbit horseradish peroxidase-conjugated secondary antibody diluted at 1:3000 at room temperature in a slow shaker.

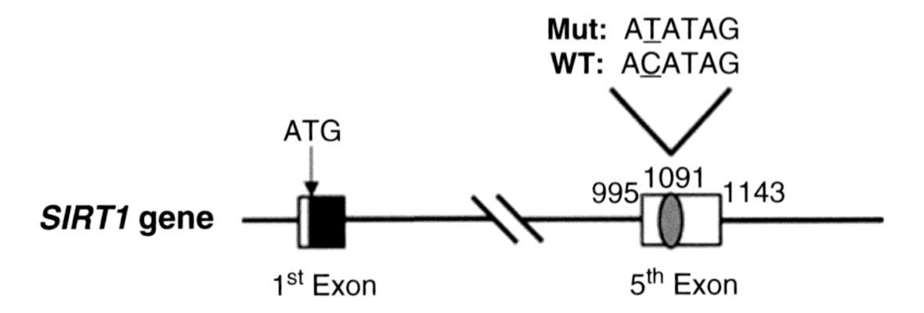

Fig. 2 Schematic representation of *SIRT1* gene and the point mutation introduced by site-directed mutagenesis

5. Wash the membrane five times, 5 min each wash with TBS-T, to remove unbound secondary antibodies.
6. Prepare the membranes for autoradiography signal detection by mixing 1:1 ratio of Western Lighting® Enhanced Chemiluminescence (ECL) substrate kit. Place the membrane on cling film and cover it with substrate. Fold down the cling film and with some tissue ensure that no air bubbles are on the membrane and remove the excess of ECL. Transfer the membrane into a developing cassette for film development (Fig. 1).

3.3 Site-Directed Mutagenesis

Subcloning is the technique used to isolate and transfer a gene of interest from its parent vector into a destination vector. Some genes are not naturally expressed at sufficiently high levels for functional studies. Therefore in many cases, subcloning allows the gene of interest to be amplified with a vector containing a strong promoter.

Site-directed mutagenesis is used in subcloning to introduce specific changes to the sequence of DNA—usually point mutations for deletions and additions—in order to study the structure and function of the DNA, RNA, or protein molecules (Fig. 2). In order to carry out site-directed mutagenesis, the gene of interest needs to be amplified by polymerase chain reaction (PCR) using primers containing the targeted mutation. The process of primer design involves the optimization of several parameters such as melting temperature, primer length, and GC base pairs content. Primers that are poorly designed can lead to poor annealing to the target DNA sequence, resulting in inefficient gene amplification. The manufacturer provides a website tool for primer design (http://www.genomics.agilent.com/primerDesignProgram.jsp).

3.3.1 Mutant Strand Synthesis

1. Synthesize two complementary oligonucleotides containing the desired mutation, flanked by unmodified nucleotide sequence. Purify these oligonucleotide primers prior to use in the following steps (*see* Mutagenic Primer Design).

Table 3
Preparation of chemicals and reagents for mutant strand synthesis

	Control reaction	Sample reaction
10× Reaction buffer	5 μl	5 μl
10 ng pWhitescriot 4.5 kb control plasmid (5 ng/ μl)	2 μl	
125 ng Oligonucleotide primer #1	1.25 μl control primer [34-mer (100 ng/ μl)]	Xμl
125 ng Oligonucleotide primer #2	1.25 μl control primer [34-mer (100 ng/ μl)]	Xμl
dNTP mix	1 μl	1 μl
QuikSolution reagent	3 μl	3 μl
Double-distilled water (ddH_2O)	35 μl	Xμl
Total volume	50 μl	50 μl

2. Prepare the control reaction as indicated in Subheading 2.2 (Table 3).

   5 μl of 10× reaction buffer.

   2 μl (10 ng) of pWhitescript 4.5 kb control plasmid (5 ng/μl).

   1.25 μl (125 ng) of oligonucleotide control primer #1 [34-mer (100 ng/μl)].

   1.25 μl (125 ng) of oligonucleotide control primer #2 [34-mer (100 ng/μl)].

   1 μl of dNTP mix.

   3 μl of QuikSolution reagent.

   36.5 μl of double-distilled water (ddH_2O) to a final volume of 50 μl.

   Then add 1 μl of PfuUltra HF DNA polymerase (2.5 U/μl).

3. Prepare the sample reaction(s) as indicated in Subheading 2.2 (Table 3) (*see* **Note 5**).

   5 μl of 10× reaction buffer.

   X μl (10 ng) of dsDNA template.

   X μl (12 ng) of oligonucleotide primer #1.

   X μl (125 ng) of oligonucleotide primer #2.

   1 μl of dNTP mix.

   3 μl of QuikSolution.

   ddH_2O to a final volume of 50 μl.

   Then add 1 μl of *PfuUltra* HF DNA polymerase (2.5 U/μl).

Table 4
Cycling parameters for the QuikChange II XL method

Segment	Cycles	Temperature (°C)	Time
1	1	95	1 min
2	18	95	50 s
		60	50 s
		68	1 min/kb of plasmid length
3	1	68	7 min

* For example, a 5 kb plasmid requires 5 min at 68 °C per cycle

4. Cycle each reaction using the cycling parameters outlined in Table 4 (for the control reaction, use a 5-min extension time and run the reaction for 18 cycles) (*see* **Note 6**).
5. Following temperature cycling, place the reaction tubes on ice for 2 min to cool the reactions to ≤37 °C (*see* **Note 7**).

3.3.2 DpnI Digestion of the Amplification Products

1. Add 1 μl of the *Dpn* I restriction enzyme (10 U/μl) directly to each amplification reaction.
2. Gently and thoroughly mix each reaction mixture by pipetting the solution up and down several times. Spin down the reaction mixtures in a microcentrifuge for 1 min, and then immediately incubate the reactions at 37 °C for 1 h to digest the parental (i.e., the nonmutated) supercoiled dsDNA.

3.3.3 Transformation of XL10-Gold Ultracompetent Cells (See Note 8*)*

1. Gently thaw the XL10-Gold ultracompetent cells on ice. For each control and sample reaction to be transformed, aliquot 45 μl of the ultracompetent cells to a pre-chilled 14 ml BD Falcon polypropylene round-bottom tube.
2. Add 2 μl of the β-ME mix provided with the kit to the 45 μl of cells (*see* **Note 9**).
3. Swirl the contents of the tube gently. Incubate the cells on ice for 10 min, swirling gently every 2 min.
4. Transfer 2 μl of the *Dpn* I-treated DNA from each control and sample reaction to separate aliquots of the ultracompetent cells.

   As an optional control, verify the transformation efficiency of the XL10-Gold ultracompetent cells by adding 1 μl of 0.01 ng/μl pUC18 control plasmid (dilute the control provided 1:10 in high-quality water) to another 45 μl aliquot of cells.

   Swirl the transformation reactions gently to mix and incubate the reactions on ice for 30 min.

Table 5
Transformation reaction plating volumes

Reaction type	Volume to plate
pWhitescript mutagenesis control	250 μl
pUC 18 transformation control	5 μl (in 200 μl of NZY + broth)[a]
Sample mutagenesis	250 μl on each of the two plates (entire transformation reaction)

[a]Place a 200 μl pool of NZY$^+$ broth on the agar plate, pipet the 5 μl of the transformation reaction into the pool, then spread the mixture

Table 6
Preparation of media and reagents (QuikChange II XL site-directed mutagenesis kit, Agilent Technologies, Inc.)

LB agar (per liter)	LB–ampicillin agar (per liter)
10 g of NaCl	1 l of LB agar
10 g of tryptone	Autoclave
5 g of yeast extract	Cool to 55 °C
20 g of agar	Add 100 mg of filter-sterilized ampicillin
Add deionized H_2O to a final volume of 1 l	Pour into petri dishes (~25 ml/100 mm plate)
Adjust pH to 7.0 with 5 N NaOH Autoclave Pour into petri dishes (~25 ml/100 mm plate)	

NZY$^+$ broth (per liter)	TE buffer
10 g of NZ amine (casein hydrolysate)	10 mM Tris-HCl (pH 7.5)
5 g of yeast extract	1 mM EDTA
5 g of NaCl Add deionized H_2O to a final volume of 1 l Adjust to pH 7.5 using NaOH Autoclave Add the following filter-sterilized supplements prior to use: 12.5 ml of 1 M $MgCl_2$ 12.5 ml of 1 M $MgSO_4$ 20 ml of 20% (w/v) glucose (or 10 ml of 2 M glucose)	

5. Preheat NZY+ broth (Table 6) in a 42 °C water bath for use in step 8 (*see* **Note 10**).
6. Heat-pulse the tubes in a 42 °C water bath for 30 s. The duration of the heat pulse is critical for obtaining the highest efficiencies. Do not exceed 42 °C (*see* **Note 11**).

7. Incubate the tubes on ice for 2 min.
8. Add 0.5 ml of preheated (42 °C) NZY+ broth to each tube and then incubate the tubes at 37 °C for 1 h with shaking at 7–10 × *g*.
9. Plate the appropriate volume of each transformation reaction, as indicated in Table 5, on agar plates containing the appropriate antibiotic for the plasmid vector.

For the mutagenesis and transformation controls, spread cells on LB–ampicillin agar plates containing 8 μg/ml X-gal and 20 mM IPTG.

10. Incubate the transformation plates at 37 °C for >16 h.

4 Notes

1. Equal amounts of IP protein concentration should be prepared for each pull down of proteins of interest as well as for the IgG sample, which is used as a negative control; the ratio between the IP lysate samples and the input control used is 1:10. Once prepared, the input control can be stored at −20 °C until the end when IP samples are also ready for final step, i.e., denaturation.
2. All steps should be performed on ice or in a refrigerated microcentrifuge.
3. Both the amount of cell lysates and antibody used to immunoprecipitate the protein of interest need to be optimized for each cell model and chosen depending on the abundance of the protein and the affinity of the antibody for the protein.
4. Depending on the IP antibody species used to pull down the protein of interest, 20 μl of protein A/G Dynabead slurry is equilibrated in phosphate-buffered saline (PBS) before use by washing three times with 500 μl PBS and discarding the excess of PBS every time by using a magnetic stand.
5. Set up an initial sample reaction using 10 ng of dsDNA template. If this initial reaction is unsuccessful, set up a series of sample reactions using a titration of various concentrations of dsDNA template ranging from 5 to 50 ng (e.g., 5, 10, 20, and 50 ng of dsDNA template) while keeping the primer concentration constant.
6. It is important to adhere to the 18-cycle limit when cycling the mutagenesis reactions. More than 18 cycles can have deleterious effects on the reaction efficiency.
7. If desired, amplification may be checked by electrophoresis of 10 μl of the product on a 1% agarose gel. A band may or may

not be visualized at this stage. In either case proceed with Dpn I digestion and transformation.

8. XL10-Gold cells are resistant to tetracycline and chloramphenicol. If the mutagenized plasmid contains only the TetR or CamR resistance marker, an alternative strain of competent cells must be used.
9. Using an alternative source of β-ME may reduce transformation efficiency.
10. Transformation of XL10-Gold ultracompetent cells has been optimized using NZY+ broth.
11. This heat pulse has been optimized for transformation in 14 ml BD Falcon polypropylene round-bottom tubes.

References

1. Shore D, Squire M, Nasmyth KA (1984) Characterization of two genes required for the position-effect control of yeast mating-type genes. EMBO J 3(12):2817–2823
2. Kaeberlein M, McVey M, Guarente L (1999) The SIR2/3/4 complex and SIR2 alone promote longevity in Saccharomyces cerevisiae by two different mechanisms. Genes Dev 13(19): 2570–2580
3. Liszt G et al (2005) Mouse Sir2 homolog SIRT6 is a nuclear ADP-ribosyltransferase. J Biol Chem 280(22):21313–21320
4. Vaquero A et al (2004) Human SirT1 interacts with histone H1 and promotes formation of facultative heterochromatin. Mol Cell 16(1): 93–105
5. Luo J et al (2000) Deacetylation of p53 modulates its effect on cell growth and apoptosis. Nature 408(6810):377–381
6. Brunet A et al (2004) Stress-dependent regulation of FOXO transcription factors by the SIRT1 deacetylase. Science 303(5666):2011–2015
7. Motta MC et al (2004) Mammalian SIRT1 represses forkhead transcription factors. Cell 116(4):551–563
8. Huffman DM et al (2007) SIRT1 is significantly elevated in mouse and human prostate cancer. Cancer Res 67(14):6612–6618
9. Eades G et al (2011) miR-200a regulates SIRT1 expression and epithelial to mesenchymal transition (EMT)-like transformation in mammary epithelial cells. J Biol Chem 286(29):25992–26002
10. Stunkel W et al (2007) Function of the SIRT1 protein deacetylase in cancer. Biotechnol J 2(11):1360–1368
11. Bradbury CA et al (2005) Histone deacetylases in acute myeloid leukaemia show a distinctive pattern of expression that changes selectively in response to deacetylase inhibitors. Leukemia 19(10):1751–1759
12. Vaziri H et al (2001) hSIR2(SIRT1) functions as an NAD-dependent p53 deacetylase. Cell 107(2):149–159
13. Luo J et al (2001) Negative control of p53 by Sir2alpha promotes cell survival under stress. Cell 107(2):137–148
14. Lain S et al (2008) Discovery, in vivo activity, and mechanism of action of a small-molecule p53 activator. Cancer Cell 13(5):454–463
15. Peck B et al (2010) SIRT inhibitors induce cell death and p53 acetylation through targeting both SIRT1 and SIRT2. Mol Cancer Ther 9(4):844–855
16. Wang RH et al (2008) Impaired DNA damage response, genome instability, and tumorigenesis in SIRT1 mutant mice. Cancer Cell 14(4): 312–323
17. Herranz D et al (2010) Sirt1 improves healthy ageing and protects from metabolic syndrome-associated cancer. Nat Commun 1:3
18. Wang RH et al (2008) Interplay among BRCA1, SIRT1, and Survivin during BRCA1-associated tumorigenesis. Mol Cell 32(1):11–20
19. Brunet A et al (2004) Stress-dependent regulation of FOXO transcription factors by the SIRT1 deacetylase. Sci Aging Knowl Environ 2004(8):2
20. Khongkow M et al (2013) SIRT6 modulates paclitaxel and epirubicin resistance and survival in breast cancer. Carcinogenesis 34(7): 1476–1486

21. Zhang T, Kraus WL (2010) SIRT1-dependent regulation of chromatin and transcription: linking NAD+ metabolism and signaling to the control of cellular functions. Biochim Biophys Acta 1804(8):1666–1675
22. Yuan J et al (2009) Histone H3-K56 acetylation is important for genomic stability in mammals. Cell Cycle 8(11):1747–1753
23. Daitoku H et al (2004) Silent information regulator 2 potentiates Foxo1-mediated transcription through its deacetylase activity. PNAS 101(27):10042–10047
24. Lin Z, Fang D (2013) The roles of SIRT1 in cancer. Genes Cancer 421(2):384–388
25. Bouras T et al (2005) SIRT1 deacetylation and repression of p300 involves lysine residues 1020/1024 within the cell cycle regulatory domain 1. J Biol Chem 280(11):10264–10276
26. Qiang L et al (2012) Brown remodeling of white adipose tissue by SIRT1-dependent deacetylation of Ppary. Cell 150(3):620–632
27. Pestell R et al (2013) Ppary deacetylation by SIRT1 determines breast tumour lipid synthesis and growth. Cancer Res 73:2-06-02
28. Yeung F et al (2004) Modulation of NF-κB-dependent transcription and cell survival by the SIRT1 deacetylase. EMBOJ 23(12):2369–2380
29. Pickard A, Wong PP, McCance DJ (2010) Acetylation of Rb by PCAF is required for nuclear localization and keratinocyte differentiation. J Cell Sci 123:3718–3726
30. Menssen A et al (2012) The c-MYC oncoprotein, the NAMPT enzyme, the SIRT1-inhibitor DBC1, and the SIRT1-inhibitor DBC1, and the SIRT1 deacetylase form a positive feedback loop. PNAS 109(4):187–196
31. Bharathy N, Taneja R (2012) Methylation muscles into transcription factor silencing. Transcription 3(5):215–220
32. Zhao X et al (2005) Regulation of MEF2 by histone deacetylase 4- and SIRT1 deacetylase-mediated lysine modifications. Mol Cell Biol 25(19):8456–8464
33. Cheng HL et al (2003) Developmental defects and p53 hyperacetylation in Sir2 homolog (SIRT1)-deficient mice. PNAS 100(19):10794–10799
34. Dehennaut V et al (2012) Molecular dissection of the interaction between HIC1 and SIRT1. Biochem Biophys Res Commun 421(2):384–388
35. Cohen HY et al (2004) Calorie restriction promotes mammalian cell survival by inducing the SIRT1 deacetylase. Sci Express 10(1126):1–4
36. Fan W, Luo J (2010) SIRT1 regulates UV-induced DNA repair through deacetylating XPA. Mol Cell 39(2):247–258
37. Yuan Z et al (2007) SIRT1 regulates the function of the Nijmegen breakage syndrome protein. Mol Cell 27(1):149–162
38. Westerheide SD et al (2009) Stress-inducible regulation of heat shock factor 1 by the deacetylase SIRT1. Science 323(5917):1063–1066
39. Tiberi L et al (2012) BCL6 controls neurogenesis through SIRT1-dependent epigenetic repression of selective notch targets. Nat Neurosci 15(12):1627–1635
40. Inoue Y et al (2007) Smad3 is acetylated by p300/CBP to regulate its transactivation activity. Oncogene 26:500–508
41. Chen Y et al (2012) Quantitative acetylome analysis reveals the roles of SIRT1 in regulating diverse substrates and cellular pathways. Am Soc Biochem Mol Biol 11(10):1048–1062
42. Nakagawa T, Guarente L (2011) Sirtuins at a glance. J Cell Sci 124:833–838
43. Ikenoue T, Inoki K, Zhao B (2008) PTEN acetylation modulates its interaction with PDZ domain. Cancer Res 68:6908–6912
44. Montie HL, Pestell RG, Merry DE (2011) SIRT1 modulates aggregation and toxicity through deacetylation of the androgen receptor in cell models of SBMA. J Neurosci 21(48):17425–17436
45. Akieda-Asai S et al (2010) SIRT1 regulates thyroid-stimulating hormone release by enhancing PIP5Kγ activity through deacetylation of specific lysine residues in mammals. PLoS One 5(7)
46. Chen IY et al (2006) Histone H2A.z is essential for cardiac myocyte hypertrophy but opposed by silent information regulator 2alpha. J Biol Chem 281(8):19369–19377
47. Yu J, Auwerx J (2010) Protein deacetylation by SIRT1: an emerging key post-translational modification in metabolic regulation. Pharmacol Res 62(1):35–41
48. Peng L et al (2011) SIRT1 deacetylates the DNA Methyltransferase 1 (DNMT1) protein and alters its activities. Mol Cell Biol 31:4720–4734
49. Hallows WC, Lee S, Denu JM (2006) Sirtuins deacetylate and activate mammalian acetyl-CoA synthetases. PNAS 103(27):10230–10235
50. Fusco S, Maulucci G, Pani G (2012) Sirt1: Def-eating senescence? Cell Cycle 11(22):4135–4146

Chapter 13

Protocols for Cloning, Expression, and Functional Analysis of Sirtuin2 (SIRT2)

Shaoping Ji, J. Ronald Doucette, and Adil J. Nazarali

Abstract

SIRT2 is a NAD$^+$-dependent deacetylase that belongs to the sirtuin family, which is comprised of seven members (SIRT1-SIRT7) in humans. Furthermore, recent study shows that the *Sirt2* gene has three transcript variants in mice. Several diverse proteins have been identified as SIRT2 substrates. SIRT2 activity involves multiple cell processes including growth, differentiation, and energy metabolism. However, little is known of SIRT2's role in oligodendrocytes or in the myelin sheath, where it is an important component. Here we describe procedures that detail *Sirt2* gene cloning, identification, expression, and biological analysis in cultured cells.

Key words SIRT2, Sirtuins, Cloning, Expression, cDNA, Knockdown

1 Introduction

Sirtuin family proteins are evolutionarily conserved and widely expressed from prokaryotes to eukaryotes including humans (1, 2). The yeast Sir2 (SIRT1 in human) protein epigenetically remodels genome structure and regulates gene expression (3, 4). SIRT2 belongs to sirtuin protein family that comprises of seven members (SIRT1–SIRT7) in humans. SIRT2 is a class III histone deacetylase, and is predominantly expressed in oligodendroglia in the central nervous system. SIRT2 protein is mainly localized in cytoplasm and is an important component of myelin sheath. SIRT2 enhances differentiation of oligodendroglia cells (CG4 cell) in vitro (5). Diverse substrates of SIRT2 have been identified, including α-tubulin, histone, CDK9, and glycolytic enzyme phosphoglycerate mutase (GEPM) (6–9). SIRT2 is also involved in the regulation of Akt activity (10).

The Sir2 gene was the first member of sirtuin family to be identified, and its gene was fully sequenced from yeast (11). Members of this family are referred to as SIRT in humans. Using yeast sir2 amino acid sequence, a protein BLAST was used to obtain partial EST sequences of human sirtuin, and five human

Sibaji Sarkar (ed.), *Histone Deacetylases: Methods and Protocols*, Methods in Molecular Biology, vol. 1436, DOI 10.1007/978-1-4939-3667-0_13, © Springer Science+Business Media New York 2016

sirtuin full-length cDNAs were cloned and characterized (12). Analysis of the expression pattern revealed that SIRT1–SIRT5 were ubiquitously expressed in humans from the fetus to the adult (12). Subsequently, SIRT6 and SIRT7 were identified by BLAST analyses of human EST with human SIRT4 (13). SIRT2 includes three sub-isoforms SIRT2.1, SIRT2.2, and SIRT2.3 (14). Here we describe a series of methods used in our laboratory to identify rat *Sirt2* gene, reverse transcription of mRNA to cDNA, and cloning into recombinant vectors, as well as protocols used for biological function analysis of *Sirt2* gene in cultured mammalian cells.

2 Materials

All solutions were prepared with analytical grade reagents and DEPC-treated water (RNase free).

CG4 cells (derived from rat glial precursors) are cultured as described previously (5). When required, freshly cultured cells were used in this protocol. Any waste generated was regularly disposed of following the manufacturer's guide from material safety data sheet (MSDS).

2.1 Cloning of the Sirt2 Gene

1. TRIzol® reagent (contains phenol and guanidine isothiocyanate).
2. Analytical grade chloroform.
3. Analytical grade isopropyl alcohol.
4. 1.5 mL Eppendorf tubes.
5. Ice-cold 75 % ethanol: Pour 75 mL of 99.5 % ethanol into a 100 mL cylinder and make up to 100 mL with water. Mix and pour into glass bottle with a sealing ring in the lid (*see* **Note 1**). Store at 4 °C (*see* **Note 2**).
6. 250 μg /mL of Oligo (dT) 18-mer.
7. RNase-free water.
8. 0.1 M of DTT solution.
9. AMV reverse transcriptase (200 U/μL).
10. 40 U/μL of RNase inhibitor.
11. 10 mM of dNTP mix (2.5 mM/each dNTP).
12. 5 U/μL of Taq DNA polymerase.
13. TBE: 40 mM Tris–Cl, pH 8.3; 45 mM boric acid; 1 mM EDTA.
14. 5 U/μL of Eco RI restriction enzyme

2.2 Sirt2 Expression in E. coli

1. LB medium (1 % of tryptone, 0.5 % of yeast extract, and 1 % of NaCl).
2. 1.0 M of isopropyl-beta-D-thiogalactopyranoside (IPTG) filter sterilized.

3. Coomassie staining solution (40 % methanol, 10 % acetic acid, and 0.025 % Coomassie Brilliant Blue R-250) filtered through Whatman #1 paper.
4. Destaining solution (40 % methanol and 10 % acetic acid).
5. Competent BL21 cells.
6. 5 U/μL of Eco RI restriction enzyme.
7. 5 U/μL of Xho I restriction enzyme.
8. pGEX-6P-1 (prokaryotic expression vector).

2.3 *Sirt2* Overexpression and Knockdown in Mammalian Cells

1. Vector pEGFP-C2.
2. HEK293 cells (does not express *Sirt2*).
3. CG4 cells (expresses *Sirt2*).
4. Lipofectamine® 2000 (liposome-mediated transfection reagent).
5. Fetal bovine serum (FBS).
6. 5 U/μL of Eco RI restriction enzyme.
7. 5 U/μL of Kpn I restriction enzyme.

3 Methods

3.1 Cloning of the Sirt2 Gene

1. For isolation of total RNA, retrieve cultured CG4 cells with harvest medium (5) and collect cell pellet by centrifugation at 400 × *g* (*see* **Note 3**) for 3 min at room temperature.
2. Add 1.0 mL of TRIzol® to pelleted 10^6 cells (or 50 mg tissue), homogenize, and gently pipette until the cell pellet is thoroughly dissolved. Leave the sample standing still for 5 min at room temperature. Add 200 μL chloroform to the sample and mix thoroughly by shaking the tube vigorously for 15 s. Centrifuge the sample at 12,000 × *g* at 4 °C for 15 min.
3. Carefully transfer aqueous phase (top layer, about 50 % of total volume) into a new tube (*see* **Note 4**), and add an equal volume of isopropyl alcohol to it. After mixing, leave the sample standing still for 10 min at room temperature. Centrifuge the sample at 12,000 × *g* for 10 min at 4 °C.
4. Remove and discard supernatant (*see* **Note 5**). Add 1.0 mL of ice-cold 75 % ethanol, close lid, and gently invert the tube a few times. Centrifuge the sample at 12,000 × *g* at 4 °C for 5 min.
5. Remove and discard supernatant (as completely as possible) and leave sample at room temperature for 7 min, allowing the sample to dry completely (*see* **Note 6**).
6. Dissolve the dried pellet (it may not be visible) in 20 μL of RNase-free water including 40 units of RNase inhibitor (*see* **Note 7**). Quantify concentration of the RNA sample using a spectrophotometer (UV at 260 and 280 nm wavelength).

7. For cDNA synthesis, prepare a reaction of reverse transcription as indicated in the following procedure:

Total RNA	2 μg
Oligo (dT) 18-mer	2 μL
dNTP mix	2 μL
Add RNase-free water	~8 μL
Total	12 μL

Mix and incubate sample at 65 °C for 5 min, centrifuge (quick spin) the sample, and leave on ice (*see* **Note 8**).

RNase inhibitor	1 μL
5× cDNA synthesis buffer	4 μL
0.1 M of DTT	2 μL
AMV reverse transcriptase	1 μL
Total	20 μL

Gently mix reaction volume by pipetting and incubate at 42 °C for 60 min. Inactivate reverse transcriptase by heating at 85 °C for 10 min. Store at -20 °C (*see* **Note 9**).

8. Design primers (*see* **Note 10**) for cloning *Sirt2* full-length coding sequence, including start codon and stop codon (NCBI accession number: NM_001008368). Forward primers: 5′-AAGCTTG CAGAG ATGGACT TCCTACGG-3′ including *Hin*d III site; reverse primer: 5′-GAATTCCTAGTGTTCCTCTTTCTC TTTG-3′ including *Eco*R I site. This *Sirt2* cDNA is used to clone *Sirt2* into pIRES2-EGFP vector.
9. Perform polymerase chain reaction (PCR) as described below:

10× PCR reaction buffer	5 μL
dNTP (10 mM or 2.5 mM/each dNTP)	1 μL (*see* **Note 11**)
Forward primer (10 μM)	2 μL
Reverse primer (10 μM)	2 μL (*see* **Note 12**)
cDNA (described above)	1 μL
Taq DNA polymerase (5 U/μL)	1 μL (*see* **Note 13**)

PCR program cycle is as follows: (1) 94 °C, 3 min; (2) 94 °C, 1 min; (3) 55 °C, 1 min; (4) 72 °C, 2 min; (5) 72 °C, 10 min. PCR reaction is repeated (**steps 2-4**) for 28 cycles on a thermal cycler (Bio-Rad) with heated lid (*see* **Note 14**).

10. Separate the PCR product by regular agarose gel electrophoresis (1% of agarose in 0.5 × TBE). Excise the correct band size

of 1071 bp from the gel (*see* **Note 15**). Isolate the DNA from the agarose gel using DNA gel extraction kit (Qiagen) according to the manufacturer's protocol.

11. Clone the DNA fragment extracted from the agarose gel into T-easy vector (Invitrogen) by T-A cloning according to the manufacturer's protocol (*see* **Note 16**).
12. The correct colonies with *Sirt2* cDNA insert will yield an approximately 1080 bp segment when digested with EcoRI restriction endonuclease (*see* **Note 17**). *Sirt2* coding region can be verified by DNA sequencing.

3.2 *Sirt2* Expression in *E. coli*

To investigate its biological activity and function, Sirt2 was expressed in *E. coli* as a fused SIRT2-GST (glutathione S-transferase) protein.

1. Amplify the Sirt2 cDNA cloned in T-easy vector by PCR as described above (Subheading 3.1) with the following primers: forward: 5′-gaattcgagatggacttcctacgg-3′ (EcoRI) and reverse: 5′-ctcgagctagtgttcctctttctcttg-3′ (XhoI). Use PCR reaction conditions as described above. The cDNA is cloned into pGEX-6P-1 for expression of SIRT2–GST fusion protein (*see* **Note 18**).
2. Separate and purify the PCR product and clone into T-easy vector as described above.
3. Verify the correct clone by DNA sequencing and digestion with EcoRI and XhoI according to the manufacturer's protocol (*see* **Note 19**).
4. Digest and purify the expression vector pGEX-6P-1 as described (*see* **Note 20**).
5. Purify the correct insert fragment and the digested vector from agarose gel as described above.
6. Perform the ligation reaction as described below:

Insert *Sirt2* cDNA	1–4.0 μL
Vector	1–4.0 μL

Mix and incubate at 65 °C for 5 min (*see* **Note 21**), then place on ice, and add:

10× Ligation buffer :	1.0 μL
T4 ligase	3.0 U
Total	10 μL (*see* **Note 22**)

Mix well and incubate ligation reaction mixture overnight at 16 °C.

7. To promote efficiency of transformation, inactivate the T4 ligase by heating at 70 °C for 5 min before transforming the sample into competent cells BL21. Spread the sample onto a LB/agar petri dish with antibiotics and culture overnight at 37 °C (following the regular protocol).
8. Pick single colonies into 2 mL LB medium and culture overnight at 37 °C with antibiotics (*see* **Note 23**).
9. Enlarge the culture sample into fresh LB by 1–200. After culturing for 2–3 h remove flask from incubator and leave at room temperature. Do not add IPTG until the medium comes to room temperature (the fusion protein may precipitate at high induction temperatures). Induce the sample for 2 h at 26 °C with 1.0 mM of IPTG and shaking.
10. Subject the sample to analysis by SDS-PAGE (10%).
11. After electrophoresis, stain gel with Coomassie staining solution for 4 h at room temperature with gentle shaking.
12. Destain gel in destaining solution at room temperature with gentle shaking, and renew destaining solution every 30 min and repeat five times. Take picture of the gel (Fig. 1).

3.3 Sirt2 Overexpression and Knockdown in Mammalian Cells

Overexpression or knockdown of a gene of interest in target cells is a normal strategy used to explore biological function of the gene.

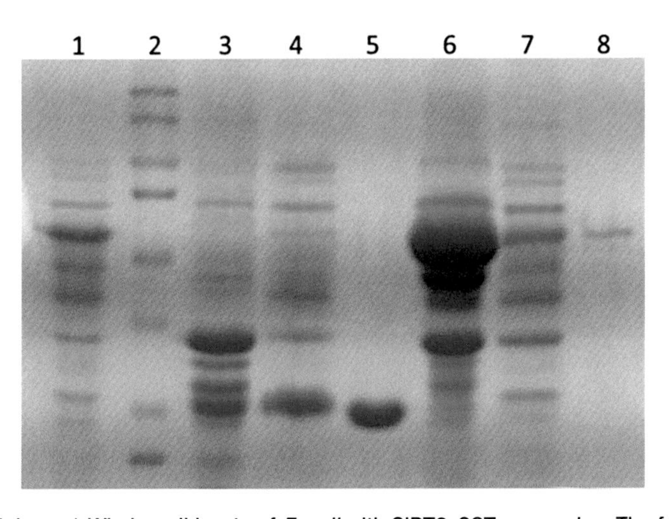

Fig. 1 *Lane 1*: Whole-cell lysate of *E. coli* with SIRT2-GST expression. The fusion protein has low solubility in *E. coli* and major portion exists as inclusion bodies; *lane 2*: protein ladder; *lane 3*: lysate pellet of *E. coli* with GST expression; *lane 4*: lysate supernatant of *E. coli* with GST expression; *lane 5*: purified GST; *lane 6*: lysate pellet of *E. coli* with SIRT2-GST expression; *lane 7*: lysate supernatant of *E. coli* with SIRT2-GST expression; *lane 8*: purified SIRT2-GST

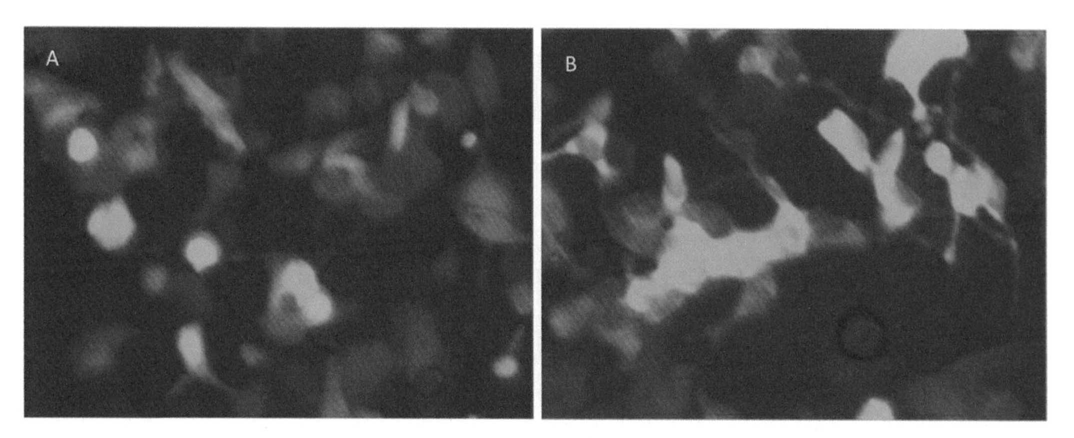

Fig. 2 SIRT2 protein is mainly localized in the cytoplasm of HEK293 cells. (**a**) HEK293 cells with only GFP expression that is distributed everywhere within the cells; (**b**) HEK293 cells with SIRT2-GFP expression that is primarily located within the cytoplasm

Expression product of SIRT2 is a cytoplasmic protein probably with some shuttling into nucleus and exporting back to cytoplasm. Here, we describe how to transfect the *Sirt2* cDNA into HEK293 cells (Fig. 2) or siRNA in CG4 cells and observe the subcellular localization of SIRT2 protein or analyze the biological function of SIRT2. The cells should be kept healthy in fresh medium and changed into non-antibiotic medium 4–12 h before transfection. The recombinant vector should be freshly purified (within 2 weeks).

1. Amplify *Sirt2* cDNA (in T-easy vector described above) with PCR using the following primers: forward: 5′-gaattcgagatg gacttcctacgg-3′ (EcoRI) and reverse: 5′-ggtaccctagtgttcct ctttctctttg-3′(KpnI), and clone into eukaryotic expression vector pEGFP-C2 (this recombinant vector is referred to as pEGFP-Sirt2) with restriction enzymes EcoRI and KpnI, respectively (as described above).
2. Transform the vector pEGFP-C2 and pEGFP-Sirt2 into competent cells and culture overnight (as described above). Isolate and purify both vectors. Quantify and adjust the concentration to 200 mg/mL and store at -20 °C.
3. siRNA against rat *Sirt2* (as sequence: 5′-AGGGAGCAUG CCAACAUAGAU-3′) and random RNA as control were synthesized by Qiagen.
4. Culture HEK293 cells in DMEM with 10 % of FBS, 100 units/ mL of penicillin, and 100 mg/mL of streptomycin and incubate cell culture at 37 °C, 5 % of CO_2 with sufficient moisture. Seed the cells into 6 mm dishes before transfection.
5. When HEK293 cells reach 80 % confluency, culture in antibiotic-free medium for 4 h before transfection. Transfection procedure is performed as indicated below.

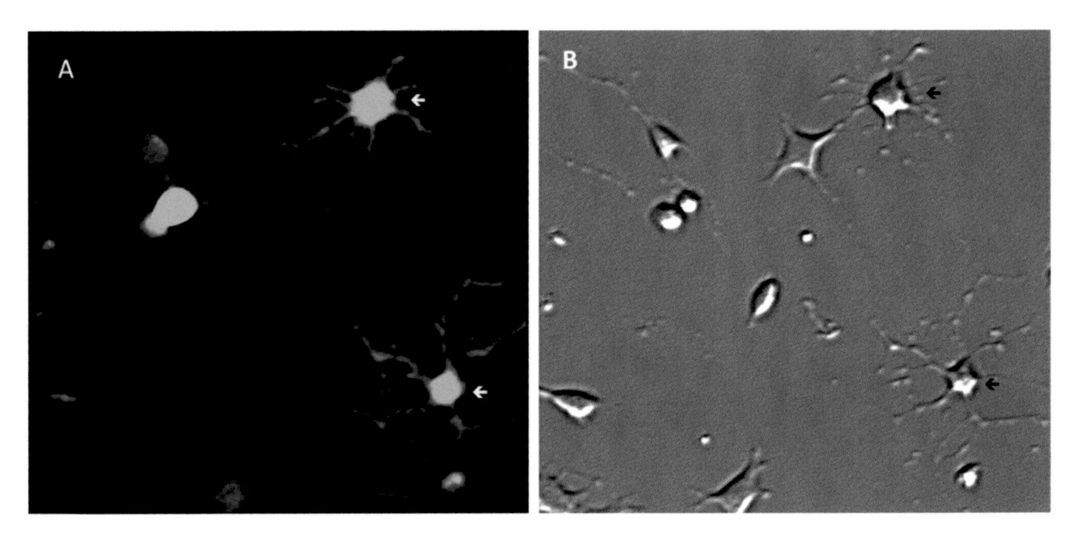

Fig. 3 SIRT2 overexpression in CG4 cells promotes process outgrowth [(**a**) fluorescence microscopy and (**b**) light microscopy, arrows] compared to untransfected cells (no GFP expression) (5)

6. Culture CG4 cells in serum-free medium as described previously (5). When cell confluency reaches 50–60 % (*see* **Note 24**), carry out the transfection procedure.
7. Add 10 μL of lipofectamine® 2000 to 300 μL of DMEM, mix and stand for 5 min. Add 10 μg of pEGFP-Sirt2 or pEGFP-C2 to separate 300 μL of DMEM and mix. Combine diluted lipofectamine® 2000 and vectors, gently mix, and allow to incubate for 20 min. Transfer mixture into culture medium and mix gently (for detailed protocol, see the manufacturer's recommendation). Change cell culture medium to normal fresh medium after transfection for 12 h and continue to culture the cells as previously.
8. Observe and take pictures of cells under a fluorescence microscope within 24–48 h after transfection (Fig. 3, **Note 25**).
9. To check expression level of the fusion protein, lyse the cell and analyze the sample by western blot analysis (using usual protocols).

4 Notes

1. Keep the ice-cold 75 % ethanol bottle lid tightly sealed to prevent ethanol from evaporating. Lower ethanol concentration will not be adequate to precipitate RNA.
2. The 75 % ethanol bottle should be kept at 4 °C in fridge and be placed inside another container to prevent evaporation and avoid causing a potential explosion.

3. The centrifugation (400 × *g*, for 3 min) to precipitate cells is sufficient. Do not use higher centrifugation as this may rupture the cells, leading to loss of cytoplasm.
4. After centrifugation, sample is divided into three parts, the top layer is the aqueous phase, the middle layer is protein, and the bottom layer is the phenolic phase. Do not touch middle and bottom phases when pipetting out aqueous phase (reverse transcription in the following steps may be blocked by phenol).
5. The pellet is visible if you have sufficient cells lysed at beginning.
6. Ensure that no droplets of ethanol are visible in the tube. Any remaining ethanol should be completely evaporated or it may inhibit the activity of reverse transcriptase.
7. In the event there is trace RNase contamination of the sample, add RNase inhibitor into the sample, for best results.
8. To facilitate the process of reverse transcription, after the elimination of RNA secondary structures at 65 °C for 5 min, the sample should immediately be put on ice to avoid re-annealing.
9. We find that cDNA stored at -20 °C should be used within 2 months, or yield efficiency of PCR will reduce.
10. Primers designed for cloning a gene should be aligned using BLAST analysis and similar sequence interference in the PCR reaction should be avoided.
11. We find that a low concentration of dNTP can reduce mutation rate in the PCR process.
12. We used a low concentration of primers to reduce nonspecific reactions in the PCR process.
13. We find that *Taq* DNA polymerase works well instead of high-fidelity DNA polymerase with lower concentration of dNTP and a smaller cycle number in the PCR process. It is convenient for the subsequent direct T-A cloning of the PCR product. We rarely use high-fidelity DNA polymerase in our laboratory.
14. Hot start for PCR program is not necessary because the PCR product will be separated by agarose gel electrophoresis, and only the correct band size is isolated for cloning.
15. Dissect the correct DNA band from the stained agarose gel within the band width and as small as possible.
16. Do not use T-A clone if PCR is not performed with regular *Taq* DNA polymerase (or other similar enzymes) which adds an extra "A" at 3′ end of PCR product. PCR product processed with high-fidelity DNA polymerase should run a program of adding "A" with *Taq* DNA polymerase for cloning into T-vector.

17. Any restriction sites flanking the insert can be used to identify the length of the insert. We recommend selecting the enzymes that share maximum efficiency in the same buffer condition.
18. Since Hind III and Eco RI do not match with vector pGEX-6P-1, this PCR step is conducted to engineer restriction endonuclease sites of Eco RI and Xho I.
19. Xho I may be blocked by methylation of DNA. Hence, we use an *E. coli* strain that is methyltransferase deficient to clone the plasmids for Xho I digestion.
20. For efficient cloning it is very important that the plasmid pGEX-6P-1 is completely digested by EcoR I and Xho I. To ensure this we use low concentration of the plasmids in a recommended volume of digestion reaction.
21. Incubation at 65 °C reduces self-annealing of insert or of plasmid and facilitates ligation between insert and plasmid.
22. Keep molar ratio of insert to plasmids at 3:1 in ligation reaction. First, both the insert and plasmid need to be quantified after purification from agarose gel. Second, calculate molecular concentration of the DNA from the molecular weight and number of DNA base pairs.
23. Do not store the transformed petri dishes at 4 °C. Instead leave them at room temperature until the colonies are ready to be picked up for expression. Some fusion proteins will not express or will have decreased expression if the bacterial colonies are cooled down.
24. Higher confluence of CG4 will cause differentiated-like changes of the cell morphology. We recommend keeping CG4 cells in a lower density in the culture.
25. SIRT2 is expressed via a bicistronic vector (pIRES-EGFP2) that gives rise to the SIRT2 protein separately from GFP, avoiding any possible impact of GFP on SIRT2 function.

Acknowledgement

This work was supported by grant from the Canadian Institutes of Health Research and the Saskatchewan Regional Partnership program.

References

1. Sherman JM, Stone EM, Freeman Cook LL, Brachmann CB, Boeke JD, Pillus L (1999) The conserved core of a human SIR2 homologue functions in yeast silencing. Mol Biol Cell 10:3045–3059
2. Brachmann CB, Sherman JM, Devine SE, Cameron EE, Pillus L, Boeke JD (1995) The SIR2 gene family, conserved from bacteria to humans, functions in silencing, cell cycle progression, and chromosome stability. Genes Dev 9:2888–2902

3. Tanny JC, Dowd GJ, Huang J, Hilz H, Moazed D (1999) An enzymatic activity in the yeast Sir2 protein that is essential for gene silencing. Cell 99:735–745
4. Fritze CE, Verschueren K, Strich R, Easton Esposito R (1997) Direct evidence for SIR2 modulation of chromatin structure in yeast rDNA. EMBO J 16:6495–6509
5. Ji S, Doucette JR, Nazarali AJ (2011) Sirt2 is a novel in vivo downstream target of Nkx2.2 and enhances oligodendroglial cell differentiation. J Mol Cell Biol 3:351–359
6. Li W, Zhang B, Tang J, Cao Q, Wu Y, Wu C, Guo J, Ling EA, Liang F (2007) Sirtuin 2, a mammalian homolog of yeast silent information regulator-2 longevity regulator, is an oligodendroglial protein that decelerates cell differentiation through deacetylating alpha-tubulin. J Neurosci 27:2606–2616
7. Xu Y, Li F, Lv L, Li T, Zhou X, Deng CX, Guan KL, Lei QY, Xiong Y (2014) Oxidative stress activates SIRT2 to deacetylate and stimulate phosphoglycerate mutase. Cancer Res 74:3630–3642
8. Eskandarian HA, Impens F, Nahori MA, Soubigou G, Coppée JY, Cossart P, Hamon MA (2013) A role for SIRT2-dependent histone H3K18 deacetylation in bacterial infection. Science 341:1238858
9. Zhang H, Park SH, Pantazides BG, Karpiuk O, Warren MD, Hardy CW, Duong DM, Park SJ, Kim HS, Vassilopoulos A, Seyfried NT, Johnsen SA, Gius D, Yu DS (2013) SIRT2 directs the replication stress response through CDK9 deacetylation. Proc Natl Acad Sci USA 110:13546–13551
10. Ramakrishnan G, Davaakhuu G, Kaplun L, Chung WC, Rana A, Atfi A, Miele L, Tzivion G (2014) Sirt2 deacetylase is a novel AKT binding partner critical for AKT activation by insulin. J Biol Chem 289:6054–6066
11. Sarén AM, Laamanen P, Lejarcegui JB, Paulin L (1997) The sequence of a 36.7 kb segment on the left arm of chromosome IV from *Saccharomyces cerevisiae* reveals 20 non-overlapping open reading frames (ORFs) including SIT4, FAD1, NAM1, RNA11, SIR2, NAT1, PRP9, ACT2 and MPS1 and 11 new ORFs. Yeast 13:65–71
12. Frye RA (1999) Characterization of five human cDNAs with homology to the yeast SIR2 gene: Sir2-like proteins (sirtuins) metabolize NAD and may have protein ADP-ribosyltransferase activity. Biochem Biophys Res Commun 260:273–279
13. Frye RA (2000) Phylogenetic classification of prokaryotic and eukaryotic Sir2-like proteins. Biochem Biophys Res Commun 273:793–798
14. Maxwell MM, Tomkinson EM, Nobles J, Wizeman JW, Amore AM, Quinti L, Chopra V, Hersch SM, Kazantsev AG (2011) The Sirtuin 2 microtubule deacetylase is an abundant neuronal protein that accumulates in the aging CNS. Hum Mol Genet 20:3986–3996

Chapter 14

Cloning and Characterization of Sirtuin3 (SIRT3)

Andy W.C. Man, Bo Bai, and Yu Wang

Abstract

Mitochondria play a pivotal role in maintaining cellular homeostasis and regulating longevity. SIRT3 is a mitochondrial sirtuin mediating the deacetylation of various metabolic and antioxidant enzymes, in turn controlling energy metabolism, stress resistance, and the pace of ageing. To study the function of SIRT3, a proteomics-based approach is employed for identifying the protein-binding partners of this enzyme in mitochondria.

Key words Sirtuin, Mitochondria, Acetylation, Protein–protein interaction

1 Introduction

Sirtuins are a family of nicotinamide adenine dinucleotide (NAD)-dependent enzymes, which catalyze mono-ADP-ribosylation and/or the removal of acetyl, acyl, malonyl, and succinyl groups from lysines of various protein substrates [1, 2]. Based on phylogenetic analysis of the conserved core domain, mammalian sirtuins are subdivided into four classes, including class I (SIRT1, SIRT2, and SIRT3), class II (SIRT4), class III (SIRT5), and class IV (SIRT6 and SIRT7) [3]. They localize in and shuttle between different subcellular compartments, such as the cytoplasm (SIRT1 and 2), nucleus (SIRT1, 6, and 7), and mitochondria (SIRT1, 3, 4, and 5) [4–6]. Class I shares high homology to Sir2, a yeast sirtuin with potent lysine-deacetylase activity [7]. In mammals, sirtuins are implicated in a wide range of cellular functions to maintain energy homeostasis, enhance stress resistance, and increase longevity [8].

SIRT3 is a major sirtuin modulating the metabolic function and oxidative stress response in mitochondria, especially under nutrient-stress conditions [9–11]. The gene of human SIRT3 is located within the 11p15.5 chromosomal region [11]. Genetic variations of SIRT3 are associated with increased human life-span [12–14]. The full-length SIRT3 protein (44 kDa) contains a targeting signal at the NH_2-terminus that is processed to produce a

Sibaji Sarkar (ed.), *Histone Deacetylases: Methods and Protocols*, Methods in Molecular Biology, vol. 1436,
DOI 10.1007/978-1-4939-3667-0_14, © Springer Science+Business Media New York 2016

28 kDa polypeptide upon import into the mitochondrial matrix [15, 16], where it catalyzes the deacetylation of acetyl-CoA acetyltransferase 1, acetyl-CoA synthase 2, glutamate dehydrogenase, long-chain acyl-CoA dehydrogenase, 3-hydroxy-3-methylglutaryl CoA synthase 2, isocitrate dehydrogenase 2, ornithine transcarbamoylase, pyruvate dehydrogenase, manganese superoxide dismutase, and components of electron transport chain complexes, including NDUFA9, succinate dehydrogenase flavoprotein, and ATP synthases [9, 17–24]. Thus, by promoting the global deacetylation of lysine residues, SIRT3 modulates the fatty acid/amino acid metabolism, antioxidant defense, oxidative phosphorylation, and energy production in mitochondria [25].

In addition to its role as a protein deacetylase, SIRT3 regulates mitochondrial functions via indirect mechanisms, such as enhancing the activities of the adenosine monophosphate-activated protein kinase (AMPK) and peroxisome proliferator-activated receptor gamma co-activator 1-alpha (PGC1α) or increasing the daf-16 homolog FOXO3a-dependent gene expressions [26, 27]. Moreover, SIRT3 prevents apoptosis by lowering reactive oxygen species and inhibiting components of the mitochondrial permeability transition pore [28–30]. Since mitochondrial dysfunction contributes to various ageing-related diseases, including diabetes, cancer, cardiac dysfunction, and neural degeneration, SIRT3 activation could be a promising target for developing novel therapies.

A proteomic approach is adopted for studying the dynamic molecular interactions between SIRT3 and other mitochondrial proteins [31]. The results have provided important insights into how SIRT3 modulates mitochondrial functions, in turn communicating to the rest of the cell and/or other organs [27, 29, 32–35].

2 Materials

Prepare all solutions and buffers using ultrapure water (deionized to attain a sensitivity of 18 MΩ-cm at 25 °C). Sterilize by autoclaving at 121 °C/15 psi for 20 min or by filtration through a 0.2 μm membrane. Determine the pH values at room temperature. All chemicals are of analytical grade. Prepare sterile plasticwares including Eppendorf and Falcon tubes. Disposable gloves should be worn when performing the experiment. Waste materials are disposed by strictly following the institutional regulations.

2.1 Chemicals and Reagents for Gene Cloning

1. G-418, ampicillin, and isopropyl β-D-1-thiogalactopyranoside (IPTG).
2. TRIZOL Reagent, pcDNA3.1, and pGEX-2 T expression vectors, transfection reagent, DH5α, and BL21 competent cells.
3. Restriction enzymes (BamHI, EcoRI, and NotI) and T4 DNA ligase.

4. Mass-Ruler high- and low-range DNA ladders.
5. Kits for DNA extraction after gel electrophoresis, reverse transcription, plasmid preparation, and high-fidelity PCR reactions.

2.2 Media and Buffers for Mammalian Cell Culture and Subcellular Fractionation

1. HepG2 (HB-8065) and HEK293 (CRL-1573) cells.
2. Dulbecco's modified Eagle medium (DMEM), fetal bovine serum (FBS), antibiotics (penicillin, streptomycin, and fungizone), and trypsin–EDTA.
3. Phosphate-buffered saline (PBS) containing 137 mM NaCl, 2.7 mM KCl, 4.3 mM Na_2HPO_4, and 1.47 mM KH_2PO_4, pH 7.4.
4. Mitochondrial isolation buffer containing 0.25 M metal-free sucrose, 1 mM EDTA, and 10 mM Tris–HCl, pH 7.2.

2.3 Media, Buffers, and Reagents for Affinity Purification

1. Bacterial culture medium.
2. Lysis buffer containing 150 mM NaCl, 50 mM Tris–HCl (pH7.4), 1 % Triton-X 100, and protease inhibitor cocktail.
3. Binding buffer containing 150 mM NaCl, 50 mM Tris–HCl (pH7.4), and 0.1 % Triton-X 100.
4. Elution buffer containing 50 mM Tris–HCl (pH 8.0) and 10 mM reduced glutathione.
5. RIPA buffer containing 25 mM Tris–HCl, 150 mM NaCl, 5 mM NaF, 1 % Na deoxycholate, 1 % NP40, 0.1 % Triton-X 100, pH 7.4.
6. Cleavage buffer containing thrombin protease (100 U/ml) in PBS.
7. Glutathione affinity column and sepharose beads.
8. Centrifugal filters.
9. Anti-FLAG M2 agarose gel and anti-FLAG antibody.

3 Methods

3.1 Cloning of Human SIRT3

1. Culture HepG2 cells in DMEM with 10 % FBS at 37 °C in humidified 5 % CO2 and 95 % air (sub-cultivation ratio of 1:4–1:6, change media twice per week). Isolate total RNA from HepG2 cells with TRIZOL reagent (*see* **Note 1**) and reverse transcribe to obtain total cDNA (*see* **Note 2**). Store products from reverse-transcriptase reactions at −20 °C.
2. Design the PCR primers by referring to the mRNA sequence of human SIRT3 (NM_012239) (*see* **Note 3**). Clone the cDNA from 35 to 1234 bp, a region encoding the full-length human SIRT3 protein that contains 399 amino acids with a predicted molecular weight of ~43.6 kDa (*see* **Note 4**).

3. Use the pGEX-2 T vector for constructing the prokaryotic expressing vector GST-hSIRT3 which encodes human SIRT3 with an NH_2-terminal GST tag (hSIRT3/GST). Use the pcDNA3.1 (+) vector to construct the mammalian expression vector hSIRT3-FLAG which encodes human SIRT3 with a COOH-terminal FLAG tag (hSIRT3/FLAG).
4. Include BamHI and EcoRI cutting sites in the forward (5′-CCG CGG ATC CAT GGC GTT CTG GGG TTG-3′) and reverse (5′-CCG CGA ATT CCT ATT TGT CTG GTC CAT CAA GC-3′) primers, respectively, to clone human SIRT3 from the HepG2 total cDNA by PCR for constructing the GST-hSIRT3 vector.
5. Modify the reverse PCR primer to introduce a FLAG peptide coding sequence, stop codon, and NotI restriction enzyme digestion site (5′-AAG GAA AAA AGC GGC CGC CTA CTT ATC GTC GTC ATC CTT GTA ATC TTT GTC TGG TCC ATC AAG CT-3′) to clone human SIRT3 from the HepG2 total cDNA by PCR for constructing the hSIRT3-FLAG vector (*see* **Note 5**).
6. Perform PCR with high-fidelity PCR kit, separate the reaction mixtures by agarose gel electrophoresis, and visualize the DNA fragments by staining with ethidium bromide and using an ultraviolet transilluminator. Purify the PCR products from the agarose gel using a DNA extraction kit (*see* **Note 6**).
7. Digest the purified human SIRT3 cDNA fragments and the cloning vectors with restriction enzymes at 37 °C for 16 h (*see* **Note 7**), purify the reaction products using a DNA extraction kit, and quantify the amounts of enzyme-digested vectors and PCR products (*see* **Note 8**) for subsequent ligation by T4 DNA ligase at room temperature for 6 h (*see* **Note 9**).
8. Add no more than 10 % of the ligation reactions into the DH5α competent cells (100 μl) for transformation. Incubate the mixture for 30 min on ice, 1 min at 42 °C, and then 2 min on ice. Add 1 ml of bacterial culture media and incubate the bacteria cultures at 37 °C for 1 h with shaking. Inoculate the bacteria on an ampicillin selection agar plate for cultivation at 37 °C for 16 h (*see* **Note 10**).
9. Select colonies from the agar plates and inoculate individual colony into bacterial culture media containing ampicillin. After incubation at 37 °C for 12–16 h with vigorous shaking, purify the plasmid using a mini-preparation kit (*see* **Note 11**).
10. Verify the purified plasmids by single or double digestions with restriction enzyme(s) to check the sizes of the vector and DNA insert (*see* **Note 12**). Confirm the authenticity of all vectors by DNA sequencing.

3.2 Expression and Purification of Recombinant SIRT3

1. Transform the prokaryotic expressing vector GST-SIRT3 into BL21 competent cells for obtaining individual colonies on agar plates as described above. Inoculate single clones into 1 ml of bacteria culture media containing ampicillin. After cultivation to the log phase, collect 700 μl of cultures to prepare the glycerol stock.
2. Expand the rest of the bacterial cultures progressively into 100 ml of bacterial culture media containing ampicillin for the induction of recombinant protein expression (*see* **Note 13**). Incubate the cultures at 37 °C with shaking at 220 rpm for about 16 h, until the absorbance of the bacterial suspension reached 0.6–0.8 at a wavelength of 600 nm. Add IPTG to a final concentration of 0.2 mM to induce the hSIRT3/GST protein expression. Incubate the culture at 37 °C with shaking for another 4 h (*see* **Note 14**).
3. Collect the bacterial pellets by centrifugation at 1800 × *g*, 4 °C, for 30 min. Resuspend the bacterial pellets with lysis buffer for sonication (*see* **Note 15**). Centrifuge the lysates at 18,000 × *g* at 4 °C for 30 min and filter the supernatant through a 0.45 μm filter (*see* **Note 16**).
4. Purify the hSIRT3/GST protein using a glutathione affinity column. Equilibrate the column with at least five column volumes of binding buffer. Dilute the bacterial lysates into binding buffer and apply to the column (*see* **Note 17**). Wash the column with at least five column volumes of binding buffer to remove nonspecifically bound bacterial proteins (*see* **Note 18**).
5. Elute the bound hSIRT3/GST recombinant protein with one column volume of elution buffer containing glutathione five times. Separate the eluted fractions by SDS-PAGE. Check the purity and quantity of the eluted recombinant protein by staining with Coomassie Brilliant Blue (*see* **Note 19**).
6. Change the solvent background of the eluted proteins to PBS using centrifugal filters (*see* **Note 20**). Store protein aliquots at −80 °C after quantifying the protein concentrations.

3.3 Pull-Down Purification of SIRT3-Interacting Proteins

1. Use the subcellular fractions enriched with mitochondria for pull-down experiments in order to identify the specific interacting proteins of SIRT3 (*see* **Note 21**).
2. Harvest HEK293 cells by mechanical scraping in cold PBS. Centrifuge and then resuspend the cells in cold isolation buffer. Homogenize the cells and remove the nuclei-enriched fractions by centrifugation of the cell homogenates at 400 × *g* for 10 min at 4 °C. Centrifuge the remaining supernatants at 10,000 × *g* for 10 min at 4 °C to collect the mitochondria-enriched fractions.

3. Wash the glutathione sepharose beads (100 μl resin) twice with cold PBS. Resuspend the beads in 200 μl PBS. Incubate the beads with purified hSIRT3/GST recombinant protein at 4 °C overnight with agitation. Discard the supernatant and wash the beads with cold PBS to remove unbound proteins (*see* **Note 22**).
4. Resuspend the mitochondria-enriched fractions from HEK 293 cells in cold RIPA buffer for sonication on ice. Incubate the lysates containing 500 μg proteins with 100 μl glutathione beads at room temperature for 2 h to remove nonspecific proteins binding to the beads. Separate the supernatants from the beads by centrifugation. Mix the pre-cleared lysates with either GST-bound or hSIRT3/GST-bound sepharose beads for a 2-h incubation at room temperature with agitation. Centrifuge the mixture at $120 \times g$ for 5 min and remove the supernatants. Wash the beads five times in PBS. Elute the protein complexes with glutathione elution buffer or cleavage buffer containing thrombin protease for subsequent mass spectrometric analysis (*see* **Note 23**).

3.4 Overexpression and Co-immunoprecipitation in Mammalian Cells

1. Perform co-immunoprecipitation to confirm the specific interactions between human SIRT3 and its binding proteins in mammalian cells (*see* **Note 24**).
2. Use the mammalian expression vector hSIRT3-FLAG for establishing stably transfected HEK293 cells. Seed the cells to reach a confluence of 60–70% in a six-well culture dish. Transfect the cells with an appropriate amount of hSIRT3-FLAG. After 48 h, sub-culture cells into 10 cm culture plates at a ratio of 1:60 for antibiotic selection in culture media containing G-418. Change fresh media every 2 days until the formation of single-cell clones, which are then collected for expanding and verifying the overexpression of FLAG-tagged human SIRT3 (*see* **Note 25**).
3. Wash the stably transfected cell cultures with cold PBS. Harvest the cells by mechanical scraping in ice-cold RIPA buffer. After homogenization on ice, incubate the cell lysates containing 500 μg proteins with sepharose beads to remove nonspecific protein bindings. After centrifugation, add 40 μl anti-FLAG M2 agarose gel (50% slurry) into the supernatant and incubate overnight at 4 °C on a shaking platform. Collect the beads by centrifugation at $120 \times g$ for 5 min.
4. Wash the beads extensively with ice-cold RIPA buffer to remove the unbound proteins. Elute the immune complexes by adding SDS-PAGE loading buffer for subsequent separation and analysis by Western blotting (*see* **Note 26**).

4 Notes

1. After washing the cells three times with cold PBS, TRIZOL reagent is applied directly to the surface of the cultures. The amount of TRIZOL needs to be adjusted according to the surface area of the cultures. The precipitated RNA is often invisible or forming a gel-like pellet, which should not be completely dried out.
2. For reverse transcription reaction, RNA samples are first mixed with DNA Wipeout Buffer for removing the trace amount of genomic DNA. The optimized blend of oligo-dT and random primers are used to obtain high yield of cDNA from all regions of the RNA transcripts.
3. Restriction enzyme cleavage sites within the cDNA sequence of human SIRT3 are identified using the online tool NEB cutter V2.0 (http://nc2.neb.com/NEBcutter2/) and excluded when designing the cloning primers.
4. The catalytic domain of human SIRT3 is from amino acid 137–346. The positively charged residues and the α-helical structure of the NH_2-terminal 25 amino acid residues are necessary for the import of SIRT3 into mitochondria. In mitochondrial matrix, SIRT3 is processed to yield a 28 kDa cleavage product [15, 16].
5. Alternatively, the pCMV-3Tag-3 mammalian expression vector can be used for expressing the protein product with three copies of FLAG epitope tag.
6. The gel slices should be completely dissolved in the extraction buffer. If necessary, adjust pH to 5.2 with 3 M sodium acetate to bring the color of the mixture to yellow.
7. The Interactive Tools (https://www.neb.com/tools-and-resources/interactive-tools) such as Double Digest Finder and NEBcloner can be referred for reaction buffer selection when setting up double digests.
8. For DNA quantification, same volumes of the DNA sample and the DNA ladder are loaded in the agarose gel. The intensity of bands will be compared with those of DNA ladder that have similar molecular weights for semi-quantification.
9. The temperature required for ligation reaction varies for blunt- and sticky-end ligations. T4 DNA ligase buffer should not be thawed at 37 °C to prevent the breakdown of ATP. Before ligation, alkaline phosphatase can be used to remove 5′ phosphate groups from DNA and prevent self-ligation.
10. Antibiotics are included for selection of bacterial colonies containing the properly assembled plasmids and should be added in the media during cooling process (usually below 55 °C).

The agar plates are labeled with antibiotics information, date, and sealed with parafilm for storage at 4 °C. Competent cells transformed with the improperly ligated plasmids cannot produce any colonies in selective media or agar plates. Thus, competent cells transformed with the ligation mixture without DNA insert can be used as negative control.

11. The procedure of cell lysis should not exceed 5 min. Mix gently by inverting the tubes until the solution becomes viscous with no aggregates. Do not vortex during extraction to avoid shearing of genomic DNA. The maximal elution efficiency is achieved between pH 7.0 and 8.5. Thus, when using ddH_2O for elution, make sure that the pH value is within this range.
12. Digestion with one restriction enzyme produces a linearized vector for verifying the size by agarose gel electrophoresis. Digestion with two restriction enzymes produces an inserted DNA fragment and a linearized vector backbone that can be verified by agarose gel electrophoresis. The lengths of linearized hSIRT3-FLAG and GST-hSIRT3 are about 6.6 kb and 6.1 kb, respectively. The 1.2 kb human SIRT3 cDNA inert and the cloning vector [4.9 kb for pGEX-2 T and 5.4 kb for pcDNA3.1 (+)] are separated after digestion with double enzymes and can be verified by DNA agarose gel electrophoresis.
13. After reaching the appropriate cell density, the induction time for recombinant protein expression varies, usually between 3 and 5 h. Do not incubate the bacterial culture too long, which will affect the quality and quantity of the recombinant protein expression. Aliquots of cultures before and after induction are collected for subsequent confirmation of recombinant protein expression.
14. After removing the supernatant, the bacterial pellet can be stored at −80 °C prior to further processing. However, it is recommended to process the pellet immediately to prevent protein degradation.
15. Sonication in short burst on ice is recommended to prevent heating up the lysates. Frothing should be avoided during sonication. The duration of sonication can be adjusted until the lysate solution is clear.
16. The cell lysates are filtered to remove cell debris which may affect the binding of proteins to the affinity column. Aliquots of the filtered lysate are collected for checking the purification efficiency.
17. To improve binding efficiency, it is recommended to use a flow rate between 1 and 5 ml/min for applying the bacterial lysates to the column. The affinity of the GST-tagged protein depends on both pH of the binding buffer and the flow speed.

18. Flow rate of 5–10 ml/min is recommended for the washing procedure. The flow-through solution from the column can be collected for checking the binding efficiency and troubleshooting the protein loss, if any.
19. The protein elutes should not be combined prior to analysis by SDS-PAGE. The fractions containing the eluted proteins (a 60 kDa recombinant hSIRT3/GST protein) are selected for further processing.
20. The solvent background is completely replaced by PBS in order to remove glutathione. The recombinant protein solution should be concentrated to 1–5 mg/ml for long-term storage.
21. Since SIRT3 is localized mainly in mitochondria, subcellular fractionation to enrich this organelle followed by pull-down experiment allows the identification of specific protein-binding partners of this enzyme.
22. The agarose beads and protein can be incubated at room temperature with agitation for up to 2 h. Incubation at 4 °C overnight is recommended to increase the binding specificity and prevent protein degradation. It is optional to block the agarose beads by incubation with 5 % BSA at room temperature for 1 h. Glutathione sepharose beads are prepared by following the same procedure.
23. A thrombin protease cleavage site is introduced between human SIRT3 and GST tag. Thrombin cleavage is performed by incubation at room temperature for 6 h. After mixing with protein-bound glutathione sepharose beads, the supernatants are collected for subsequent analysis.
24. Co-immunoprecipitation is performed with the total cell lysates, but not the subcellular fractions. Compared to pull-down experiment, the interactions between proteins from different subcellular compartments during co-immunoprecipitation can be a limiting factor for identifying true interactions. Thus, it is mainly used for the verification of protein–protein interactions.
25. Anti-FLAG antibody is used to detect the FLAG-tagged human SIRT3 by Western blotting, which shows two protein bands with molecular weights of 44 kDa and 28 kDa, respectively. In the stably transfected HEK293 cells, the majority of SIRT3 is presented as a truncated 28 kDa protein.
26. Endogenous SIRT3 can be immunoprecipitated from mitochondrial organelles isolated from human and mouse tissues for verifying its protein-binding partners.

References

1. Giblin W, Skinner ME, Lombard DB (2014) Sirtuins: guardians of mammalian healthspan. Trends Genet 30:271–286
2. Choudhary C, Weinert BT, Nishida Y, Verdin E, Mann M (2014) The growing landscape of lysine acetylation links metabolism and cell signalling. Nat Rev Mol Cell Biol 15:536–550
3. Frye RA (2000) Phylogenetic classification of prokaryotic and eukaryotic Sir2-like proteins. Biochem Biophys Res Commun 273:793–798
4. Finkel T, Deng CX, Mostoslavsky R (2009) Recent progress in the biology and physiology of sirtuins. Nature 460:587–591
5. Bai B, Liang Y, Xu C, Lee MY, Xu A, Wu D, Vanhoutte PM, Wang Y (2012) Cyclin-dependent kinase 5-mediated hyperphosphorylation of sirtuin-1 contributes to the development of endothelial senescence and atherosclerosis. Circulation 126:729–740
6. Xu C, Cai Y, Fan P, Bai B, Chen J, Deng HB, Che CM, Xu A, Vanhoutte PM, Wang Y (2015) Calorie restriction prevents metabolic ageing caused by abnormal SIRT1 function in adipose tissues. Diabetes 64:1576–1590
7. Wang Y, Xu C, Liang Y, Vanhoutte PM (2012) SIRT1 in metabolic syndrome: where to target matters. Pharmacol Ther 136:305–318
8. Imai S, Guarente L (2014) NAD+ and sirtuins in aging and disease. Trends Cell Biol 24:464–471
9. Hirschey MD, Shimazu T, Huang JY, Schwer B, Verdin E (2011) SIRT3 regulates mitochondrial protein acetylation and intermediary metabolism. Cold Spring Harb Symp Quant Biol 76:267–277
10. Cooper HM, Spelbrink JN (2008) The human SIRT3 protein deacetylase is exclusively mitochondrial. Biochem J 411:279–285
11. Onyango P, Celic I, McCaffery JM, Boeke JD, Feinberg AP (2002) SIRT3, a human SIR2 homologue, is an NAD-dependent deacetylase localized to mitochondria. Proc Natl Acad Sci U S A 99:13653–13658
12. Albani D, Ateri E, Mazzuco S, Ghilardi A, Rodilossi S, Biella G, Ongaro F, Antuono P, Boldrini P, Di Giorgi E, Frigato A, Durante E, Caberlotto L, Zanardo A, Siculi M, Gallucci M, Forloni G (2014) Modulation of human longevity by SIRT3 single nucleotide polymorphisms in the prospective study "Treviso Longeva (TRELONG)". Age 36:469–478
13. Bellizzi D, Dato S, Cavalcante P, Covello G, Di Cianni F, Passarino G, Rose G, De Benedictis G (2007) Characterization of a bidirectional promoter shared between two human genes related to aging: SIRT3 and PSMD13. Genomics 89:143–150
14. Rose G, Dato S, Altomare K, Bellizzi D, Garasto S, Greco V, Passarino G, Feraco E, Mari V, Barbi C, BonaFe M, Franceschi C, Tan Q, Boiko S, Yashin AI, De Benedictis G (2003) Variability of the SIRT3 gene, human silent information regulator Sir2 homologue, and survivorship in the elderly. Exp Gerontol 38:1065–1070
15. Schwer B, North BJ, Frye RA, Ott M, Verdin E (2002) The human silent information regulator (Sir)2 homologue hSIRT3 is a mitochondrial nicotinamide adenine dinucleotide-dependent deacetylase. J Cell Biol 158:647–657
16. Hallows WC, Albaugh BN, Denu JM (2008) Where in the cell is SIRT3? – functional localization of an NAD+-dependent protein deacetylase. Biochem J 411:e11–e13
17. Cimen H, Han MJ, Yang Y, Tong Q, Koc H, Koc EC (2010) Regulation of succinate dehydrogenase activity by SIRT3 in mammalian mitochondria. Biochemistry 49:304–311
18. Lombard DB, Alt FW, Cheng HL, Bunkenborg J, Streeper RS, Mostoslavsky R, Kim J, Yancopoulos G, Valenzuela D, Murphy A, Yang Y, Chen Y, Hirschey MD, Bronson RT, Haigis M, Guarente LP, Farese RV Jr, Weissman S, Verdin E, Schwer B (2007) Mammalian Sir2 homolog SIRT3 regulates global mitochondrial lysine acetylation. Mol Cell Biol 27:8807–8814
19. Hirschey MD, Shimazu T, Goetzman E, Jing E, Schwer B, Lombard DB, Grueter CA, Harris C, Biddinger S, Ilkayeva OR, Stevens RD, Li Y, Saha AK, Ruderman NB, Bain JR, Newgard CB, Farese RV Jr, Alt FW, Kahn CR, Verdin E (2010) SIRT3 regulates mitochondrial fatty-acid oxidation by reversible enzyme deacetylation. Nature 464:121–125
20. Qiu X, Brown K, Hirschey MD, Verdin E, Chen D (2010) Calorie restriction reduces oxidative stress by SIRT3-mediated SOD2 activation. Cell Metab 12:662–667
21. Sebastian C, Mostoslavsky R (2010) SIRT3 in calorie restriction: can you hear me now? Cell 143:667–668
22. Shimazu T, Hirschey MD, Hua L, Dittenhafer-Reed KE, Schwer B, Lombard DB, Li Y, Bunkenborg J, Alt FW, Denu JM, Jacobson MP, Verdin E (2010) SIRT3 deacetylates mitochondrial 3-hydroxy-3-methylglutaryl CoA synthase 2 and regulates ketone body production. Cell Metab 12:654–661

23. Hallows WC, Yu W, Smith BC, Devries MK, Ellinger JJ, Someya S, Shortreed MR, Prolla T, Markley JL, Smith LM, Zhao S, Guan KL, Denu JM (2011) Sirt3 promotes the urea cycle and fatty acid oxidation during dietary restriction. Mol Cell 41:139–149
24. Ahn BH, Kim HS, Song S, Lee IH, Liu J, Vassilopoulos A, Deng CX, Finkel T (2008) A role for the mitochondrial deacetylase Sirt3 in regulating energy homeostasis. Proc Natl Acad Sci U S A 105:14447–14452
25. Rardin MJ, Newman JC, Held JM, Cusack MP, Sorensen DJ, Li B, Schilling B, Mooney SD, Kahn CR, Verdin E, Gibson BW (2013) Label-free quantitative proteomics of the lysine acetylome in mitochondria identifies substrates of SIRT3 in metabolic pathways. Proc Natl Acad Sci U S A 110:6601–6606
26. Jacobs KM, Pennington JD, Bisht KS, Aykin-Burns N, Kim HS, Mishra M, Sun L, Nguyen P, Ahn BH, Leclerc J, Deng CX, Spitz DR, Gius D (2008) SIRT3 interacts with the daf-16 homolog FOXO3a in the mitochondria, as well as increases FOXO3a dependent gene expression. Int J Biol Sci 4:291–299
27. Lombard DB, Zwaans BM (2014) SIRT3: as simple as it seems? Gerontology 60:56–64
28. Verdin E, Hirschey MD, Finley LW, Haigis MC (2010) Sirtuin regulation of mitochondria: energy production, apoptosis, and signaling. Trends Biochem Sci 35:669–675
29. Kincaid B, Bossy-Wetzel E (2013) Forever young: SIRT3 a shield against mitochondrial meltdown, aging, and neurodegeneration. Front Aging Neurosci 5:48
30. Bause AS, Haigis MC (2013) SIRT3 regulation of mitochondrial oxidative stress. Exp Gerontol 48:634–639
31. Law IK, Liu L, Xu A, Lam KS, Vanhoutte PM, Che CM, Leung PT, Wang Y (2009) Identification and characterization of proteins interacting with SIRT1 and SIRT3: implications in the anti-aging and metabolic effects of sirtuins. Proteomics 9:2444–2456
32. Giralt A, Villarroya F (2012) SIRT3, a pivotal actor in mitochondrial functions: metabolism, cell death and aging. Biochem J 444:1–10
33. Shih J, Donmez G (2013) Mitochondrial sirtuins as therapeutic targets for age-related disorders. Genes Cancer 4:91–96
34. Osborne B, Cooney GJ, Turner N (2014) Are sirtuin deacylase enzymes important modulators of mitochondrial energy metabolism? Biochim Biophys Acta 1840:1295–1302
35. Parihar P, Solanki I, Mansuri ML, Parihar MS (2015) Mitochondrial sirtuins: emerging roles in metabolic regulations, energy homeostasis and diseases. Exp Gerontol 61:130–141

Chapter 15

Identification of Sirtuin4 (SIRT4) Protein Interactions: Uncovering Candidate Acyl-Modified Mitochondrial Substrates and Enzymatic Regulators

Rommel A. Mathias, Todd M. Greco, and Ileana M. Cristea

Abstract

Recent studies have highlighted the three mitochondrial human sirtuins (SIRT3, SIRT4, and SIRT5) as critical regulators of a wide range of cellular metabolic pathways. A key factor to understanding their impact on metabolism has been the discovery that, in addition to their ability to deacetylate substrates, mitochondrial sirtuins can have other prominent enzymatic activities. SIRT4, one of the least characterized mitochondrial sirtuins, was shown to be the first known cellular lipoamidase, removing lipoyl modifications from lysine residues of substrates. Specifically, SIRT4 was found to delipoylate and modulate the activity of the pyruvate dehydrogenase complex (PDH), a protein complex critical for the production of acetyl-CoA. Furthermore, SIRT4 is well known to have ADP-ribosyltransferase activity and to regulate the activity of the glutamate dehydrogenase complex (GDH). Adding to its impressive range of enzymatic activities are its ability to deacetylate malonyl-CoA decarboxylase (MCD) to regulate lipid catabolism, and its newly recognized ability to remove biotinyl groups from substrates that remain to be defined. Given the wide range of enzymatic activities and the still limited knowledge of its substrates, further studies are needed to characterize its protein interactions and its impact on metabolic pathways. Here, we present several proven protocols for identifying SIRT4 protein interaction networks within the mitochondria. Specifically, we describe methods for generating human cell lines expressing SIRT4, purifying mitochondria from crude organelles, and effectively capturing SIRT4 with its interactions and substrates.

Key words Sirtuin 4, Lipoic acid, Lipoyl, Lipoamide, PDH, Protein–protein interactions

1 Introduction

Sirtuins (SIRTs) comprise a family of seven mammalian nicotinamide adenine dinucleotide (NAD^+)-dependent enzymes that regulate genome expression, stress response, and aging [1, 2]. Defining the functional diversity of SIRTs is of immense interest, given their roles in the regulation of critical cellular homeostasis, and involvement with conditions such as cancer, and cardiovascular and neurodegenerative diseases [3–8]. Significant progress towards characterizing individual SIRTs has begun to elucidate

Sibaji Sarkar (ed.), *Histone Deacetylases: Methods and Protocols*, Methods in Molecular Biology, vol. 1436,
DOI 10.1007/978-1-4939-3667-0_15, © Springer Science+Business Media New York 2016

their specificity in the context of cellular localization, protein interactions, enzymatic function, and substrates, shedding light on these important molecular effectors.

SIRTs display diverse subcellular localizations and enzymatic functions. For example, SIRT1, SIRT6, and SIRT7 are nuclear, SIRT3–5 are predominantly mitochondrial, and SIRT2 is cytoplasmic [9, 10]. Although SIRT3–5 are housed within mitochondria, they have different deacyl catalytic functions and impact diverse metabolic pathways. SIRT3 exerts robust deacetylation activity for numerous mitochondrial substrates [11, 12] and regulates energy homeostasis [13]. SIRT4 deacetylates malonyl-CoA decarboxylase (MCD) to regulate lipid catabolism [14], and has been shown to delipoylate pyruvate dehydrogenase (PDH) to inhibit acetyl-CoA production [15]. SIRT4 has also been shown to perform ADP-ribosylation on glutamate dehydrogenase (GLUD1) [9]. SIRT5 desuccinylates, demalonylates, and deglutarylates protein substrates such as carbamoyl phosphate synthase 1 (CPS1) to regulate the urea cycle [16–18]. Thus, mitochondrial SIRTs are emerging as important enzymes, intrinsic to pivotal metabolic reactions that dictate mitochondrial regulation and homeostasis.

The conventional approach to discover new sirtuin enzymatic activities has been to screen various acyl-modified histone peptides with recombinant SIRT proteins in *in vitro* reactions [19]. While this approach can be used to directly compare the catalytic efficiency of SIRTs to various substrates, it is somewhat limited by the number of candidates tested in the screen. Moreover, it is often difficult to translate the *in vitro* findings *in vivo*, depending on the subcellular locations of the enzyme and potential substrate in cells. Furthermore, identification of the biological protein substrates in the cell remains challenging, particularly if there are several possible proteins that contain the acyl modification, or if there are as yet unknown substrates. An alternative strategy to this systematic screening approach is provided by proteomic profiling. The interacting SIRT protein partners are identified by mass spectrometry, and statistical bioinformatic analysis are employed to discover enriched functional categories.

We recently utilized this proteomics-based strategy to discover SIRT4 as the first known cellular lipoamidase (or delipoylase), enzymatically removing lipoyl modifications from lysine residues of substrate proteins [15]. Immunoaffinity purification (IP) of SIRT4 from mitochondrial fractions revealed its interactions with lipoyl-modified dehydrogenases and biotin-dependent carboxylases. The value of defining SIRT4 protein interactions within mitochondria in order to discover novel substrates was demonstrated by a combination of follow-up experiments. In vitro steady-state enzyme kinetic assays confirmed the catalytic efficiency of SIRT4 for lipoyl- and biotin-modified peptides. Additionally, experiments performed in human cells (i.e., fibroblasts) and in livers from SIRT4 knockout mice demonstrated that SIRT4 modulates the lipoyl-lysine levels of dihydrolipoyl acetyltransferase (DLAT), the E2 component of PDH, and the overall PDH activity. The impact of SIRT4 on other dehydrogenase complexes and on biotin-dependent carboxylases remains to be investigated. Furthermore, the contribution of the SIRT4 lipoamidase activity to various conditions of health and disease remains to be determined. Therefore, future studies of SIRT4 interactions with substrates and functional protein complexes in different cells and tissues promises to significantly expand the knowledge regarding its impact on cellular metabolism in health and disease.

In this chapter, we describe protocols that have proven effective for the identification of SIRT4 protein interactions. Firstly, retroviral transduction is used to generate stable cell lines overexpressing SIRT4-EGFP. This approach is commonly implemented when high-affinity, high-specificity antibodies required for the IP of the protein of interest are not available. While studying the interactions of the endogenous protein is always preferred, a tag can offer higher affinity (and sometimes specificity) of isolation. In addition to the studies of SIRT4 and other human sirtuins [20, 21], we have previously used this approach to discover and validate novel functional protein interactions for the eleven human histone deacetylases [22], as well as of other relevant families of proteins, such as the PYHIN proteins involved in transcriptional regulation and immune response [23]. In the second section, we describe a protocol for cellular fractionation and mitochondrial isolation. This is used to deplete the high abundance cellular contaminants from the IP, whilst enriching for trace SIRT4 interactions within the mitochondria. The third section contains methods for characterization of SIRT4 protein interactions. We have included protocols that describe conjugation of antibodies to magnetic beads, and the lysis conditions for achieving effective SIRT4 isolations. We also provide protocols for gel-based protein separation and tryptic digestion into peptides for mass spectrometry interrogation. The chapter concludes with instructions for data analysis, and determination of the specificity of the observed protein interactions based on the SAINT algorithm [24]. This integrated pipeline can be applied to any mitochondrial protein of interest to discover novel potential substrates, interacting protein partners, and expand the knowledge of its molecular function.

2 Materials

2.1 Generation of Fibroblasts Stably Expressing SIRT4-EGFP

1. Cell lines: MRC5 and Phoenix (ATCC). Cells are maintained at 37 °C with 5% CO_2.
2. Culture medium: DMEM containing 10% (v/v) fetal bovine serum and 1% (v/v) penicillin–streptomycin solution.
3. Opti-MEM reduced serum medium.

4. Lipofectamine 2000.
5. 1.5 mL Eppendorf tubes (Sterile).
6. 0.45 μm syringe filters.
7. 10 mL syringes.
8. Tissue culture plates: 6, 10, and 15 cm.
9. Geneticin (G418).
10. Selection medium: Culture medium containing 400 μg / mL G418.

2.2 Conjugation of Anti-GFP Antibody to Magnetic Beads

1. Dynabeads M-270 Epoxy (Life Technologies).
2. Affinity purified antibodies against an epitope tag or, if available, the protein of interest (e.g., anti-GFP antibodies described below for the isolation of GFP-tagged SIRT4). Store at −80 °C.
3. 0.1 M sodium phosphate buffer, pH 7.4 (4 °C, filter sterilized). Prepare as 19 mM NaH_2PO_4, 81 mM Na_2HPO_4. Adjust pH to 7.4, if necessary.
4. 3 M ammonium sulfate (filter sterilized). Prepare in 0.1 M sodium phosphate buffer, pH 7.4.
5. 100 mM glycine–HCl, pH 2.5 (4 °C, filter sterilized). Prepare in water and adjust to pH 2.5 with HCl.
6. 10 mM Tris–HCl, pH 8.8 (4 °C, filter sterilized). Prepare in water and adjust to pH 8.8 with HCl.
7. 100 mM triethylamine: Prepare fresh in water. CAUTION: Triethylamine is toxic and extremely flammable, and must be handled in a chemical hood and disposed appropriately.
8. DPBS, pH 7.4 (Dulbecco's phosphate-buffered saline (1×), liquid).
9. DPBS containing 0.5 % Triton X-100. Prepare fresh in DPBS.
10. DPBS containing 0.02% sodium azide. Prepare fresh in DPBS. CAUTION: Sodium azide is a toxic solid compound and must be handled in a chemical hood and disposed appropriately.
11. Rotator (at 30 °C).
12. Magnetic separation tube rack.
13. Tube shaker, e.g., TOMY micro tube mixer.
14. Safe-Lock tubes, 2 mL round bottom.
15. Ultrapure water (e.g., from a Milli-Q Integral water purification system).

2.3 Isolation of Mitochondria from Cultured Human Cells

1. Polycarbonate membrane filters, 14 μm (STERILTECH Corporation).
2. Swinney filter holders (Maine Manufacturing).
3. 5 mL syringes.

4. Trypsin–EDTA (0.5 %), no phenol red.
5. Dulbecco's phosphate-buffered saline (DPBS).
6. Homogenization buffer: 0.25 M sucrose, 1 mM EDTA, 20 mM HEPES–NaOH, pH 7.4.
7. Dilution buffer: 0.25 M sucrose, 6 mM EDTA, 120 mM HEPES–NaOH, pH 7.4.
8. OptiPrep™ Density Gradient Medium.
9. 50 % OptiPrep solution: Prepare fresh just before use by mixing 4 mL OptiPrep stock with 0.8 mL dilution buffer.
10. OptiPrep gradient solutions (Prepare fresh and store on ice) (Table 1).
11. 15 mL conical tubes.
12. Benchtop centrifuge.
13. Ultracentrifuge (capable of 100,000 × *g*).
14. Ultracentrifuge rotor and respective buckets.
15. Thin-wall polypropylene 4 mL ultracentrifuge tubes.
16. Safe-Lock microcentrifuge tubes, 1.5 mL.

2.4 Characterization of SIRT4 Interacting Proteins by Immunoaffinity Isolation

1. Mitochondrial-enriched fraction. It is preferred to isolate mitochondria fresh and to use it immediately.
2. 10× IP buffer: 0.2 M HEPES–KOH, pH 7.4, 1.1 M KOAc, 20 mM $MgCl_2$, 1 % Tween 20, 10 μM $ZnCl_2$, 10 μM $CaCl_2$. Filter-sterilize, and store at 4 °C for up to 6 months.
3. 10 % Triton X-100.
4. 4 M NaCl.
5. Protease inhibitor cocktail.
6. Lysis buffer: 20 mM HEPES–KOH, pH 7.4, 0.11 M KOAc, 2 mM $MgCl_2$, 0.1 % Tween 20, 1 μM $ZnCl_2$, 1 μM $CaCl_2$ 0.6% Triton X-100, 200 mM NaCl, protease inhibitors. The final composition and amount of the lysis buffer has been optimized for

Table 1 Optiprep density mixtures

OptiPrep solutions (%)	50 % OptiPrep solution (mL)	Homogenization buffer (mL)
10	0.4	1.6
15	0.6	1.4
20	0.8	1.3
25	1.0	1.0
30	1.2	0.8

analyzing SIRT4 interactions within mitochondria isolated from five 15 cm culture dishes. Prepare 2 mL of buffer per IP sample by combining 1.58 mL of Milli-Q grade water, 0.2 mL of 10× IP buffer, 0.12 mL of 10% Triton X-100, and 0.1 mL of 4 M NaCl. Prepare fresh prior to each experiment and store on ice. Add 1/100 (v/v) protease inhibitor cocktail immediately before use.

7. Wash buffer: Same composition as lysis buffer, except without protease inhibitors. Prepare 10 mL of buffer per IP sample.
8. DPBS, pH 7.4 (Dulbecco's phosphate-buffered saline (1×), liquid).
9. Magnetic beads conjugated with antibodies. Stored at 4 °C and used for IP within 2 weeks after conjugation.
10. Polytron for tissue homogenization (e.g., PT 10–35 Polytron from Kinematica).
11. Centrifuge and rotor, capable of 8000 × *g* at 4 °C.
12. Safe-Lock microcentrifuge tubes, 2 mL round bottom.
13. Magnetic separation rack.
14. 4× LDS sample buffer.
15. 10× reducing agent: Bond-breaker TCEP solution, neutral pH (Pierce).
16. 10× alkylating agent: 0.5 M chloroacetamide in water. Aliquot and store at ≤−20 °C.
17. Heat block at 70 °C.

2.5 SDS-PAGE and In-Gel Digestion of SIRT4 Interacting Proteins

1. Primary eluate from immunoaffinity purification.
2. 4–12% Bis–Tris pre-cast SDS-PAGE gel, 10 well.
3. SDS-PAGE electrophoresis system.
4. 20× MOPS SDS running buffer.
5. 1× running buffer: Dilute 20× MOPS SDS Running Buffer to 1×.
6. Molecular weight standards.
7. 4× LDS sample buffer.
8. Coomassie blue stain.
9. Ultrapure dH_2O (e.g., from a Milli-Q purification system).
10. Rocking platform.
11. Sheet protector and scanner.
12. Solution basins.
13. Safe-Lock microcentrifuge tubes, 2 mL round bottom.
14. Axygen 96-well plates with sealing mat.
15. Multichannel pipet.
16. Pipet tips, 200 μL (Low retention tips are highly recommended).

17. Ceramic plate, forceps, razor blade, Windex.
18. Mickle Gel Slicer or equivalent tool that cuts gel lanes into slices.
19. Lens paper.
20. Water, minimum HPLC grade.
21. Acetonitrile (ACN), HPLC grade.
22. Formic acid stock (FA), LC-MS grade, 99+%.
23. 1% FA solution: Mix 1.0 mL of FA stock in 99.0 mL ultrapure water. Store at RT.
24. 0.1 M ammonium bicarbonate (ABC): Dissolve 0.80 g of ammonium bicarbonate solid in 0.1 L of HPLC grade water. Store at RT and use within 1 month, or sterile filter for long term storage.
25. Destain solution: Mix 2 mL of ACN with 2 mL of 0.1 M ammonium bicarbonate. Prepare fresh before use in a solution basin.
26. Rehydration solution: Mix 2.5 mL of ultrapure water with 2.5 mL of 0.1 M ammonium bicarbonate. Prepare fresh before use in solution basin.
27. 0.5 μg/μl trypsin stock, sequencing grade. Store at −80 °C, limit to <5 freeze–thaw cycles.
28. 12.5 ng/μl trypsin solution: Mix 8.5 μL of trypsin stock with 331.5 μL of rehydration solution. Prepare fresh immediately before use.
29. Vortex device with an adaptor for 96-well plates.
30. Glass autosampler vials, MS Certified with 200 μL fused inserts and pre-slit caps.
31. 50% ACN–0.5% FA solution. Mix 0.5 mL of ACN and 0.5 mL of FA solution (1%) in microfuge tube.
32. Vacuum concentrator.

2.6 Peptide Desalting Using StageTips

1. Microcentrifuge.
2. Pipet tips, 200 μL (Low retention tips are highly recommended).
3. 16 G needle (Hamilton).
4. Syringe plunger, 100 μL (Hamilton).
5. Empore SDB-RPS disks (3 M).
6. 10% trifluoroacetic acid (TFA) in MS grade water.
7. 0.2% TFA in MS grade water.
8. Elution buffer: 5% ammonium hydroxide and 80% acetonitrile in water.
9. FA solution: 1% formic acid and 4% acetonitrile in water.
10. Autosampler vials.

2.7 Nanoliquid Chromatography Tandem Mass Spectrometry Analysis

1. Nanoflow HPLC system.
2. Mobile phase A (MPA): 0.1 % FA/99.9 % water. Store in amber bottle for up to 6 months.
3. Mobile phase B (MPB): 0.1 % FA/97 % ACN/2.9 % water. Store in amber bottle for up to 6 months.
4. Analytical column, e.g., Acclaim PepMap RSLC 75 μm ID × 25 cm.
5. Mass spectrometer.
6. Nanospray ESI source.
7. SilicaTip Emitter, Tubing (OD × ID) 360 μm × 20 μm; Tip (ID) 10 μm.

2.8 Mass Spectrometry Data Analysis and Interpretation

1. Multi-core/multi-CPU 64-bit PC workstation with at least 12 GB of RAM and 2 TB of storage.
2. Software for generating peaklists and scoring PSMs, with support for precursor ion quantification e.g., Proteome Discoverer 1.4, Mascot 2.3, Scaffold 4.
3. SAINT (http://www.crapome.org/).
4. Spreadsheet software.

3 Methods

3.1 Generation of Fibroblasts Stably Expressing SIRT4-EGFP

Highly specific antibodies with strong SIRT4 affinity can be used to directly study endogenous SIRT4 interactions. However, when these are not available, an alternative strategy can be to express a tagged version of the protein, such as SIRT4-EGFP. Different vectors can be used to express the tagged protein; however, it is preferable to select one that leads only to moderate levels of overexpression. For the protocol below, the retroviral vector pLXSN is used to express SIRT4-EGFP together with a neomycin-resistance marker in human fibroblasts (specifically, MRC5 cells). This approach was proven effective for discovering a novel enzymatic activity of SIRT4, as well as its substrates within the mitochondria [15]. The process involves transfection of Phoenix cells to generate retroviral particles in the cellular supernatant, followed by transduction of MRC5 cells (Fig. 1a). The protocol below describes reagent amounts required to generate one MRC5 cell line. For construction of additional cell lines, adjust reagents accordingly. For instance, MRC5 cells expressing EGFP should be generated in parallel to use as a control for nonspecific protein associations to magnetic beads or the tag.

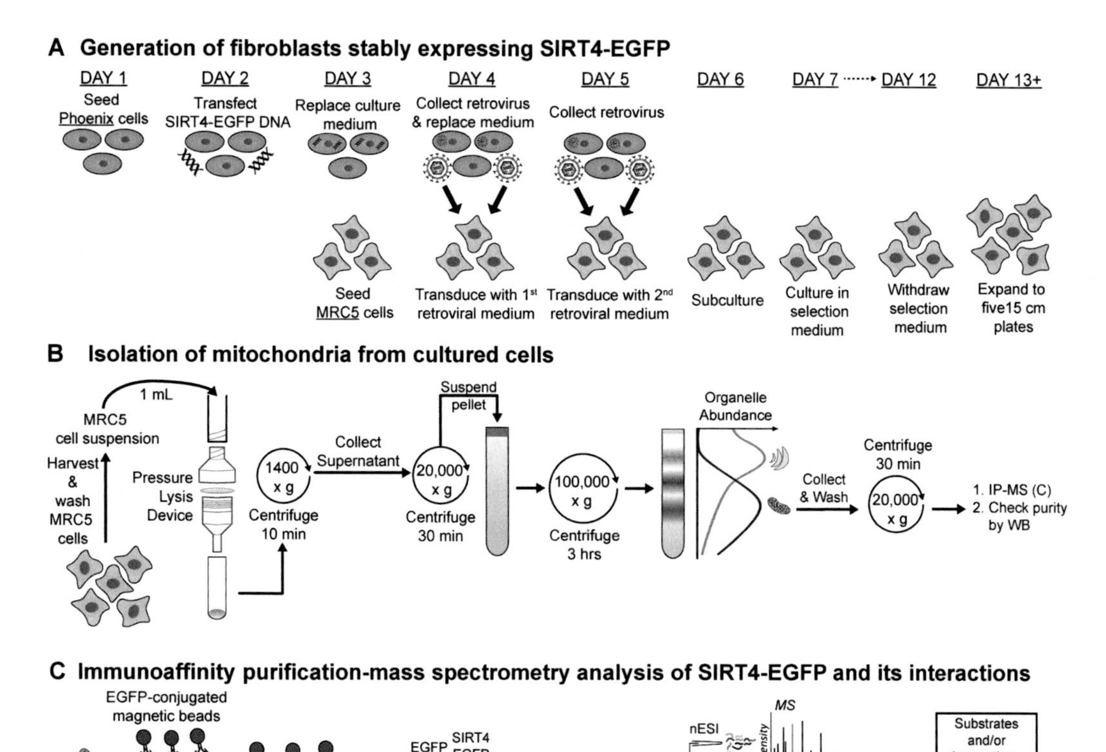

Fig. 1 Experimental approach for studying interactions and substrates of SIRT4. (**a**) Timeline for the generation of MRC5 fibroblasts (*bottom row*) stably expressing SIRT4-EGFP (*orange cells*) by transduction of retrovirus generated from Phoenix cells (*top row*). Two rounds of transduction are performed, followed by G418 selection of EGFP-tagged SIRT4 expressing cell. MRC5 cells that survive the selection are expanded to 5 × 15 cm plates in normal growth medium. (**b**) Scheme for isolation of mitochondria from EGFP-tagged SIRT4 expressing MRC5 cells. A crude organelle fraction is first obtained by 3× passage of MRC5 cells through a pressure lysis apparatus containing two polycarbonate filter disks. Organelles are then resolved by ultracentrifugation through a self-prepared OptiPrep density gradient. Western blot analysis should be used to confirm mitochondria are enriched in fractions 3 and 4, and also to check purity. (**c**) Graphic workflow for proteomic analysis of SIRT4-containing complexes isolated from purified mitochondria. Using optimized lysis buffer, SIRT4-EGFP and associated proteins are extracted from mitochondria and mixed with EGFP-conjugated magnetic beads. Affinity purified proteins are then eluted from the beads, resolved by SDS-PAGE, and digested in-gel with trypsin. Tryptic peptide mixtures are then analyzed by reverse-phase nanoliquid chromatography (RP-nLC) coupled to nanoelectrospray (nESI) tandem mass spectrometry. The total number of spectra collected and assigned to each protein (spectral counts) between control and SIRT4 isolations are used to determine specific interactions, which are subsequently analyzed by bioinformatics

3.1.1 Day 1

1. For each unique MRC5 cell line that will be transduced, seed one 6 cm plate with 1.7×10^6 Phoenix cells in 5 mL culture medium. Place in a humidified incubator set at 5 % CO_2 and 37 °C (unless otherwise stated, all subsequent overnight cell culture incubations are performed under these conditions).

3.1.2 Day 2

1. In the afternoon, ensure Phoenix cells are 85–90 % confluent.
2. Label two 1.5 mL microfuge tubes, A and B.
3. In Tube A, add 500 μL Opti-MEM and 7 μL Lipofectamine.
4. In Tube B, add 500 μL Opti-MEM and 2 μg DNA (e.g., SIRT4-EGFP or EGFP).
5. Incubate for 5 min at room temperature (RT).
6. Gently add contents of Tube B to Tube A, and incubate for 20 min at RT.
7. Replace culture medium of Phoenix cells seeded on Day 1 with 4 mL of fresh medium (without penicillin–streptomycin).
8. Add Lipofectamine–DNA mixture dropwise to Phoenix cells, swirl gently to mix, and incubate overnight.

3.1.3 Day 3

1. In the morning, aspirate culture medium from Phoenix cells and replace with 5 mL of fresh medium. Incubate overnight. This supernatant will be used for first retroviral transduction on Day 4 (**steps 1–6**).
2. In the afternoon, seed 1.5×10^6 MRC5 cells in a 10 cm plate in culture medium. Incubate overnight.

3.1.4 Day 4

1. Gently collect first retrovirus-containing supernatant from Phoenix cells in 15 mL conical tube (from Day 3, **step 1**).
2. Add 5 mL fresh culture medium to Phoenix cells and incubate overnight. This supernatant will be used for second retroviral transduction on Day 5.
3. Centrifuge Phoenix cell supernatant (first retrovirus collection) at $500 \times g$ for 5 min at RT to pellet lifted cells.
4. Pass supernatant through 0.45 μm syringe filter and collect in a 15 mL conical tube.
5. Add 5 mL fresh culture medium to the filtered Phoenix cell supernatant containing retrovirus (*see* **Note 1**).
6. Remove culture medium from MRC5 cells (seeded on Day 3, **step 2**) and replace with retrovirus-containing medium. Incubate overnight.

3.1.5 Day 5

1. Gently collect second retrovirus-containing supernatant from Phoenix cells in 15 mL conical tube (from Day 4, **step 2**).
2. Perform a second round of retroviral transduction, following the same procedure as described above (Day 4, **steps 3–6**).

3.1.6 Day 6

1. Passage MRC5 cells, and re-seed to have 60–70% confluence in 10 mL culture medium. Incubate overnight.

3.1.7 Day 7

1. Aspirate MRC5 cell culture medium and replace with 10 mL selection medium (*see* **Note 2**).
2. Culture in selection medium for 3–6 days. During this time, monitor the amount of cell detachment/lysis and change medium as needed (*see* **Note 3**). Expression of SIRT4-EGFP can be evaluated by direct fluorescence microscopy and/or western blotting.
3. Withdraw selection medium when cell lysis has stopped (6 days or less). Continue to passage cells in culture medium (no selection) for 2–3 passages (*see* **Note 4**) and expand into five 15 cm plates.
4. At ~95% confluence, cells should be harvested for mitochondria isolation (*see* Subheading 3.3). It is recommended to perform mitochondria isolation from freshly harvested cells and then proceed immediately to immunoaffinity purification of SIRT4-EGFP (*see* Subheading 3.4). Therefore, preparation of the EGFP antibody-conjugated magnetic beads (*see* Subheading 3.2) should be started 2 days prior to the collection of cells for mitochondria isolation.

3.2 Conjugation of Antibodies to Magnetic Bead for SIRT4 Immunoaffinity Purifications

This protocol has been optimized for the conjugation of M-270 Epoxy Dynabeads with affinity purified antibodies [25]. The use of other magnetic beads (e.g., M-450 or MyOne Dynabeads) is possible, but the amount of antibody used for conjugation should be adjusted based on the binding capacity of the bead. Commercially available antibodies can be used with the M-270 Epoxy Dynabeads. However, if the antibody is stored in buffer containing free amines (e.g., Tris), the amount of antibody covalently conjugated to the surface epoxy groups will be limited; it is best to avoid such buffers. It is recommended to begin this section of the protocol in the afternoon and resume (**step 7**) in the morning of the following day.

1. For each immunoisolation to be performed, weigh out 7 mg of magnetic Dynabeads in a round-bottom tube (*see* **Note 5**).
2. Add 1 mL sodium phosphate buffer (pH 7.4) to the beads. Mix by vortexing for 30 s, followed by 15 min on a tube shaker (vigorous setting).

3. Place the tube on a magnetic rack. Once the beads have attached to the magnet and the supernatant has a clear appearance (~30 s), aspirate the wash buffer.
4. Remove the tube from the rack. Add 1 mL sodium phosphate buffer (pH 7.4) and mix by vortexing for 30 s. Aspirate the wash buffer as above.
5. Remove the tube from the rack. In the following order, add the necessary amount of (1) antibodies, (2) sodium phosphate buffer, and (3) ammonium sulfate solution.
   (a) The optimal total volume for bead conjugation is a 20 μL reaction volume per mg of M-270 epoxy beads.
   (b) The amount of antibody that should be used during conjugation is 4–5 μg Ab per mg M-270 epoxy beads.
   (c) The 3 M ammonium sulfate solution is added last to give a final concentration of 1 M.
   (d) For example, for 14 mg of washed beads, add 42 μg of antibody. Second, add 0.1 M sodium phosphate buffer such that its volume equals 280 μL, minus the volume of antibody and the volume of 3 M ammonium sulfate. Finally, add 93 μL, of 3 M ammonium sulfate.
6. Wrap the tube with parafilm and incubate the bead suspension overnight on a rotator at 30 °C (*see* **Note 6**).
7. The following morning, place the tube against the magnetic rack.
8. OPTIONAL: Retain the supernatant to assess the efficiency of bead conjugation by SDS-PAGE.
9. Wash the beads sequentially with the following buffers (1 mL each): sodium phosphate buffer; 100 mM glycine–HCl, pH 2.5 (FAST WASH); 10 mM Tris–HCl, pH 8.8; 100 mM triethylamine (FAST WASH); D-PBS (4 WASHES); DPBS containing 0.5 % Triton X-100 (15 min wash with gentle agitation); DPBS (*see* **Note 7**).
10. Resuspend washed beads in 12.5 μL DPBS + 0.02 % NaN_3 per mg of beads. Measure the final volume of the bead slurry to determine the bead concentration (mg of beads/μL DPBS).
11. Beads can be used immediately or stored for up to 2 weeks at 4 °C. After 1 month of storage, their efficiency for isolation decreases by approximately 40 %.

3.3 Isolation of Mitochondria from Cultured Cells

Given that SIRT4 is known to be primarily mitochondrial [10], subcellular enrichment of mitochondria prior to immunoaffinity purification is advantageous. This approach will deplete highly abundant cytoplasmic and nuclear proteins that are unlikely to be physiological substrates of SIRT4, improving the detection of trace and/or transient protein interactions that may be in lower abundance (Fig. 1b).

1. Precool ultracentrifuge rotor and buckets at 4 °C.
2. Prepare five pressure filtration apparatuses. Unscrew and open swinney filter holder, and place two polycarbonate membrane filters inside (shiny side facing towards syringe attachment side of swinney). Screw swinney closed extremely tight, to ensure sample does not leak out during cell lysis (*see* **Note 8**). Attach syringe to swinney.
3. Remove culture medium, and rinse cells with 10 mL DPBS.
4. Lift cells with 3 mL/plate Trypsin–EDTA solution. Briefly incubate at 37 °C, if necessary.
5. Add 12 mL culture medium to inactivate trypsin, and centrifuge at 500 × *g* for 5 min to pellet cells.
6. Aspirate and discard medium, and wash cells with 10 mL DPBS. Centrifuge at 500 × *g* for 5 min to pellet cells.
7. Resuspend cells in 5 mL homogenization buffer (*see* **Note 9**).
8. Transfer 1 mL to the each syringe of each pressure filtration apparatus, and gently push syringe plunger down to force cell suspension through filtration apparatus, collecting filtrate in a clean 15 mL tube in ice.
9. Unscrew syringe from swinney, and separate plunger from syringe. Reconnect syringe to swinney.
10. Repeat **steps 8–9** twice, using the filtrate from the first lysis (*see* **Note 10**).
11. Pool filtrates from the same cell lines and centrifuge at 1400 × *g* for 10 min at 4 °C to pellet nuclei.
12. Carefully aspirate the supernatant, and transfer to new 15 mL tube. Centrifuge at 20,000 × *g* for 30 min at 4 °C to pellet crude organelles.
13. During centrifugation, prepare two 10–30% discontinuous OptiPrep gradients (*see* Subheading 2.3) in thin wall ultracentrifuge tubes by top-layering 0.7 mL of each OptiPrep solution, starting with the densest solution (30%) at the bottom of the tube (*see* **Note 11**). Store gradients on ice.
14. After centrifugation, aspirate and discard the supernatant.
15. Gently resuspend the crude organelle pellet in 0.7 mL homogenization buffer by pipetting until the pellet has been dispersed to a homogenous suspension.
16. To one OptiPrep gradient, carefully overlay the 0.7 mL crude organelle suspension on top of the 10% OptiPrep solution. To the other OptiPrep gradient, layer 0.7 mL of Homogenization Buffer to serve as the balance during ultracentrifugation.
17. Transfer tubes to opposing ultracentrifuge buckets.

18. Attach all six buckets to their respective position on the SW60 rotor and perform ultracentrifugation at 100,000 × *g* for 3 h at 4 °C.
19. Carefully aspirate six 0.7 mL fractions from the top of the gradient. Collect Mitochondria should be primarily enriched in fraction 4, and to a lesser extent, in fraction 3 (*see* **Note 12**).
20. Wash mitochondria by adding 0.8 mL DPBS, and mixing gently.
21. Centrifuge at 20,000 × *g* for 30 min at 4 °C to pellet mitochondria.
22. Assess purity by western blotting (*see* **Note 13**).
23. Ideally, as mentioned above, the immunoaffinity isolation of SIRT4-EGFP (*see* Subheading 3.4) should be performed on freshly isolated mitochondria. This will produce the most consistent isolation of SIRT4-EGFP and its interacting proteins, particularly for less stable interactions. However, if experimental limitations require collection of mitochondria on different days, freshly isolated mitochondria should be immediately flash frozen in liquid nitrogen and stored at −80 °C until ready to proceed with the immunoaffinity isolation section.

3.4 Immunoaffinity Isolation of SIRT4 Interacting Proteins

SIRT4 enzymatic activity towards various acyl-modified peptides has been traditionally screened by systematically testing candidate peptides using in vitro enzymatic assays, e.g., as in Ref. [19]. This approach can be laborious and is often prohibited by the number of candidate substrates in the screen. Proteomics can be employed for a less biased approach for identifying potential substrates, which can then be validated by in vitro enzyme kinetic assays, as in [15]. Using optimized mitochondrial lysis conditions and stringency, SIRT4 and its interactions are isolated and subjected to mass spectrometry-based identification (Fig. 1c). Following immunoaffinity isolation, there are several time points when the protocol can be paused and continued at a later time (as indicated below).

1. Prepare lysis buffer (containing protease inhibitors) and wash buffer.
2. Add 1 mL of lysis buffer to purified mitochondria (isolated from five 15 cm plates, Subheading 3.3) and vortex 3 × 20 s, cooling in between on ice. Mix by rotation for 10 min at 4 °C.
3. Centrifuge the lysate at 5000 × *g* for 10 min at 4 °C.
4. During centrifugation step, place tube of antibody-conjugated magnetic beads against a magnetic rack for 30–60 s. Discard the storage buffer and wash 3 × 1 mL with wash buffer by gently pipetting to resuspend the beads. <u>Do not vortex</u>.
5. Resuspend the beads in 100 μL of wash buffer and place on ice.

6. Carefully transfer the clarified mitochondrial lysates (supernatant) into a clean round-bottom microfuge tube and store on ice. Retain (1) the insoluble cell/tissue pellet and (2) 40 μL of the clarified lysates as the input fraction for isolation efficiency analysis by western blot.
7. Mix the antibody-conjugated beads to a homogenous suspension by gently pipetting. Pipette a 7 mg equivalent of beads into each sample of clarified mitochondrial lysates.
8. Rotate the lysate–bead suspension on a rotator at 4 °C for 1 h (*see* **Note 14**).
9. During the incubation, prepare 1× NuPAGE sample buffer and three clean round-bottom microfuge tubes per IP sample. Ensure that wash buffer and DPBS are cooled on ice.
10. After incubation, place samples on magnetic rack and allow the beads to adhere to the tube wall (~20 s). Transfer the flow-through (unbound) fraction to a clean tube and retain for isolation efficiency analysis by western blot.
11. Gently resuspend the beads in 1 mL of wash buffer and transfer the bead suspension to a clean round-bottom tube.
12. Separate the beads from the buffer wash buffer using the magnetic rack as above. Aspirate the wash buffer. *Perform this step between all subsequent wash steps.*
13. Wash the beads 3 × 1 mL with wash buffer. On the third wash, transfer the bead slurry to a second clean round-bottom tube.
14. Wash the beads 2 × 1 mL with wash buffer.
15. Wash the beads with 1 mL DPBS and transfer slurry to a third clean round-bottom tube.
16. Repeat DPBS wash and ensure buffer is completely removed.
17. To elute the protein complexes from the beads, add 40 μL of 1× sample buffer to beads.
18. Incubate for 10 min at 70 °C, then 10 min at RT with agitation (*see* **Note 15**).
19. Isolate beads on the magnetic rack and transfer the primary eluate to a clean microcentrifuge tube.
20. For assessment of elution efficiency, resuspend the beads in 1× sample buffer and repeat **steps 18** and **19**, except incubate the beads for 5 min at 95 °C.
21. Isolate beads on the magnetic rack and transfer the secondary eluate to a clean microcentrifuge tube.
22. Add 5 μL of 10× reducing agent and 5 μL of 10× alkylating agent to primary and secondary eluates. Heat at 70 °C for 10 min. Retain 10% aliquots of primary and secondary eluates for western analysis of isolation efficiency.

23. At this point, samples can either be stored at ≤−20 °C, or the remaining 90 % of the primary eluate can be processed immediately by SDS-PAGE and in-gel digestion to identify the co-isolated proteins by mass spectrometry (*see* Subheading 3.5).
24. To assess the efficiency of immunoisolation, analyze equal percentages (e.g., 5 %) of the following samples by western blotting: cell pellet and input (**step 6**), flow-through (**step 10**), primary and secondary eluates (**step 22**).

3.5 SDS-PAGE and In-Gel Digestion of SIRT4 Interacting Proteins

The proteins co-isolated with SIRT4 are next identified by mass spectrometry. Samples can be prepared for mass spectrometry analysis by digesting the proteins with an enzyme using either an in-gel or in-solution protocol. An example of an in-solution digestion protocol was described in detail in [26]. Here, we provide an in-gel digestion protocol that was successfully implemented when studying SIRT4 interactions [15]. This protocol is performed over the course of 3 days. To minimize keratin and other environmental contaminants, it is recommended to wear a lab coat and hair protection, to avoid close contact and limit environmental exposure of the pre-cast gel during sample loading and protein staining. While many SDS-PAGE systems can be employed for proteomic analysis, a system that has pre-cast gels available is highly recommended to further reduce environmental contaminants. Here we use the NuPAGE system as we have found that using the MOPS running buffer in combination with a 4–12 % Bis–Tris pre-cast acrylamide gels excels at resolving larger molecular weight proteins, which usually generate more peptide complexity (*see* **Note 16**).

3.5.1 Day 1

1. If necessary, thaw primary eluate sample(s) at 70 °C.
2. Set up the electrophoresis system.
3. Load 30 μL of primary eluates, leave 1–2 empty lanes between samples (*see* **Note 17**).
4. Load 20 μL of 1× sample buffer into empty wells to ensure even running of the samples.
5. Electrophorese briefly (2–3 min) at 100 V, then turn off system and load remaining sample volume (~15 μL).
6. Continue electrophoresis at 150 V until the dye front has migrated approximately 1/3 of the way down the gel (*see* **Note 18**).
7. Open gel cassette to expose gel and discard the wells. Working with wet gloves, transfer gel (by the thick ridge at the bottom) into a plastic tray containing ultrapure water. Remove bottom ridge.
8. Wash the gel 3 × 5 min with ultrapure H_2O while rocking.

9. Cover the gel completely in Coomassie blue stain. Incubate for 1–3 h until the protein bands become clearly visible.
10. Dispose of the staining solution. Rinse briefly with ultrapure H_2O. Add fresh ultrapure H_2O and destain overnight at 4 °C with gentle rocking (*see* **Note 19**).

3.5.2 Day 2

Working solution volumes are calculated for 16 total gel fractions and based on 90 μl of solution per well, except where noted (*see* **Note 20**). For higher throughput, a 96-well plate with sealing mat and a multichannel pipette is recommended. A different set of tips should be used for each set of gel pieces from the sample lane, but each set can be used for all steps (Day 2 and 3).

1. Briefly wash the destained gel several times with ultrapure water to remove any excess staining reagent.
2. Place the gel in a sheet protector. Keep the gel wet with ultrapure water. Place on a bed scanner to digitize the gel image. Print out the image to use as a reference during grouping of the gel slices (**step 9**).
3. Wash the ceramic plate, forceps, blade of the gel slicer, and a razor blade with Windex, then with ultrapure water.
4. Transfer the gel to the ceramic plate and cut out the vertical stained region of one sample lane. Trim the lane to remove the lower portion just at the dye front and the top 2 mm near the well.
5. Place a small piece of lens paper on the cutting stage of the Mickle Gel Slicer, wet the lens paper with ultrapure water and flatten it out.
6. Place the excised gel lane on the lens paper and add a drop of ultrapure water.
7. Use the Mickle Gel Slicer (or equivalent) to cut the gel lane into 1 mm slices (~20–30 total slices).
8. Lift the lens paper and flip the cut gel lane onto the ceramic plate. Carefully remove the lens paper. Keep cut gel slices slightly wet for ease of sample handling.
9. Starting from the bottom of the gel, divide/group gel pieces such the amount of protein per group (estimated by the Coomassie blue stain intensity) is normalized. For example, for darkly stained bands/band clusters, group only 1–2 gel pieces, while lightly stained regions (usually upper MW region) can be grouped into 3–4 gel slices (*see* **Note 21**)
10. Once the grouping has been defined, use a razor blade to cut each group gel slices into three segments (*see* **Note 22**).
11. Transfer gel pieces using the forceps or the flat side of razor blade into one well of 96-well plate.
12. Repeat **steps 4–12** for additional sample lanes.

13. Add 90 μL of destain solution to each well, seal plate, and agitate on a vortexer (medium setting) at 4 °C for 10 min. Aspirate and discard. For all aspiration steps, make sure the gel pieces are not stuck to pipet tips when discarding waste solution as sample loss could occur.
14. Repeat the previous step with fresh destain solution. While destaining, prepare and aliquot rehydration solution and ACN (5 mL each) in separate solution basins.
15. Aspirate the destain solution, add 90 μL ACN to the gel pieces, seal plate with mat, and invert manually several times. Let it stand at RT until the gel pieces are white (1–2 min).
16. Remove ACN and add 90 μL of rehydration solution, seal the plate, and manually invert several times. Incubate at 4 °C until the gel pieces have swelled and are translucent (<5 min).
17. Aspirate rehydration solution and perform another round of dehydration–rehydration.
18. Aspirate rehydration solution and perform a final dehydration with ACN. Aspirate the ACN and let residual evaporate.
19. While ACN is evaporating, prepare the trypsin solution. Add 20 μL of trypsin solution to the dried gel pieces and incubate at RT until the gel pieces swell and become translucent (~5 min).
20. Add ~30 μL of rehydration solution to the gel pieces (use enough volume to fully cover the gel pieces). Seal the plate and incubate overnight at 37 °C.

3.5.3 Day 3

1. In the morning, add 30 μL 1 % FA to each sample and incubate for 4 h at RT.
2. **CRITICAL STEP**: Transfer the extracted peptides to separate microfuge tubes. Ensure no gel pieces are transferred to the microfuge tubes. Keep extracted peptides at 4 °C.
3. Add 30 μL of 50 % ACN–0.5 % FA solution to the gel pieces. Incubate for 2 h at RT.
4. Combine the second extraction with each respective first extraction.
5. To further normalize the amount of peptides per fraction, pool peptides from neighboring gel regions that were minimally stained (*see* **Note 23**).
6. Concentrate the samples (to ~ 25 μL) by vacuum centrifugation to remove ACN.
7. Proceed to peptide desalting procedure (Subheading 3.6).

3.6 Peptide Desalting Using StageTips

1. Prepare one StageTip for each sample by using a 16 G needle to cut and deposit a single Empore SDB-RPS disk into the bottom of a 200 μL pipette tip using the syringe plunger (*see* **Note 24**).
2. Using 10 % TFA stock, acidify samples to final concentration of 1 % TFA.
3. Apply the sample to the top of the StageTip (*see* **Note 25**) and centrifuge at 1000 × *g* until all solution has passed through the StageTip (*see* **Note 26**).
4. Wash disk with 100 μL of 0.2 % TFA.
5. Apply 50 μL of Elution buffer to the StageTip and manually collect the eluate in an autosampler vial.
6. Concentrate samples by vacuum centrifugation to near-dryness.
7. Add FA solution to achieve a final volume of 9 μL. Vortex briefly to mix.
8. Proceed immediately to nLC-MS/MS analysis (Subheading 3.7) or store at −80 °C for future analysis.

3.7 Nanoliquid Chromatography-Tandem Mass Spectrometry Analysis

Many HPLC and MS system configurations are suitable for analyzing tryptic peptide digests obtained from in-gel digestion. Here, we present an approach that is tailored for the acquisition of label-free spectral counting data, since we have effectively implemented this label-free approach [24] for determining the specificity of interactions for SIRT4 and for other proteins [15, 22, 23]. Towards this goal, the instrumentation and associated method should be optimized for the acquisition rate of MS/MS spectra, while also balancing the sensitivity and the depth of analysis, improving the detection of low abundance interactions. An LC system capable of low flow rates (<0.5 μL/min) and high pressure support (>400 bar) is highly preferable, as these capabilities allow the use of analytical columns with inner diameters ≤75 μm and lengths ≥ 25 cm. Additionally, a high sensitivity detector for MS/MS acquisition is encouraged. For example, in our study of SIRT4 interactions [15], we used a Dionex Ultimate 3000 nanoliquid chromatography system directly coupled to an LTQ Orbitrap Velos mass spectrometer.

1. Ensure that the system is properly calibrated according to the manufacturer's specifications.
2. Using MS instrument software, create an appropriate acquisition method (*see* **Note 27**).
3. Using the LC instrument software, create a reverse-phase LC method. Program the method to separate peptides over 90 min using a linear gradient of 4– 40 % mobile phase B.

4. Create a shorter length (e.g., 30 min) gradient method to use for analysis of a standard sample that serves as a quality control for instrument performance before, during, and after the analysis of the SIRT4 samples. A typical quality control sample would be a tryptic digest of a single protein (e.g., albumin) or whole cell lysates (e.g., HeLa).
5. Perform duplicate (at a minimum) injections of the standard sample to ensure that the system is performing at an acceptable level prior to injecting experimental samples.
6. For experimental samples, inject 4 μL of each fraction using the appropriate LC-MS/MS method designed above.
7. After an experimental set of injections has been complete, inject the standard peptide mixture to confirm that instrument performance has been maintained throughout the analysis.

3.8 Data Analysis and Interpretation

3.8.1 Peptide Identification and Protein Assignment

1. Extract all MS/MS spectra from raw mass spectrometry data, removing MS/MS spectra that do not contain at least ten peaks.
2. Create instrument and experiment-specific database search methods.
   (a) Define static peptide modification for cysteine carbamidomethylation.
   (b) Define variable modification for methionine oxidation (*see* **Note 28**).
3. Submit spectra to an appropriate workflow to obtain peptide spectrum matches and protein group assignments (*see* **Note 29**).
4. Select peptide and protein scoring filters to achieve a desired false discovery rate (e.g., ≤1 %).
5. Export data tables for interaction specificity analysis using the SAINT algorithm. The tables should contain, at minimum, protein group descriptions with respective accession numbers and total spectrum counts.

3.8.2 SAINT Interaction Specificity Analysis Using Label-free Spectral Counting

1. Register for a free user account at the website, www.crapome.org (*see* **Note 30**).
2. Select “Workflow 3: Analyze Your Data”.
3. OPTIONAL: Select additional negative controls from the CRAPOME database (*see* **Note 31**).
4. Using the exported data tables (*see* above), generate a compatible SAINT matrix input file, as specified in the workflow step 2, Upload Data.
5. Upload SAINT matrix file and proceed to step 3, Data Analysis.

6. Under the "Analysis Options", enable "Probability Score", choose the "SAINT" model, and increase the "n-iter" option to 10,000 (*see* **Note 32**). Run Analysis.
7. After the analysis has completed, save and open the output file, which reports the individual and average SAINT scores (AvgP) for each identified protein. SAINT scores range from 0 (lowest probability of specific interaction) to 1 (highest probability of specific interaction).
8. Evaluate the performance of SAINT to distinguish between specific interactions and nonspecific background. If many interactions are known for a particular protein of interest, the sensitivity and specificity of the analysis can be estimated by constructing ROC plots. If no prior interaction knowledge is available, then construct a histogram for the distribution of SAINT scores. Use these analyses to select a SAINT score cut-off that eliminates the majority of nonspecific interactions (false positives), while retaining the highest scoring interactions (*see* **Note 33**).

3.9 Conclusion

The protocol detailed above provides a method for isolating SIRT4 and identifying its protein interactions in the mitochondria of human fibroblasts (Fig. 1a). The approach to generate cell lines stably expressing SIRT4-EGFP using retroviral transduction is also amenable for other cell types. By performing an upstream biochemical organelle fractionation (Fig. 1b) to obtain enriched mitochondria, a major source of potential nonspecific interactions from abundant cytoplasmic and nuclear proteins during immunoaffinity isolation is reduced (Fig. 1c). This overall approach was proven effective for the identification of SIRT4 interactions and has led to the discovery of SIRT4 enzymatic delipoylase activity [15]. In addition to the identification of candidate SIRT4 substrates, the resulting interaction datasets can contain novel information about proteins that may regulate the diverse enzymatic activities of SIRT4, including its ability to remove lipoyl, biotinyl, and acetyl modifications. Furthermore, such studies can reveal protein complexes that may facilitate the targeting of SIRT4 to its substrates. New hypotheses can be generated by analyzing candidate interactions using bioinformatics and functional pathway enrichment analyses. Different bioinformatics platforms can be used for this purpose and, while the description of these computational approaches is outside the scope of this chapter, we point the reader to data workflows used in several protein interaction studies [15, 22, 23, 26, 27]. As the knowledge regarding the substrates and functions of SIRT4 still remains limited, we hope that this protocol will aid future studies to better understand its contribution to diverse metabolic pathways in health and disease states.

4 Notes

1. For some cell types, polybrene (4–8 μg/mL) can be used to enhance gene transfer efficiency, however, this is usually toxic to MRC5 cells.
2. Prior to selection, a G418 killing curve should be optimized to determine appropriate concentration of G418 required to kill non-transduced cells.
3. Un-transduced cells will begin to lift and undergo cell lysis, leading to the medium becoming cloudy with cellular debris. Replace medium daily, and continue to select in G418-containing medium until cells stop lifting, and begin to expand (usually within 6 days). As an alternative, or in addition to G418, FACS may also be used to select for transfected cells which will express EGFP.
4. MRC5 cells cultured in G418 will have reduced proliferation and can acquire enlarged morphology. Two to three passages in regular medium are usually required for them to recover and stabilize.
5. Unless otherwise indicated, all steps should be performed at room temperature. Round-bottom tubes are the preferred tube shape, which minimizes bead trapping during the conjugation. The required amount of beads is dependent on both the experimental objective and the abundance of the protein to be immunoaffinity purified. As an approximate guide, 1–2 mg beads are appropriate for small-scale optimization experiments, 5–7 mg beads are usually sufficient for single immunoaffinity purifications, and 10–20 mg beads may be required for proteins of high abundance. During the washing steps, proceed immediately from one wash step to the next and do not allow the beads to sit without a washing solution between each step.
6. For efficient conjugation, tubes should rotate end-over-end (not a lateral rocking). Also, ensure the beads remain wet during this rotation. A minimum conjugation volume of 200 μl is recommended, which equates to a minimum of 10 mg beads used for conjugation. This is achieved in the protocol by performing at least two affinity enrichments (7 mg beads each for EGFP IP and SIRT4-EGFP IP).
7. Bead washes with the acidic (glycine) and basic (trimethylamine) solutions should be performed quickly (labeled as "FAST"), so as not to denature the antibody. To perform a FAST wash, the beads should only be pipetted in these solutions enough to disperse the pellet, which can usually be achieved by adding the solution directly over the pellet, followed by one additional rinse (if necessary). The beads should be immediately placed on the magnet, then when the solution

turns clear, the wash buffer should be aspirated and the subsequent neutralization buffer added to the tube.

8. For each plate of cells, prepare one pressure filter apparatus containing two filters. Usually each device will efficiently lyse 1 mL of cell suspension after 3 rounds. After use, the filters are disposed; however, the swinneys can be reused following short treatment with bleach, and extensive washing with water.
9. 1 mL of homogenization buffer per plate of cells is recommended.
10. The efficiency of each round of lysis can be observed by viewing a few microliters of the cell lysis suspension under the microscope to determine percentage of cell lysis (cell free nuclei). Continue until 70–90% of cells have been lysed, which is expected to take approximately three rounds with MRC5 cells.
11. Samples in this protocol have been top-loaded; we have found that alternatives, such as bottom-loading, are not as reproducible. When preparing the gradient, carefully layer the fractions on top of each other to minimize the mixing of the solutions. This can be done by angling the pipette tip towards the wall of the tube, and dispensing solutions gently as the pipette moves upwards out of the tube.
12. To estimate the density of specific fractions, fractions can be collected from the empty OptiPrep gradient that was used as a balance. These fractions can then be diluted 10,000-fold with water, and the absorbance can be measured at 244 nm.
13. Location of mitochondria on gradient can be evaluated by western blotting across all 6 fractions for mitochondrial markers COXIV, SIRT4, or EGFP (for tagged SIRT4). Mitochondria should be enriched in fractions 3 and 4. These fractions can be pooled and washed together to increase yield of mitochondria. Purity estimations can be performed simultaneously, by probing fractions with markers for other organelles such as endosome (EEA1), endoplasmic reticulum (calreticulin), Golgi (GM130), and lysosome (LAMP1).
14. Longer incubation times tend to promote the accumulation of nonspecific binders and the loss of weak interacting partners [25].
15. We have found that when using high affinity antibodies (e.g., anti-GFP) stringent heat and detergent denaturing conditions are sometimes required for efficient elution of the target proteins from the beads.
16. If improved resolution of proteins in the lower molecular weight region is desired, use the 20× MES running buffer instead of the 20× MOPS running buffer.
17. Since the total volume of primary eluates is ~45 μL, each sample is loaded in two separate additions (30 and 15 μL).

18. If it is known that the sample complexity or protein load is high, then resolving proteins for entire gel length may be beneficial.
19. If necessary, the stained gel can be stored in ultrapure water at 4 °C, wrapped tightly, up to several weeks until proceeding to in-gel digestion, though some sample loss may occur.
20. The total number of gel slice fractions per lane will vary slightly depending on the number of distinct bands and the resolving distance. Working solution volumes should be adjusted accordingly.
21. When using a 96-well plate, do not place more than 4 × 1 mm gel slices per well.
22. Do not mince gel slices as this increases the likelihood of transferring them during extraction.
23. For the study of SIRT4 interactions, after pooling of neighboring gel fractions, four fractions were subjected to LC-MS/MS analysis. For samples with more complexity or greater amount of co-isolated proteins, increasing the number of fractions may be beneficial. Alternatively, if an appropriate HPLC system is available, longer reverse phase separation gradients can be used.
24. Ensure that the disk makes contact with the walls of the pipette tip and is located a few mm above the tapered end of the tip. Each disk can bind up to ~20 μg of peptides. Though rarely necessary for in-gel digestion, if greater capacity is required, an additional disk can be layered in the same StageTip to increase binding capacity. The number of washes should be increased, equal to the total number of disks.
25. In contrast to C_{18} Empore disks, SDB-RPS Empore disks do not require activation with organic solvents.
26. Binding and washing of peptides over StageTips can be performed manually by applying pressure with a small plastic syringe or by centrifugation of the StageTip in a collection tube with an adapter. Independent of method, it is important that the peptide binding is done slowly (~25 μL / min). For the centrifugation method, usually a speed of ~1000 × *g* is appropriate for sample binding and 2000 × *g* for washing.
27. Several considerations are required when designing an LC-MS/MS method, many of which are instrument-specific. However, in general the MS acquisition cycle should be designed based on the performance characteristics of the LC system. It is critical that the MS cycle time, determined largely by the number of full and tandem MS scans, permits acquisition of multiple full scans over the average LC elution peak.

For example, given LC peak widths of 15–30 s, an optimal time for a single acquisition cycle would be in the range of 2–4 s for data-dependent methods.

28. Other variable modifications may be included in the primary database search, such as phosphorylation, acetylation, or deamidation. However, as addition of modifications increases both search time and space, it is recommended to include only those modifications that are a frequent occurrence, e.g., > 5 % relative to the total number of identified peptides.
29. When selecting an analysis workflow, ensure that it incorporates the ability to control for false positive sequence matches, e.g., by performing database searching against reversed protein sequences to estimate false discovery rates. If available, it is highly recommended to use a software platform that also controls false identification rates at the protein level.
30. An alternative to the online SAINT algorithm is to download the latest version of the SAINT source files (www.sourceforge.com) and compile it for your appropriate operating system. This strategy allows the SAINT algorithm to be run locally in the command-line, but requires additional computational knowledge. For a more detailed description of the underlying SAINT algorithm and its associated parameters *see* Ref. 28.
31. To compute meaningful SAINT specificity scores at least two biological replicates of the experimental and control isolations are required. Ideally, control isolations are "user" controls performed in parallel to the experimental samples; however, user controls can be replaced and/or supplemented with negative control data from the CRAPOME database [29] to provide additional stringency. These datasets are easily added when using the online SAINT workflow #3.
32. Several user-defined options are available when running SAINT. A thorough discussion of their recommended usage can be found in [28].
33. For examples of ROC curves and histogram distributions that illustrate the distribution of SAINT scoring *see* Refs. 22 and 23.

Acknowledgements

We are grateful for funding from NIH grants R01HL127640 and R21AI102187 (I.M.C.), an NHMRC of Australia Early Career CJ Martin Fellowship #APP1037043 (R.A.M.), and an NJCCR post-doctoral fellowship (T.M.G.).

References

1. Guarente L (2000) Sir2 links chromatin silencing, metabolism, and aging. Genes Dev 14(9): 1021–1026
2. Imai S, Armstrong CM, Kaeberlein M et al (2000) Transcriptional silencing and longevity protein Sir2 is an NAD-dependent histone deacetylase. Nature 403(6771):795–800
3. Donmez G (2012) The neurobiology of sirtuins and their role in neurodegeneration. Trends Pharmacol Sci 33(9):494–501
4. Min SW, Sohn PD, Cho SH et al (2013) Sirtuins in neurodegenerative diseases: an update on potential mechanisms. Front Aging Neurosci 5:53
5. Roth M, Chen WY (2014) Sorting out functions of sirtuins in cancer. Oncogene 33(13): 1609–1620
6. Sebastian C, Satterstrom FK, Haigis MC et al (2012) From sirtuin biology to human diseases: an update. J Biol Chem 287(51): 42444–42452
7. Winnik S, Auwerx J, Sinclair DA et al (2015) Protective effects of sirtuins in cardiovascular diseases: from bench to bedside. Eur Heart J. doi:10.1093/eurheartj/ehv290
8. Yuan H, Su L, Chen WY (2013) The emerging and diverse roles of sirtuins in cancer: a clinical perspective. Onco Targets Ther 6:1399–1416
9. Haigis MC, Mostoslavsky R, Haigis KM et al (2006) SIRT4 inhibits glutamate dehydrogenase and opposes the effects of calorie restriction in pancreatic beta cells. Cell 126(5):941–954
10. Michishita E, Park JY, Burneskis JM et al (2005) Evolutionarily conserved and nonconserved cellular localizations and functions of human SIRT proteins. Mol Biol Cell 16(10):4623–4635
11. Lombard DB, Alt FW, Cheng HL et al (2007) Mammalian Sir2 homolog SIRT3 regulates global mitochondrial lysine acetylation. Mol Cell Biol 27(24):8807–8814
12. Rardin MJ, Newman JC, Held JM et al (2013) Label-free quantitative proteomics of the lysine acetylome in mitochondria identifies substrates of SIRT3 in metabolic pathways. Proc Natl Acad Sci U S A 110(16):6601–6606
13. Brautigam CA, Wynn RM, Chuang JL et al (2006) Structural insight into interactions between dihydrolipoamide dehydrogenase (E3) and E3 binding protein of human pyruvate dehydrogenase complex. Structure 14(3):611–621
14. Laurent G, German NJ, Saha AK et al (2013) SIRT4 coordinates the balance between lipid synthesis and catabolism by repressing malonyl CoA decarboxylase. Mol Cell 50(5):686–698
15. Mathias RA, Greco TM, Oberstein A et al (2014) Sirtuin 4 is a lipoamidase regulating pyruvate dehydrogenase complex activity. Cell 159(7):1615–1625
16. Du J, Zhou Y, Su X et al (2011) Sirt5 is a NAD-dependent protein lysine demalonylase and desuccinylase. Science 334(6057): 806–809
17. Peng C, Lu Z, Xie Z, et al. (2011) The first identification of lysine malonylation substrates and its regulatory enzyme. Mol Cell Proteomics 10(12), DOI:10.1074/mcp.M111.012658
18. Tan M, Peng C, Anderson KA et al (2014) Lysine glutarylation is a protein posttranslational modification regulated by SIRT5. Cell Metab 19(4):605–617
19. Feldman JL, Baeza J, Denu JM (2013) Activation of the protein deacetylase SIRT6 by long-chain fatty acids and widespread deacylation by mammalian sirtuins. J Biol Chem 288(43):31350–31356
20. Miteva YV, Cristea IM (2014) A proteomic perspective of Sirtuin 6 (SIRT6) phosphorylation and interactions and their dependence on its catalytic activity. Mol Cell Proteomics 13(1):168–183
21. Tsai YC, Greco TM, Boonmee A et al (2012) Functional proteomics establishes the interaction of SIRT7 with chromatin remodeling complexes and expands its role in regulation of RNA polymerase I transcription. Mol Cell Proteomics 11(5):60–76
22. Joshi P, Greco TM, Guise AJ et al (2013) The functional interactome landscape of the human histone deacetylase family. Mol Syst Biol 9(1):672
23. Diner BA, Li T, Greco TM et al (2015) The functional interactome of PYHIN immune regulators reveals IFIX is a sensor of viral DNA. Mol Syst Biol 11(1):787
24. Choi H, Larsen B, Lin ZY et al (2011) SAINT: probabilistic scoring of affinity purification-mass spectrometry data. Nat Methods 8(1):70–73
25. Cristea IM, Williams R, Chait BT et al (2005) Fluorescent proteins as proteomic probes. Mol Cell Proteomics 4(12):1933–1941

26. Greco TM, Miteva Y, Conlon FL et al (2012) Complementary proteomic analysis of protein complexes. Methods Mol Biol 917: 391–407
27. Greco TM, Diner BA, Cristea IM (2014) The impact of mass spectrometry-based proteomics on fundamental discoveries in virology. Annu Rev Virol 1(1):581–604
28. Choi H, Liu G, Mellacheruvu D et al (2012) Analyzing protein-protein interactions from affinity purification-mass spectrometry data with SAINT. Curr Protoc Bioinformatics 39(8.15):1–23
29. Mellacheruvu D, Wright Z, Couzens AL et al (2013) The CRAPome: a contaminant repository for affinity purification-mass spectrometry data. Nat Methods 10(8):730–736

Chapter 16

Generation and Purification of Catalytically Active Recombinant Sirtuin5 (SIRT5) Protein

Surinder Kumar and David B. Lombard

Abstract

Sirtuin-family deacylases promote health and longevity in mammals. The sirtuin SIRT5 localizes predominantly to the mitochondrial matrix. SIRT5 preferentially removes negatively charged modifications from its target lysines: succinylation, malonylation, and glutarylation. It regulates protein substrates involved in glucose oxidation, ketone body formation, ammonia detoxification, fatty acid oxidation, and ROS management. Like other sirtuins, SIRT5 has recently been linked with neoplasia. Therefore, targeting SIRT5 pharmacologically could conceivably provide new avenues for treatment of metabolic disease and cancer, necessitating development of SIRT5-selective modulators. Here we describe the generation of SIRT5 bacterial expression plasmids, and their use to express and purify catalytically active and inactive forms of SIRT5 protein from *E. coli*. Additionally, we describe an approach to assay the catalytic activity of purified SIRT5, potentially useful for identification and validation of SIRT5-specific modulators.

Key words Sirtuins, Site directed mutagenesis, Desuccinylation, PDC, E1α, PDHA1

1 Introduction

Sirtuins are NAD$^+$-dependent enzymes that regulate diverse cellular processes, thereby maintaining metabolic homeostasis and genomic integrity [1]. Mammals possess seven sirtuin family members (SIRT1-SIRT7) [2, 3], which display diverse subcellular localization patterns, catalytic activities, protein targets, and biological functions [1, 2, 4]. Owing to their ability to catalyze removal of acetyl moieties from lysine residues, sirtuins were initially described as class III deacetylases. However, certain members of the mammalian sirtuin family carry out a range of enzymatic activities other than deacetylation. SIRT3, the predominant mitochondrial deacetylase [5], has recently been shown to possess decrotonylase activity [6]. Likewise, SIRT4 can catalyze both deacetylation and ADP-ribosylation reactions [7–9]. The least well-characterized sirtuin, SIRT5, displays very low deacetylase activity [10]; instead SIRT5 preferentially catalyzes removal of

Sibaji Sarkar (ed.), *Histone Deacetylases: Methods and Protocols*, Methods in Molecular Biology, vol. 1436,
DOI 10.1007/978-1-4939-3667-0_16, © Springer Science+Business Media New York 2016

negatively charged modifications: malonylation, succinylation, and glutarylation [10–14]. SIRT5-mediated modification of carbamoyl phosphate synthetase 1 (CPS1) [10, 13, 15], 3-hydroxy-3-methylglutaryl CoA synthase 2 (HMGCS2) [12], Cu/Zn superoxide dismutase (SOD1) [16], and urate oxidase [17] increases enzymatic activity, while desuccinylation of pyruvate dehydrogenase complex (PDC) and succinate dehydrogenase (SDH) by SIRT5 reduces their activities [11]. PDC catalyzes oxidative decarboxylation of pyruvate into acetyl-CoA, which is subsequently used in the Krebs cycle to generate ATP [18, 19]. In other contexts, altered activities of the SIRT5 targets, PDC and SDH, have been linked with neoplasia and cancer cell metabolic reprogramming [2, 20]. Additionally, SIRT5-mediated desuccinylation and activation of SOD1 suppresses ROS levels and promotes the growth of lung cancer cells [16]. Recently, it has been shown that SIRT5 is overexpressed in advanced non-small cell lung cancer (NSCLC) and promotes chemoresistance in NSCLC via enhancement of NRF2 activity. Consistently, SIRT5 knockdown represses the growth of NSCLC cell lines and increases their susceptibility to genotoxic drugs [21]. These findings suggest a potential oncogenic function of SIRT5 in NSCLC.

Despite the fact that SIRT5 is broadly expressed [15, 22], SIRT5-deficient mice are healthy, fertile, and without major clinical phenotype [5, 12, 23]. This suggests that SIRT5 is largely dispensable for metabolic homeostasis at least under basal conditions, rendering it an attractive potential drug target. The production of catalytically active SIRT5 protein is first step towards identification of SIRT5-specific modulators.

Alternative splicing of the human *SIRT5* mRNA results in two distinct SIRT5 protein isoforms, differing at their C-termini [24, 25]. Full-length isoform-1 comprises 310 amino acids, containing a 36-amino acid N-terminal mitochondrial localization signal (MLS) [26]. The MLS is proteolytically removed during mitochondrial import, which results in a 274 amino acid mature SIRT5 protein [25–27]. The second isoform comprises 299 amino acids. Whereas ectopically expressed isoform-1 is present throughout the cell, isoform-2 appears to be strictly mitochondrial [25]. In mouse liver, SIRT5 protein is predominantly mitochondrial, but also present in the cytosol and the nucleus [11].

In this chapter, we describe a method for generation of catalytically active and inactive forms of mature, SIRT5 isoform-1 protein, utilizing the basic techniques of molecular cloning and biochemistry. Replacing a catalytic Histidine (H) residue with Tyrosine (Y) at position 158 of full length SIRT5 attenuates catalytic activity [11, 13, 15]. Initially we describe site directed mutagenesis to replace C472 with T in the coding sequence of SIRT5. This single nucleotide change converts a Histidine (H) to a Tyrosine (Y) at position 158 of full-length, unprocessed SIRT5 protein.

Then we discuss the cloning of *SIRT5* and *SIRT5C472T* sequences, encoding the mature, processed forms (lacking the mitochondrial targeting signal) of SIRT5 and SIRT5H158Y proteins, respectively, into the bacterial expression plasmid pET15b. The pET15b plasmids harboring *SIRT5* and *SIRT5C472T* sequences are then used for overexpression and purification of catalytically active and inactive (H158Y) forms of SIRT5 protein. Finally, a desuccinylation assay is described to evaluate the catalytic activity of purified proteins. The desuccinylation activity assay can further be exploited for the validation of SIRT5-specific modulators (inhibitors or activators), identified in large scale screening of compound libraries.

2 Materials

2.1 Insertion of H158Y Mutation into SIRT5 CodingSequence and Preparation of SIRT5 and SIRT5-H158Y Bacterial Expression Plasmids with N-Terminal His-Tag

1. Human SIRT5-pBABE-puro plasmid DNA.
2. Custom-synthesized oligonucleotides.
3. 10 mM dNTP mix.
4. PfuUltra High-Fidelity DNA polymerase and 10× PfuUltra High-Fidelity reaction buffer.
5. Thermocycler.
6. Dpn I restriction enzyme.
7. Chemically competent DH5α bacterial cells.
8. Water bath.
9. LB medium (1 L): 10 g tryptone, 5 g yeast extract, 10 g NaCl.
10. LB-agar (1 L): 10 g tryptone, 5 g yeast extract, 10 g NaCl, 2 g agar.
11. 50 mg/ml carbenicillin or 100 mg/ml ampicillin: Dissolve the required amount in autoclaved double distilled water and filter through 0.2 μm syringe filter. Aliquot and store at −20 °C.
12. Miniprep Plasmid Purification Kit.
13. 50× TAE Stock (1 L): 242 g Tris base (MW = 121.1), 57.1 mL glacial acetic acid, and 100 mL 0.5 M EDTA. Dissolve Tris in about 600 mL of ddH_2O. Then add EDTA and acetic acid and bring the final volume to 1 L with ddH_2O. Store at room temperature. This 50× stock solution can be diluted with water to make a 1× working solution.
14. Agarose.
15. PCR cleanup/purification kit.
16. pET15b plasmid.
17. NdeI and XhoI restriction enzymes.
18. Gel extraction kit.
19. T4 DNA ligase and 10× DNA ligase buffer.
20. NanoDrop spectrophotometer.

2.2 Bacterial Expression, Solubility Determination, and Purification of His Tagged SIRT5 and SIRT5-H158Y Proteins

1. Chemical competent *E. coli* BL21 (DE3).
2. 1 M IPTG: Dissolve 2.38 g of IPTG in 10 ml of autoclaved double distilled water and filter through 0.2 μm syringe filter. Aliquot and store at −20 °C.
3. 4× Laemmli buffer: 40% glycerol, 240 mM Tris–HCl pH 6.8, 8% SDS, 0.04% bromophenol blue, 5% beta-mercaptoethanol.
4. 4–20% polyacrylamide Gel.
5. 1× running buffer: 0.1% (w/v) SDS, 192 mM glycine, 400 mM Tris base.
6. Coomassie brilliant blue stain.
7. Sonicator equipped with a microtip probe.
8. Lysozyme solution: Dissolve 10 mg/ml in 10 mM Tris–HCl, pH 8.0.
9. 50% Ni-NTA slurry.
10. Lysis buffer: 50 mM NaH_2PO_4, 300 mM NaCl, 10 mM imidazole. Adjust to pH 8.0 using NaOH.
11. Wash buffer: 50 mM NaH_2PO_4, 300 mM NaCl, 20 mM imidazole. Adjust to pH 8.0 using NaOH.
12. Elution buffer: 50 mM NaH_2PO_4, 300 mM NaCl, 250 mM imidazole, 5% glycerol. Adjust pH 8.0 using NaOH.
13. Protein concentration assay kit.

2.3 In Vitro Desuccinylation Assay

1. Porcine heart PDC.
2. PDC centrifugation buffer: 100 mM KH_2PO_4 (pH 7.5), 0.05% lauryl maltoside, 2.5 mM EDTA, 30% glycerol.
3. 3× reaction buffer: 75 mM Tris–HCl, pH 8.0, 600 mM NaCl, 15 mM KCl, 3 mM $MgCl_2$, 0.3% PEG 8000, 9.375 mM NAD^+.
4. 1× running buffer: 0.1% (w/v) SDS, 192 mM glycine, 400 mM Tris base.
5. 1× transfer buffer: 192 mM glycine, 400 mM Tris base, 20% methanol. Store transfer buffer at 4 °C.
6. PVDF membrane.
7. Bovine serum albumin (BSA).
8. Anti-lysine-succinyl antibody.
9. Anti-pyruvate dehydrogenase E1-alpha subunit antibody.
10. 1× TBST (Tris buffered saline containing 0.01% Tween 20): 50 mM Tris–HCl, pH 7.6, 150 mM NaCl, 0.01% (v/v) Tween 20.
11. Ponceau S staining solution: 0.1% (w/v) Ponceau S, 5% (v/v) acetic acid.
12. Chemiluminescent HRP substrate.

13. Stripping buffer (1 L): 15 g glycine, 1 g SDS, 10 ml Tween 20. Adjust pH to 2.2 using HCl.
14. 1× PBS: 137 mM NaCl, 2.7 mM KCl, 10 mM Na_2HPO_4, 2 mM KH_2PO_4. Adjust pH to 7.4 using HCl.

3 Methods

3.1 Engineering H158Y Mutation into the SIRT5 Coding Sequence

1. Obtain plasmid encoding human SIRT5 (*see* **Notes 1** and **2**).
2. Custom synthesize the following two complementary oligonucleotides containing the desired mutation (underlined), flanked by unmodified nucleotide sequence.
   (a) SIRT5 H158Y sense oligo: 5′GAACCTTCTGGAGATC TATGGTAGCTTATTTAAAAC 3′
   (b) SIRT5 H158Y anti-sense oligo: 5′ GTTTTAAATAAGCT ACCATAGATCTCCAGAAGGTTC 3′
3. Resuspend oligos in appropriate amount of ddH_2O to a final concentration of 100 μM. Dilute further by adding ddH_2O to a concentration of 10 μM.
4. Set up a series of reactions using different amounts (5, 10, 20, and 50 ng) of template DNA (human SIRT5-pBABE-puro plasmid) as below:

   5 μl of 10× PfuUltra HF reaction buffer.

   1 μl of template DNA (5–50 ng).

   1.14 μl of 10 μM SIRT5 H158Y Top Oligo (125 ng).

   1.14 μl of 10 μM SIRT5 H158Y Bottom Oligo (125 ng).

   1 μl of 10 mM dNTP mix.

   1 μl of PfuUltra HF DNA Polymerase (2.5 U/μl).

   39.72 μl of ddH_2O.
5. Perform amplification reaction in a thermocycler, set to the following parameters:
   (a) 95 °C, 30 s.
   (b) 95 °C, 30 s.
   (c) 55 °C, 1 min.
   (d) 68 °C, 7.5 min.
   (e) Repeat **steps** (**b**–**d**) for a total of 16 cycles
   (f) 68 °C, 10 min.
   (g) 4 °C, Hold.
6. Add 1 μl of Dpn I restriction enzyme (10 U/μl) to each reaction tube. Mix the contents gently and thoroughly by pipetting up and down multiple times. Spin down briefly, and incubate at 37 °C for 1 h to digest parental template DNA.

7. Transform Dpn I treated DNA into chemical competent DH5α bacteria via heat shock as below:
   (a) Thaw chemical competent DH5α bacterial cells on ice, and aliquot 50 μl in separate fresh autoclaved tubes.
   (b) Transfer 1 μl of Dpn I treated DNA from each sample tube to separate aliquots of chemical competent DH5α bacteria. Mix the content of tubes by gentle tapping and incubate on ice for 5 min.
   (c) Heat shock the transformation reactions for 90 s in a water bath set to 42 °C, and immediately transfer the tubes to ice again for 5 min.
   (d) Add 1 ml of LB medium to all tubes aseptically, and incubate at 37 °C for 1 h with shaking at 220 rpm. Plate 50–100 μl of the transformation mix on LB agar plates containing 50 μg/ml carbenicillin or 100 μg/ml ampicillin, and incubate at 37 °C for ~16 h.
8. Inoculate few colonies (~4–6) from each plate into separate tubes containing 5 ml LB medium with 50 μg/ml carbenicillin or 100 μg/ml ampicillin. Incubate overnight at 37 °C with shaking at 220 rpm.
9. Harvest the cells by centrifugation at 6500 × *g* for 5 min at room temperature, and purify the plasmid DNA using a standard miniprep plasmid purification protocol or kit.
10. Confirm the insertion of desired mutation by nucleotide sequencing using pBABE sequencing primers (Weinberg Lab): 5′ CTTTATCCAGCCCTCAC 3′ and 5′ ACCCTAACTG ACACACATTCC 3′

3.2 Preparation of SIRT5 and SIRT5-H158Y Bacterial Expression Plasmids with N-Terminal His-Tag

1. Synthesize the following primers to amplify the SIRT5 coding sequence lacking the mitochondrial targeting signal, for in vitro studies. These oligos are designed to introduce appropriate restriction enzyme sites at each end of the amplified PCR product for subsequent cloning into the bacterial expression plasmid pET15b, which provides an in-frame His-tag followed by a thrombin site at the SIRT5 N-terminus.
   (a) SIRT5 forward primer with NdeI restriction site (underlined):
   5′ CCGG<u>CATATGA</u>GTTCAAGTATGGCAGATTTTCG 3′
   (b) SIRT5 reverse primer with XhoI restriction site (underlined):
   5′ CCGG<u>CTCGAGT</u>TAAGAAACAGTTTCATTTTC 3′
2. Make 10 μM working solutions of both primers as described in Subheading 3.1, **step 3**.

3. Using these primers and human SIRT5-pBABE-puro and SIRT5-H158Y-pBABE-puro plasmids as templates, perform PCR as described below:
   (a) Set up following reaction (per tube):
       2.5 μl of 10× PfuUltra HF reaction buffer.
       1 μl of template DNA (50 ng).
       1 μl of 10 μM SIRT5 forward primer with NdeI restriction site.
       1 μl of 10 μM SIRT5 reverse primer with XhoI restriction site.
       1 μl of 10 mM dNTP mix.
       0.5 μl of PfuUltra HF DNA Polymerase (2.5U/μl).
       18 μl of ddH_2O.
   (b) Perform amplification reaction in a thermocycler set to the following parameters:
       (i) 95 °C, 3 min.
       (ii) 95 °C, 1 min.
       (iii) 55 °C, 1 min.
       (iv) 68 °C, 1.5 min.
       (v) Repeat **steps** (**ii**) to (**iv**) for 34 times.
       (vi) 68 °C, 10 min.
       (vii) 4 °C, Hold.
4. Verify the PCR fragment size on an 1 % agarose gel, and purify the amplified products from the PCR reactions using PCR cleanup/purification kit following manufacturer's instructions.
5. Perform restriction digestions of purified PCR products and pET15b with NdeI and XhoI restriction enzymes following manufacturer's instructions (*see* **Note 3**).
6. Perform Agarose Gel Electrophoresis using restriction enzymes digested PCR products and pET15b. Excise the DNA bands from the gel, and extract DNA from gel pieces using gel extraction kit following manufacturer instructions. Measure the concentrations of extracted DNA using NanoDrop spectrophotometer.
7. Set the ligation reactions using restriction enzymes digested and gel extracted pET15b and SIRT5/SIRT5H158Y PCR products in plasmid to insert ratio of 1:3 as follows (*see* **Note 4**):
   2 μl of 10× DNA ligase buffer.
   50 ng of pET15b.
   22 ng of SIRT5/SIRT5HY.
   1 μl of T4 DNA Ligase.
   ddH_2O to a final volume of 20 μl.

8. Incubate the ligation reactions at 16 °C for 16 h—overnight.
9. Transform 10 μl of ligation products into chemical competent DH5α bacterial cells by the heat shock method as described in Subheading 3.1, **step** 7.
10. Inoculate a few colonies (~6–8) from each plate into separate tubes containing 5 ml LB medium with 50 μg/ml carbenicillin or 100 μg/ml ampicillin. Incubate overnight at 37 °C under shaking at 220 rpm.
11. Harvest the cells by centrifugation at 6500 × *g* for 5 min at room temperature, and purify the plasmid DNA using a mini-prep plasmid purification kit following manufacturer's instructions.
12. Confirm the cloning of SIRT5 and SIRT5H158Y inserts into pET15b by double restriction digestion and/or PCR amplification.
13. Sequence the entire PCR fragment to confirm that unintended mutations are not present using pET15b sequencing primer:

    5′ TAATACGACTCACTATAGGG 3′

3.3 Bacterial Expression of His Tagged SIRT5 and SIRT5-H158Y Proteins

1. For expression of recombinant His tagged SIRT5 and SIRT5-H158Y proteins in *E. coli*, transform SIRT5-pET15b and SIRT5-H158Y-pET15b plasmids into chemical competent *E. coli* BL21 (DE3) by heat shock method as described in Subheading 3.1, **step** 7.
2. Pick single colonies from each transformation, and inoculate in 2.5 ml of LB medium supplemented with 50 μg/ml carbenicillin or 100 μg/ml ampicillin. Also inoculate one 2.5 ml culture with a colony transformed with the control plasmid pET15b. Inoculate one extra set of cultures to serve as an uninduced control. Grow the cultures at 37 °C until OD_{600} reaches ~0.45–0.5.
3. Induce the expression of recombinant SIRT5 by adding IPTG to a final concentration of 1 mM, and grow the cultures for an additional 6–8 h. Do not add IPTG to uninduced controls.
4. Transfer 1 ml from all uninduced and induced cultures to separate microcentrifuge tubes. Harvest the cells by centrifugation for 1 min at 15,000 × *g*, and discard supernatants.
5. Resuspend pellets in 25 μl of ddH_2O, and mix with 25 μl of 4× Laemmli sample buffer. Heat the samples at 95–100 °C for 10 min, and centrifuge at 15,000 × *g* for 10 min.
6. Monitor the expression of recombinant proteins by running 10 μl supernatant from each tube from **step 5** (i.e., cell lysates from un-induced and IPTG induced cultures) on 4–20 % poly-

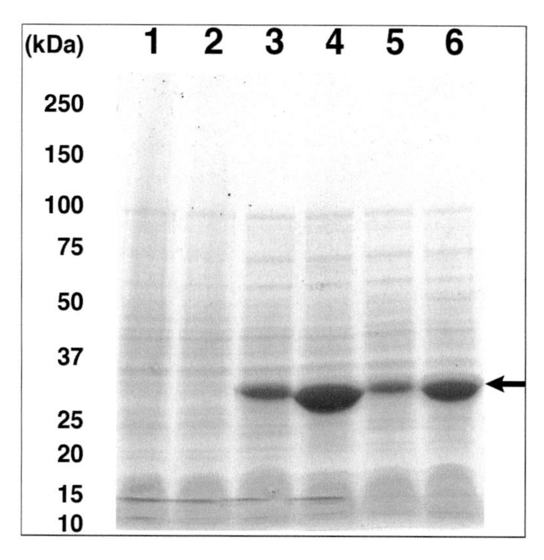

Fig. 1 SDS-PAGE analysis of recombinant SIRT5 and SIRT5H158Y expression. Expression of SIRT5 (*lane 4*) and SIRT5H158Y (*lane 6*) in *E. coli* following induction with 1 mM IPTG at 37 °C. *Lane 3* and *lane 5* show the expression of SIRT5 and SIRT5H158Y respectively, in the absence of IPTG. *Lane 1* and *lane 2* show the vector control bacterial lysate in the absence and presence of 1 mM IPTG, respectively

acrylamide gel in 1× running buffer at 200 V for 1 h, followed by Coomassie brilliant blue staining. The SDS-PAGE analysis of BL21 (DE3) cells, transformed with recombinant vectors having SIRT5 and SIRT5-H158Y coding sequences, shows overexpression of 32 kDa proteins (Fig. 1, lanes 3 and 4 for SIRT5, lanes 5 and 6 for SIRT5-H158Y) corresponding to expected size of SIRT5 and SIRT5-H158Y with a 6× His-tag and a thrombin site (*see* **Notes 5** and **6**).

3.4 Assessment of Solubility of Tagged SIRT5 and SIRT5-H158Y Proteins

1. Inoculate single bacterial colony harboring SIRT5-pET15b or SIRT5-H158Y-pET15b plasmid into 5 ml LB medium containing 50 μg/ml carbenicillin or 100 μg/ml ampicillin, and incubate overnight at 37 °C while shaking at 220 rpm.
2. The next day, inoculate 50 ml of LB medium containing 50 μg/ml carbenicillin or 100 μg/ml ampicillin with 2.5 ml of the overnight cultures, and incubate at 37 °C, with shaking, until the OD_{600} reaches ~0.5. This will take ~45 min.
3. Transfer 1 ml of each culture to separate microcentrifuge tubes. Pellet the cells by centrifugation at 15,000 ×g for 1 min. Resuspend the pellets in 25 μl of ddH_2O, mix with 25 μl of 4× Laemmli sample buffer, and save at −20 °C. These will serve as uninduced controls.

4. Induce the remaining cultures for the expression of recombinant proteins by adding IPTG to a final concentration of 1 mM, and grow the cultures for an additional 6–8 h at 37 °C with shaking.
5. Harvest bacterial cells by centrifugation at 10,000 × *g* for 10 min at 4 °C, and resuspend cell pellet in 5 ml of lysis buffer for native purification.
6. Add lysozyme to a final concentration of 1 mg/ml, and incubate on ice for 30 min (*see* **Note 7**).
7. Sonicate the cell suspension at 250 W with 6–8 bursts of 10 s followed by 20 s intervals for cooling. Keep samples on ice at all times.
8. Transfer 10 μl of each lysate to separate microcentrifuge tubes, mix with 10 μl of 4× Laemmli sample buffer, and freeze at −20 °C. These are induced controls.
9. Centrifuge remaining lysates at 10,000 × *g* for 30 min at 4 °C. Transfer the supernatants containing soluble proteins to fresh tubes, and keep on ice. Resuspend the pellets into 5 ml of lysis buffer to make a suspension of insoluble proteins.
10. Mix 10 μl of soluble protein supernatant with 10 μl of 4× Laemmli sample buffer. Similarly mix 10 μl suspensions of insoluble proteins with 10 μl of 4× Laemmli sample buffer.
11. Heat these samples along with uninduced, and induced controls, from **steps 3** and **8** respectively, at 95–100 °C for 10 min.
12. Centrifuge at 15,000 × *g* for 10 min, and run 20 μl of the supernatants from the un-induced controls and 10 μl from all other samples including vector control on 4–20% polyacrylamide gel followed by staining with Coomassie brilliant blue. SDS-PAGE analysis shows that both His-tagged SIRT5 and SIRT5H158Y recombinant proteins corresponding to ~32 kDa are soluble in lysis buffer for native purification (Fig. 2) and can be purified under native conditions (*see* **Note 8**).

3.5 Purification of His Tagged SIRT5 and SIRT5-H158Y Proteins

1. Grow 5 ml each of starter cultures from single bacterial colonies as described above in the Subheading 3.4, **step 1**.
2. Inoculate 1 L each of LB medium containing 50 μg/ml carbenicillin or 100 μg/ml ampicillin with 5 ml of the starter cultures, and grow at 37 °C with shaking at 220 rpm, until the OD_{600} reaches ~0.5.
3. Induce the expression of recombinant proteins by adding IPTG to a final concentration of 1 mM, and grow the cultures for an additional 8 h.
4. Harvest bacterial cells expressing recombinant proteins by centrifugation at 10,000 × *g* for 10 min at 4 °C. Weigh the bacterial cell pellets (*see* **Note 9**).

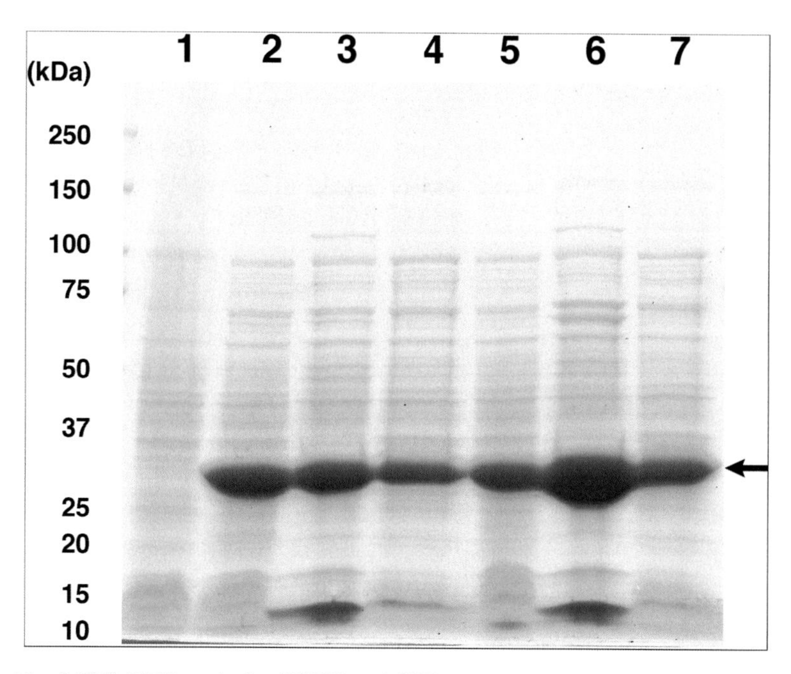

Fig. 2 SDS-PAGE analysis of SIRT5 and SIRT5H158Y solubility under native conditions. IPTG induced *E. coli* cells expressing SIRT5 or SIRT5H158Y were lysed under native conditions and centrifuged. Whole cell lysates (SIRT5; *lane 2* and SIRT5H158Y; *lane 5*), supernatants containing soluble proteins (SIRT5; *lanes 3* and SIRT5H158Y; *lane 6*) and pellets containing insoluble material (SIRT5; *lanes 4* and SIRT5H158Y; *lane 7*) were separated by SDS-PAGE and stained with Coomassie brilliant blue. *Lane 1* shows induced *E. coli* cells harboring empty control vector

5. Resuspend the cell pellets in lysis buffer at 5 ml per gram wet weight, and add lysozyme to final concentration of 1 mg/ml. Incubate on ice for 30 min (*see* **Note 10**).
6. Sonicate the cell suspension with a sonicator set to 250 W with 6–8 bursts of 10 s followed by 20 s intervals for cooling. Keep samples all time on ice (*see* **Note 11**).
7. Centrifuge lysates at 10,000 × *g* for 30 min at 4 °C to pellet the insoluble fractions and other cellular debris. Supernatants (i.e., cleared lysates), containing the soluble proteins, will be used for purification of His tagged SIRT5 and SIRT5-H158Y proteins in subsequent steps (*see* **Note 8**).
8. Transfer 1 ml 50% Ni-NTA slurry per 4 ml of cleared lysate to 50 ml Falcon tubes, and centrifuge at 700 × *g* for 2 min at 4 °C.
9. Remove and discard the supernatants, and resuspend Ni-NTA pellets in four resin-bed volumes of lysis buffer. Centrifuge again at 700 × *g* for 2 min at 4 °C, and carefully remove and discard buffer.

10. Transfer cleared lysates from **step 7** to Ni-NTA pellets from **step 9**, and mix gently by shaking on a rotary shaker (~200 rpm) at 4 °C for 45–60 min.
11. Load the lysate–Ni–NTA mixtures into separate columns, and collect the flow-through from each column. If desired, save for SDS-PAGE analysis.
12. Wash resin twice with four bed volumes of wash buffer, and collect the flow-through for SDS-PAGE analysis, if desired.
13. Elute His-tagged proteins from the resin with 1 ml elution buffer. Repeat this **step 8** ten times, collecting each fraction in a separate tube (*see* **Note 12**).
14. Monitor protein elution by measuring the absorbance of the fractions at 280 nm, and fractionate on a 4–20% polyacrylamide gel to directly analyze the purified proteins (Fig. 3).
15. Measure the concentrations of eluted protein in desired fractions by suitable Protein Assay Kit (*see* **Note 13**).

3.6 In Vitro Desuccinylation Assay to Assess the Activity of Purified SIRT5 and SIRT5-H158Y Proteins

1. Mix 0.5 ml (~7.5 mg) of PDC with 1 ml of PDC centrifugation buffer, and centrifuge at 135,000 × *g* for 2 h at 4 °C (*see* **Note 14**).
2. Remove supernatant leaving 200 μl. Resuspend pellet in residual 200 μl of supernatant by pipetting up and down several times.
3. Measure the concentration of resuspended PDC.
4. Set up three reaction tubes, each with 30 μg of purified porcine heart PDC and 20 μl of 3× desuccinylation reaction buffer. In one tube add 10 μg of purified SIRT5 protein, and in another add 10 μg of purified SIRT5H158Y protein. Do not add any SIRT5 protein to the third (negative control) tube. Bring up the total final reaction volume to 60 μl in all three tubes by adding ddH_2O. Incubate in water bath set at 37 °C for 2 h with occasional agitation via finger tap (*see* **Note 15**).
5. Analyze samples from each reaction tube by immunoblot using rabbit anti lysine-succinyl antibody as discussed below:
   (a) Add 20 μl of 4× Laemmli sample buffer to each tube, and heat at 95–100 °C for 5 min.
   (b) Load 20 μl of each sample and 8 μl of protein standard ladder on a 4–20% polyacrylamide gel, and run the gel at 200 V in 1× running buffer until the dye front reaches to the bottom of the gel.
   (c) Transfer the fractionated proteins from polyacrylamide gel to a PVDF membrane in 1× transfer buffer at 800 mA for 50 min at 4 °C.
   (d) Visualize the transfer of fractionated proteins on PVDF membrane using Ponceau S staining solution. Make a digital scanned image of the stained membrane.

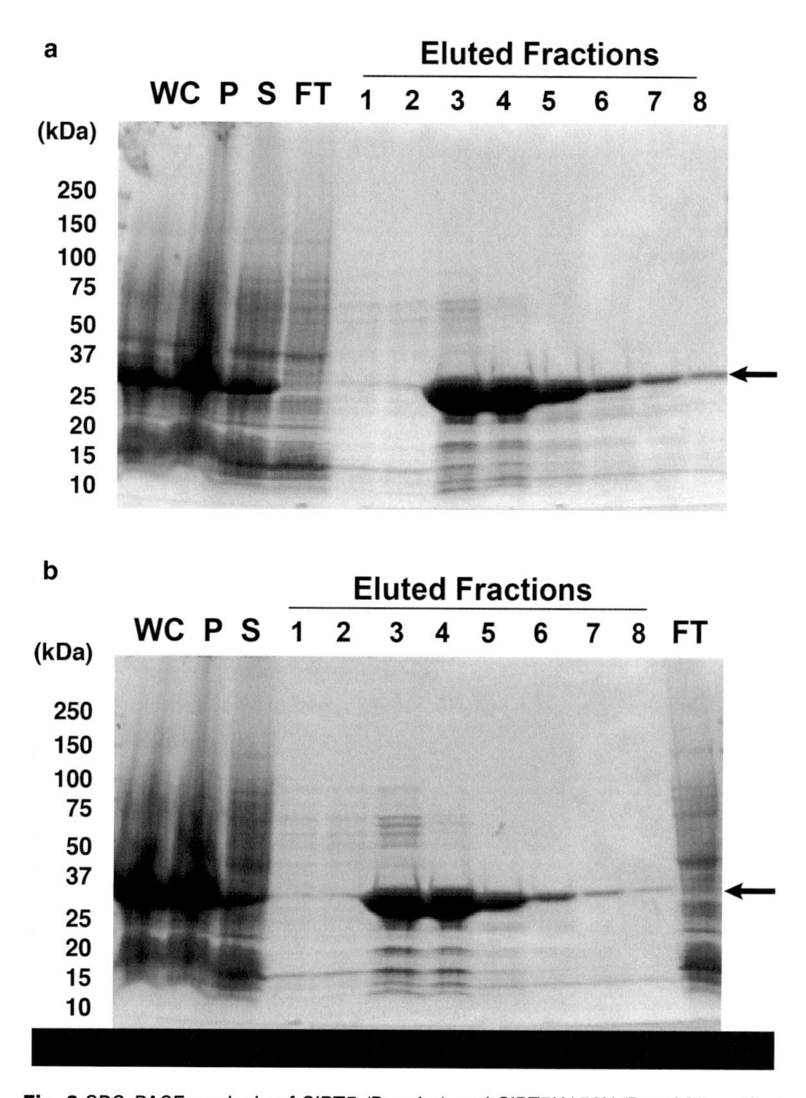

Fig. 3 SDS-PAGE analysis of SIRT5 (Panel **a**) and SIRT5H158Y (Panel **b**) purified under native conditions. IPTG-induced *E. coli* cells expressing SIRT5 and SIRT5H158Y were lysed under native conditions and centrifuged. Whole cell lysates (WC), supernatants containing soluble proteins (S), pellets containing insoluble fractions (P) and eluted protein fractions (1–8, 10 μl each) were separated electrophoretically; FT shows the flow through (supernatant following incubation with Ni–NTA) in both panels

(e) Block the nonspecific binding sites by incubating the membrane in 5 % skim milk in 1× TBST for 30 min at room temperature on a rocker.

(f) Discard blocking solution, and wash briefly with 1× TBST.

(g) Incubate overnight at 4 °C with rabbit anti lysine-succinyl antibody, diluted at 1:1000 in 5 % BSA in 1× TBST, while shaking on a rocker.

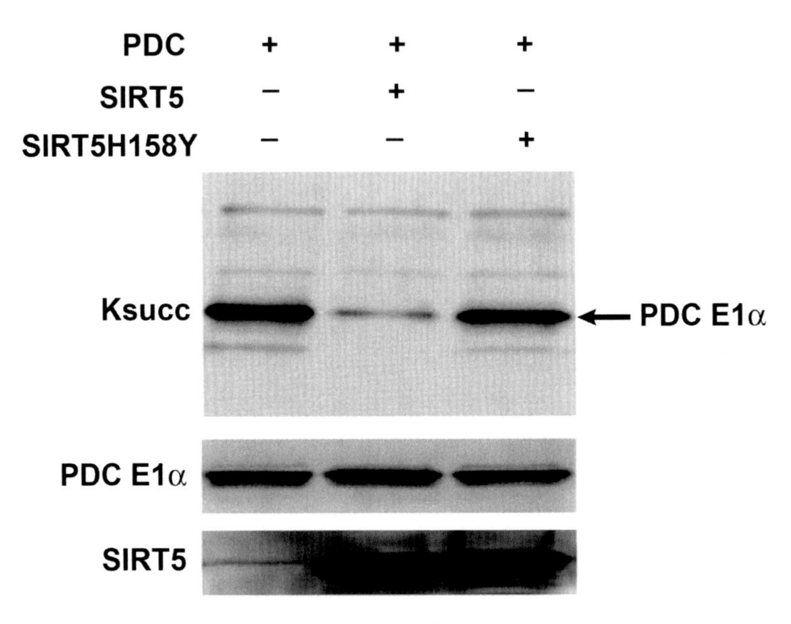

Fig. 4 Desuccinylase activity of purified SIRT5 and SIRT5H158Y. Pelleted porcine heart PDC was treated with or without purified SIRT5 and SIRT5H158Y proteins and processed for immunoblotting with anti-Ksucc antibody as described in the Subheading 3. Total PDC E1α serves as loading control

(h) Collect antibody solution, and wash membrane three times for 10 min each with 1× TBST.

(i) Incubate the membrane with HRP-conjugated secondary antibody, diluted at 1:10,000 in 5 % skim milk in 1× TBST, for 1 h at room temperature while shaking on a rocker.

(j) Wash membrane three times for 10 min each with 1× TBST.

(k) Develop signal using appropriate chemiluminescent HRP Substrate. Figure 4 shows that SIRT5-treated PDC has decreased levels of lysine succinylation on the E1α subunit compared with untreated or SIRT5H158Y treated controls.

(l) Wash membrane with distilled water for 10 min at room temperature.

(m) Strip membrane by incubating with stripping buffer while shaking at room temperature for 10 min. Discard buffer, and repeat one more time.

(n) Wash membrane twice with 1× PBS for 10 min each, followed by two washes with 1× TBST.

(o) Block and re-probe with mouse anti-pyruvate dehydrogenase E1-alpha subunit antibody, diluted at 1:1000 in 5% BSA in 1× TBST, as described above in the Subheading 3.6, **step 5** (**e-g**).

(p) Repeat steps from **h–k** (*see* **Note 16**).

4 Notes

1. Human SIRT5-pBABE-puro plasmid was a kind gift from Dr. Marcia Haigis, Department of Cell Biology, Harvard Medical School.
2. Always use plasmid DNA isolated from dam^+ strains of the *E. coli*. Plasmid DNA isolated from dam^- strains of *E. coli* (e.g., JM110 is not suitable) as DpnI, which is used to selectively digest parental plasmid DNA, cleaves only when its recognition site is methylated.
3. The pET15b plasmid should be sequentially digested.
4. The plasmid:insert ratio can be increased up to 1:5 if ligation efficiency is poor.
5. Both SIRT5 and SIRT5H158Y display leaky expression in *E. coli* BL21 (DE3), which is confirmed by immunoblotting with anti-His antibody. However, we do not observe any toxicity of SIRT5 expression to *E. coli* BL21 (DE3) cells.
6. Expression of recombinant proteins can further be confirmed by immunoblotting using anti-His antibody or anti-SIRT5 antibody.
7. Always use a freshly prepared lysozyme solution.
8. A significant proportion of SIRT5 and SIRT5H158Y proteins remain insoluble and can be found in the pellet. If desired, these can be purified from insoluble pellet under denaturing purification conditions.
9. First weigh the empty tubes. Following centrifugation, weigh again tubes containing bacterial pellets. To determine the weight of bacterial pellets, subtract the weights of empty tubes from that of bacterial pellets containing tubes.
10. To avoid protein degradation during purification, inclusion of protease inhibitors in the lysis buffer is recommended.
11. If the lysate is very viscous, add RNase A (10 μg/ml) and DNase I (5 μg/ml), and incubate on ice for 10–15 min.
12. Ni-NTA resin can be washed with phosphate buffer, and collected for reuse in subsequent purifications.
13. Eluted fractions should be stored at −70 °C.
14. Always use tubes certified for ultracentrifugation.
15. 3× reaction buffer can be stored at −20 °C for 3–4 weeks.
16. SIRT5-treated material can be used for MS analysis. Also we have found that recombinant SIRT5 is active against target specific peptides (not shown); hence, it is suitable for the screening of specific modulators (inhibitors or activators) by high-throughput methodology.

Acknowledgements

We thank members of Lombard laboratory for helpful discussions. Work in our laboratory is supported by the National Institute of Health (R01GM101171, R21CA177925), Department of Defense Grant (OC140123), the Glenn Foundation for Medical Research, and the Discovery Fund of the University of Michigan Comprehensive Cancer Center. Research reported in this publication was supported in part by the National Center for Advancing Translational Sciences of the National Institutes of Health under Award Number UL1TR000433. The content is solely the responsibility of the authors and does not necessarily represent the official views of the National Institutes of Health. Natalie German and Dr. Marcia Haigis are gratefully acknowledged for providing the SIRT5 retroviral plasmid.

References

1. Giblin W, Skinner ME, Lombard DB (2014) Sirtuins: guardians of mammalian healthspan. Trends Genet. doi:10.1016/j.tig.2014.04.007
2. Kumar S, Lombard DB (2015) Mitochondrial sirtuins and their relationships with metabolic disease and cancer. Antioxid Redox Signal. doi:10.1089/ars.2014.6213
3. Feldman JL, Dittenhafer-Reed KE, Denu JM (2012) Sirtuin catalysis and regulation. J Biol Chem 287:42419–42427. doi:10.1074/jbc.R112.378877, R112.378877 [pii]
4. Frye RA (2000) Phylogenetic classification of prokaryotic and eukaryotic Sir2-like proteins. Biochem Biophys Res Commun 273:793–798. doi:10.1006/bbrc.2000.3000, Pii: S0006-291X(00)93000-6
5. Lombard DB, Alt FW, Cheng HL, Bunkenborg J, Streeper RS, Mostoslavsky R, Kim J, Yancopoulos G, Valenzuela D, Murphy A, Yang Y, Chen Y, Hirschey MD, Bronson RT, Haigis M, Guarente LP, Farese RV Jr, Weissman S, Verdin E, Schwer B (2007) Mammalian Sir2 homolog SIRT3 regulates global mitochondrial lysine acetylation. Mol Cell Biol 27:8807–8814. doi:10.1128/MCB.01636-07
6. Bao X, Wang Y, Li X, Li XM, Liu Z, Yang T, Wong CF, Zhang J, Hao Q, Li XD (2014) Identification of 'erasers' for lysine crotonylated histone marks using a chemical proteomics approach. eLife 3. doi: 10.7554/eLife.02999
7. Ahuja N, Schwer B, Carobbio S, Waltregny D, North BJ, Castronovo V, Maechler P, Verdin E (2007) Regulation of insulin secretion by SIRT4, a mitochondrial ADP-ribosyltransferase. J Biol Chem 282:33583–33592. doi:10.1074/jbc.M705488200
8. Haigis MC, Mostoslavsky R, Haigis KM, Fahie K, Christodoulou DC, Murphy AJ, Valenzuela DM, Yancopoulos GD, Karow M, Blander G, Wolberger C, Prolla TA, Weindruch R, Alt FW, Guarente L (2006) SIRT4 inhibits glutamate dehydrogenase and opposes the effects of calorie restriction in pancreatic beta cells. Cell 126:941–954. doi:10.1016/j.cell.2006.06.057
9. Laurent G, German NJ, Saha AK, de Boer VC, Davies M, Koves TR, Dephoure N, Fischer F, Boanca G, Vaitheesvaran B, Lovitch SB, Sharpe AH, Kurland IJ, Steegborn C, Gygi SP, Muoio DM, Ruderman NB, Haigis MC (2013) SIRT4 coordinates the balance between lipid synthesis and catabolism by repressing malonyl CoA decarboxylase. Mol Cell 50:686–698. doi:10.1016/j.molcel.2013.05.012
10. Du J, Zhou Y, Su X, Yu JJ, Khan S, Jiang H, Kim J, Woo J, Kim JH, Choi BH, He B, Chen W, Zhang S, Cerione RA, Auwerx J, Hao Q, Lin H (2011) Sirt5 is a NAD-dependent protein lysine demalonylase and desuccinylase. Science 334:806–809. doi:10.1126/science.1207861
11. Park J, Chen Y, Tishkoff DX, Peng C, Tan M, Dai L, Xie Z, Zhang Y, Zwaans BM, Skinner ME, Lombard DB, Zhao Y (2013) SIRT5-mediated lysine desuccinylation impacts diverse metabolic pathways. Mol Cell 50:919–930. doi:10.1016/j.molcel.2013.06.001
12. Rardin MJ, He W, Nishida Y, Newman JC, Carrico C, Danielson SR, Guo A, Gut P, Sahu AK, Li B, Uppala R, Fitch M, Riiff T, Zhu L, Zhou J, Mulhern D, Stevens RD, Ilkayeva OR, Newgard CB, Jacobson MP, Hellerstein M, Goetzman ES, Gibson BW, Verdin E (2013)

SIRT5 regulates the mitochondrial lysine succinylome and metabolic networks. Cell Metab 18:920–933. doi:10.1016/j.cmet.2013.11.013
13. Tan M, Peng C, Anderson KA, Chhoy P, Xie Z, Dai L, Park J, Chen Y, Huang H, Zhang Y, Ro J, Wagner GR, Green MF, Madsen AS, Schmiesing J, Peterson BS, Xu G, Ilkayeva OR, Muehlbauer MJ, Braulke T, Muhlhausen C, Backos DS, Olsen CA, McGuire PJ, Pletcher SD, Lombard DB, Hirschey MD, Zhao Y (2014) Lysine glutarylation is a protein post-translational modification regulated by SIRT5. Cell Metab 19:605–617. doi:10.1016/j.cmet.2014.03.014
14. Peng C, Lu Z, Xie Z, Cheng Z, Chen Y, Tan M, Luo H, Zhang Y, He W, Yang K, Zwaans BM, Tishkoff D, Ho L, Lombard D, He TC, Dai J, Verdin E, Ye Y, Zhao Y (2011) The first identification of lysine malonylation substrates and its regulatory enzyme. Mol Cell Proteomics 10(M111):012658. doi:10.1074/mcp.M111.012658
15. Nakagawa T, Lomb DJ, Haigis MC, Guarente L (2009) SIRT5 deacetylates carbamoyl phosphate synthetase 1 and regulates the urea cycle. Cell 137:560–570. doi:10.1016/j.cell.2009.02.026
16. Lin ZF, Xu HB, Wang JY, Lin Q, Ruan Z, Liu FB, Jin W, Huang HH, Chen X (2013) SIRT5 desuccinylates and activates SOD1 to eliminate ROS. Biochem Biophys Res Commun 441: 191–195. doi:10.1016/j.bbrc.2013.10.033
17. Nakamura Y, Ogura M, Ogura K, Tanaka D, Inagaki N (2012) SIRT5 deacetylates and activates urate oxidase in liver mitochondria of mice. FEBS Lett 586:4076–4081. doi:10.1016/j.febslet.2012.10.009
18. Patel MS, Nemeria NS, Furey W, Jordan F (2014) The pyruvate dehydrogenase complexes: structure-based function and regulation. J Biol Chem 289:16615–16623. doi:10.1074/jbc.R114.563148
19. Ozden O, Park SH, Wagner BA, Song HY, Zhu Y, Vassilopoulos A, Jung B, Buettner GR, Gius D (2014) Sirt3 deacetylates and increases pyruvate dehydrogenase activity in cancer cells. Free Radic Biol Med. doi:10.1016/j.freeradbiomed.2014.08.001
20. Zwaans BM, Lombard DB (2014) Interplay between sirtuins, MYC and hypoxia-inducible factor in cancer-associated metabolic reprogramming. Dis Model Mech 7:1023–1032. doi:10.1242/dmm.016287
21. Lu W, Zuo Y, Feng Y, Zhang M (2014) SIRT5 facilitates cancer cell growth and drug resistance in non-small cell lung cancer. Tumour Biol. doi:10.1007/s13277-014-2372-4
22. Michishita E, Park JY, Burneskis JM, Barrett JC, Horikawa I (2005) Evolutionarily conserved and nonconserved cellular localizations and functions of human SIRT proteins. Mol Biol Cell 16:4623–4635. doi:10.1091/mbc.E05-01-0033, Pii: E05-01-0033
23. Yu J, Sadhukhan S, Noriega LG, Moullan N, He B, Weiss RS, Lin H, Schoonjans K, Auwerx J (2013) Metabolic characterization of a Sirt5 deficient mouse model. Sci Rep 3:2806. doi:10.1038/srep02806
24. Mahlknecht U, Ho AD, Letzel S, Voelter-Mahlknecht S (2006) Assignment of the NAD-dependent deacetylase sirtuin 5 gene (SIRT5) to human chromosome band 6p23 by in situ hybridization. Cytogenet Genome Res 112:208–212. doi:10.1159/000089872
25. Matsushita N, Yonashiro R, Ogata Y, Sugiura A, Nagashima S, Fukuda T, Inatome R, Yanagi S (2011) Distinct regulation of mitochondrial localization and stability of two human Sirt5 isoforms. Genes Cells 16:190–202. doi:10.1111/j.1365-2443.2010.01475.x
26. Gertz M, Steegborn C (2010) Function and regulation of the mitochondrial sirtuin isoform Sirt5 in mammalia. Biochim Biophys Acta 1804:1658–1665. doi:10.1016/j.bbapap.2009.09.011
27. Nakagawa T, Guarente L (2009) Urea cycle regulation by mitochondrial sirtuin, SIRT5. Aging (Albany NY) 1:578–581

Chapter 17

Sirtuin 6 (SIRT6) Activity Assays

Minna Rahnasto-Rilla, Maija Lahtela-Kakkonen, and Ruin Moaddel

Abstract

SIRT6 has been shown to possess weak deacetylation, mono-ADP-ribosyltransferase activity, and deacylation activity in vitro. SIRT6 selectively deacetylates H3K9Ac and H3K56Ac. Several SIRT6 assays have been developed including HPLC assays, fluorogenic assays, FRET, magnetic beads, in silico, and bioaffinity chromatography assays. Herein, we describe detailed protocols for the HPLC based activity/inhibition assays, magnetic beads deacetylation assays, bioaffinity chromatographic assays as well as fluorogenic and in silico assays.

Key words Deacetylation assays, Activation assays, Inhibition assays, Bioaffinity chromatography, In silico screening

1 Introduction

Sirtuins are NAD+-dependent histone deacetylase enzymes (HDACs) that function as regulators of many cellular processes [1], and are evolutionarily conserved. To date, seven sirtuins have been identified (SIRT1-7), with SIRT1-3 being the most studied. However, SIRT6 has gained interest in age-associated diseases due to its role genomic stability, oxidative stress, and glucose metabolism [2]. While other SIRTuins have been shown to be less selective in which acetylated peptides they deacetylate, SIRT6 has been shown to only deacetylate H3K9Ac and H3K56Ac. SIRT6 has also been shown to possess mono-ADP-ribosyltransferase activity [3] and deacylation activity [4]. In fact, SIRT6 preferentially hydrolyzes long-chain fatty acyl groups from lysines with much higher activity than its deacetylation activity in vitro [4].

Due to the increased interest in SIRT6, several assays have been developed as seen in Table 1, including HPLC assays [5, 6], fluorogenic assays [7–9], FRET [10], magnetic beads [11], in silico and bioaffinity chromatography [12]. Of these assays, the SIRT6 deacetylation assays coupled to HPLC-MS are the most commonly used. These assays have the added advantage in that

Sibaji Sarkar (ed.), *Histone Deacetylases: Methods and Protocols*, Methods in Molecular Biology, vol. 1436,
DOI 10.1007/978-1-4939-3667-0_17, © Springer Science+Business Media New York 2016

Table 1
The reaction conditions of different SIRT6 in vitro assays

Activation	Peptide substrate	NAD (mM)	SIRT6	Reaction volume (μL)	Inc. time (min)	Ex-/Em-wl
HPLC	QTARK(Ac)STGG: 70 μM [6]	0.5	4 μM		60	NA
	ARTKQTARK(Ac)STGGK APRKQLA 40 μM [13]	0.5	0.05 μg/μL	60	30	NA
HPLC	QTARK(Ac)STGGAc: 300 μM[14]	2	4 μM	40	360	–
	KQTARK(Ac)STGGWW: 50 μM [5]	1	10 μM	30	60	–
	KQTARK(Myr)STGGWW: 50 μM [5]	0.5	1 μM	30	15	–
	EALPK(Myr)TGGPQWW: 50 μM [5]	0.5	1 μM	30	15	–
	ARTKQTARK(Ac)STGGK APRKQLA 150 μM [13]	0.5	0.05 μg/μL	60	60	NA
Frontal chrom.						
Magnetic beads	ARTKQTARK(Ac)STGGK APRKQLA 43 μM [11]	0.2	–	50	240	–
	RYQK(Ac)-AMC[a]: 320 μM [9]	3	0.09 μg/μL	50	90	370/430 nm
Fluorogenic	EALPK(Myr)-AMC[a]: 10 μM[8]	1	1 μM	60	120	360/460 nm
	RHKK(Ac)-AMC[a, b] 400 μM	3	0.09 μg/μL	50	45–90	355–370/ 430–460 nm
FRET	(DABCYL)-ISGASE(Myr)DIVHSE-(EDANS)G: 10 μM [10]	1	0.5 μM	60	60	336/490

[a]7-amino-4-methylcoumarin
[b]Cayman commercial kit

they have been demonstrated to be robust in determining both the inhibition and activation activity of the tested compounds for SIRT6 and are discussed in greater detail below [6, 13]. The HPLC deacetylation assays, predominantly use H3K9Ac as the peptide substrate, and the activity is determined based on the quantification of the deacetylated peptide, H3K9 (Fig. 1). In order to study the deacylation activity of SIRT6, two different substrates

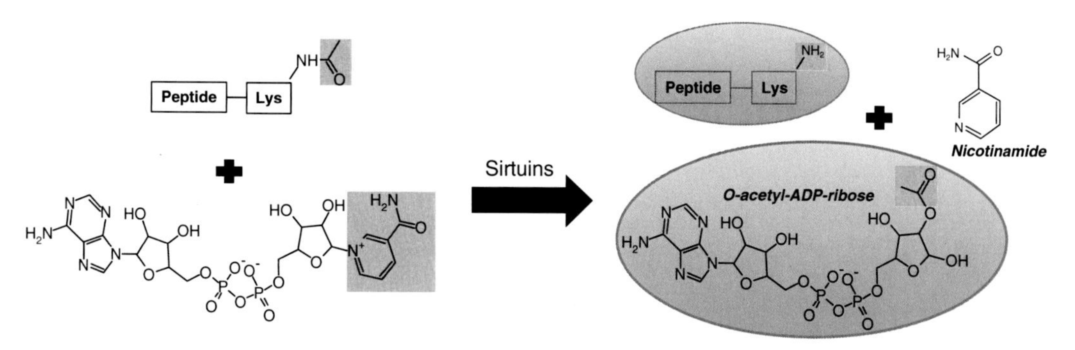

Fig. 1 Sirtuin mediated deacetylation. In most cases HPLC detection is based on the quantification of deacetylated substrate or O-acetyl-ADP-ribose (*red oval*)

were used TNFα (EALPK(Myr)TGGPQWW) [5] or H3K9myr [14] containing the acyl lysine modifications. While these assays are equally important, they are not discussed here and we refer the readers to Ref. [7] and Ref. [13] for experimental details.

Fluorogenic assays for SIRT6 differ from the HPLC assays as they produce a fluorescent signal after deacetylation/deacylation-coupled proteolytic cleavage of the substrate. A potential disadvantage with this method may arise with the use of complex mixtures, as it may result in the assay becoming susceptible to autofluorescence, fluorescence quenching, or anti-proteolytic properties of unknown compounds in the complex mixture [10]. Kokkonen et al. (2014) profiled a collection of fluorogenic substrates using SIRT6 fluorogenic deacetylation assay for the identification of novel SIRT6 inhibitors, and this is discussed in greater detail below [9]. Currently, there are two commercially available kits for measuring SIRT6 activity: Biovision and Cayman. Biovision provides a fluorometric SIRT6 inhibitor screening kit, where the substrate is deacetylated, and fluoresces after cleavage of the peptide by a proteolytic enzyme following deacetylation (please see http://www.biovision.com). Cayman's SIRT6 Direct Fluorescent Screening Assay measures both SIRT6 activation and inhibition using a fluorescence-based method where a p53 sequence Arg-His-Lys-Lys(ε-acetyl)-AMC is incubated with SIRT6 and NAD+. Deacetylation sensitizes the substrate such that subsequent treatment with the developer releases a fluorescent product (please see www.caymanchemicals.com for more details).

In another approach, SIRT6 was immobilized onto the surface of magnetic beads and the deacetylation activity was determined by measuring the production of the deacetylated H3K9Ac peptide [11]. An added advantage of this method lies in its reusability and its use in ligand fishing from complex matrices including botanical extracts [15]. In order to study the binding affinities of tested compounds for SIRT6, the SIRT6 protein was subsequently immobilized onto the surface of an open tubular capillary [15].

The immobilized SIRT6 was well characterized using frontal affinity chromatography and it was demonstrated that the SIRT6-OT column correctly predicted a compounds deacetylation activity with a single chromatographic run [16].

More recently, an in silico method for the screening of SIRT6 inhibitors has also been developed [7]. The availability of the experimentally identified ligands for SIRT6 opened up the possibility to use ligand-based approaches for virtual screening. Ligand-based virtual screening methods can be an efficient approach to find novel regulators. Usually they can be used for screening large databases including millions of compounds, which cannot be done with structure-based methods. Thus ligand-based methods are widely used for identifying novel regulators. Ligand-based virtual screening usually produces chemically and structurally similar hits than already known hits. Therefore, the main drawback of ligand-based methods is its capability to identify structurally diverse scaffolds. This method has also been discussed in greater detail below.

2 Materials

2.1 SIRT6 Deacetylation Activation Assay [13]

1. Tris buffered saline (TBS) [25 mM, pH 8.0] containing 137 mM NaCl, 1 mM $MgCl_2$.
2. 10 mM NAD+ (NAD) in TBS (*see* **Note 1**).
3. 0.96 mM ARTKQTARK(Ac)STGGKAPRKQLA [H3K9Ac: residues 1–21].
4. GST-Tag SIRT6 (SIRT6-GST) (*see* **Note 2**): The human SIRT6 expression vector hSIRT6-pGEX-6P3 was kindly provided by Prof. Katrin Chua (Stanford, USA). The recombinant GST-tagged SIRT6 was produced by fermentation in *Escherichia coli* BL21(DE3)-pRARE. The production was done at +16 °C with 0.1 mM IPTG for 20 h and the soluble overexpressed protein was affinity purified on glutathione agarose.
5. Active His-tagged SIRT6 (SIRT6-HIS), human recombinant: SIRT6 protein is a solution of 1 mg/mL in Tris buffer [50 mM, pH 8.0] containing 100 mM NaCl and 20% glycerol.
6. Zorbax Eclipse XDB-C18 column (4.6 mm × 50 mm, 1.8 μm).
7. HPLC: Shimadzu prominence system consisting of a CBM-20A, LC-20 AB binary pumps, an SIL-20 AC-HT autosampler and a DGU-20A3 degassing unit.
8. MS/MS: 5500 QTRAP from Applied Biosystems/MDS Sciex equipped with Turbo V electrospray ionization source (TIS)®.

2.2 SIRT6 Deacetylation Fluorogenic Assay [9]

1. Tris buffered saline (TBS) [50 mM, pH 8] containing 137 mM NaCl, 2.7 mM KCl, and 1 mM $MgCl_2$.
2. 50 mM NAD+ (NAD) in TBS.
3. 10 mM RYQK(Ac)-AMC (AMC) in TBS was purchased from CASLO ApS.
4. Developer solution (6 μg/μl trypsin, 4 mM NAD).
5. (SIRT6-GST): Please see **item 4** of Subheading 2.1.
6. Black 1/2 AREAPLATE-96 F.

2.3 SIRT6-Magnetic Beads Deacetylation Assay [11]

1. BcMag amine-terminated magnetic beads (50 mg/mL, 1 μm).
2. The manual magnetic separator Dynal MPC-S.
3. 2-(*N*-morpholino)ethanesulfonic acid buffer (MES) [100 mM, pH 5.5].
4. N-Hydroxysulfosuccinimide (Sulfo-NHS).
5. 1-ethyl-3-(3-methylaminopropyl)carbodiimide (EDC).
6. Storage Buffer (SB): phosphate buffer [10 mM, pH 7.4] containing 150 mM NaCl and 0.02 % sodium azide.
7. HDAC assay buffer (Tris–HCl, [50 mM,pH 8.0] containing 150 mM NaCl, 1 mM DTT, and 0.2 mM NAD+).
8. ARTKQTARK(Ac)STGGKAPRKQLA [H3K9Ac: residues 1–21] in TBS.
9. Zorbax Eclipse XDB-C18 column (4.6 mm × 50 mm, 1.8 μm).
10. HPLC: Shimadzu prominence system consisting of a CBM-20A, LC-20 AB binary pumps, an SIL-20 AC-HT autosampler and a DGU-20A3 degassing unit.
11. 5500 QTRAP from Applied Biosystems/MDS Sciex equipped with Turbo V electrospray ionization source (TIS)®.
12. Mobile phase: Eluent A: water containing 0.02 % formic acid; Eluent B: acetonitrile containing 0.02 % formic acid.
13. SIRT6 protein [11]: The expression plasmid pMBP-SIRT6 encodes a maltose-binding protein (MBP)-SIRT6 fusion protein. It was created by amplifying the full-length chicken SIRT6 cDNA from DT40 cDNA using primers SIRT6F (5′-CGCGGATCCATGGCGGTGAATTACGC-3′)/SIRT6R(5′CGCTCGAGTCAGGT GAGGAGAGGCTC-3′), digesting the PCR product with BamHI and XhoI, and ligating is into the BamHI/XhoI sites of pMH6.

2.4 SIRT6 Frontal Chromatographic Analysis [15]

1. Reagents similar to Subheading 2.3 unless otherwise specified.
2. Open tubular capillary (30 cm × 100 μm i.d.).
3. Rabbit peristaltic pump.

4. Frontal mobile phase: ammonium acetate [10 mM, pH 7.4]: methanol (90:10 v/v) containing 0.2 mM NAD+.(*see* **Notes 3** and **4**)
5. Agilent HPLC-MS: Series 1100 Liquid Chromatography/ Mass Selective Detector equipped with a vacuum de-gasser (G 1322 A), a binary pump (1312 A), an autosampler (G1313 A) with a 20 μL injection loop, a mass selective detector (G1946 B) supplied with atmospheric pressure ionization electrospra and an online nitrogen generation system.

2.5 SIRT6 In Silico-Screen [7]

1. The homology model for SIRT6 was constructed using the crystal structure of SIRT3 complexed with a substrate (PDB id: 3glr) as a template. The sequence identity between SIRT3 and SIRT6 was about 34%. The homology model for SIRT6 was constructed with ORCHESTRAR tool in SYBYL-X version 1.2.
2. The protein structure was prepared for the molecular docking with protein preparation wizard in Schrödinger's Maestro (Schrödinger's Release 2011 Maestro version 9.2 LLC, New York, 2011). This procedure includes a short minimization (with convergence criteria of RSMD 0.30 Å) for heavy atoms with OPLS_2005 force field.
3. Several known SIRT1 and SIRT2 peptidic and pseudopeptidic inhibitors were selected for docking studies. The 3D structure of compounds were constructed with Discovery Studio (Dassault Systèmes BIOVIA, Discovery Studio Modeling Environment, Release 2.5, San Diego: Dassault Systèmes, 2010) and MOE (Molecular Operating Environment (MOE) version 2010.10; Chemical Computing Group Inc., Canada, 2010) and refined with Ligprep (Schrödinger release 2011: Schrödinger, LLC, New York, 2011). All tautomeric states were determined at pH 7 ± 2.

3 Methods

3.1 SIRT6 Deacetylation Assay [13]

3.1.1 Acitvation Assay

1. 43.3 μL TBS is added to 3.6 μL NAD (final 0.6 mM), 6 μL 10 mM DTT, and 2.5 μL 0.96 mM H3K9Ac (final 40 μM).
2. 0.6 μL of DMSO is used as a control or 0.6 μL of tested compounds is dissolved in DMSO (*see* **Note 5**).
3. The reaction is started by the addition of 4 μl of SIRT6 protein (final 0.05 μg/μL) (SIRT6-GST or SIRT6-HIS).
4. The solution is incubated for 30 min at 37 °C.
5. The reaction is terminated with the addition of 6 μL cold (100%) formic acid (final concentration 10%).

6. The resulting solution is centrifuged at 16,100 × *g* on Eppendorf Microcentrifuge 5424 (SoCal BioMedical) for 15 min.
7. The samples are then collected and analyzed using a HPLC coupled to a 5500 QTRAP.
8. The chromatographic separation of H3K9 and acetylated H3K9 is achieved on a Zorbax Eclipse XDB-C18 column (4.6 mm × 50 mm, 1.8 μm) at room temperature. The mobile phase consists of Eluent A and B with the following gradient: 0–2.0 min, 0% B; 2.0–10 min, 0—8% B; 10–10.10 min, 8—80% B; 10.10–12 min, 80%; 12–12.1 min 80–0% B; 15 min, 0% B at 0.9 mL/min. The total run time is 15 min and the injection volume per sample is 20 μl.
9. Positive electrospray ionization data are acquired using multiple reaction monitoring (MRM). The TIS instrumental source settings for temperature, curtain gas, ion source gas 1 (nebulizer), ion source gas 2 (turbo ion spray), entrance potential, and ion spray voltage were 550 °C, 20 psi, 60 psi, 50 psi, 10 V, and 5500 V, respectively. The TIS compound parameter settings for declustering potential, collision energy, and collision cell exit potential were 231 V, 45 V, and 12 V, respectively, for H3K9Ac and 36 V, 43 V, and 12 V, respectively, for H3K9. The standards are characterized using the following MRM ion transitions: H3K9Ac (*m/z* 766.339 → 760.690) and H3K9 (*m/z* 752.198 → 746.717).

3.1.2 Inhibition Assay [13]

1. 36.4 μL of TBS is mixed with 3.6 μL NAD (final 0.6 mM), 6 μL 10 mM DTT, and 9.4 μL 0.96 mM H3K9Ac (final 150 μM) in a 1.5 mL Eppendorf tube.
2. 0.6 μL of DMSO is added to the control tubes and 0.6 μL of varying concentrations of tested compounds in DMSO is added to the eppendorf tube (*see* **Note 5**).
3. The reaction is initiated by the addition of 4 μL SIRT6 protein (GST or His-tagged) (final 0.05 μg/μL) and the solution is incubated for 60 min at 37 °C with 300 rpm rotation on the Thermo shaker incubator (Hangzhou Allsheng Instruments Co., Ltd).
4. The reaction is terminated by the addition of 6 μL 100% cold formic acid (final concentration 10%).
5. The samples are centrifuged at 16,100 × *g* for 15 min in Eppendorf Microcentrifuge 5424 (SoCal BioMedical) and the supernatant is analyzed as described above in **steps 7–9**.

3.2 SIRT6 Deacetylation Fluorogenic Assay [9]

1. 3.2 μL NAD (final 3.2 mM), 1.6 μL AMC (final 320 μM), and 2.5 μL inhibitor solution/DMSO are added to 38.7 μL of TBS, in a well plate (*see* **Note 6**).

2. 4 μL the SIRT6-GST is added to start the reaction.
3. The solution is incubated for 90 min at 37 °C.
4. 50 μL of developer is added to the solution for 30 min at room temperature to terminate the reaction.
5. The activity is determined by measuring the fluorescence with excitation and emission wavelengths of 380 and 440 nm using EnVision 2104 Multilabel Reader.

3.3 SIRT6-Magnetic Beads Deacetylation Assay [11]

1. 0.5 mL (25 mg) of BcMag beads is rinsed with 1 mL of MES buffer.
2. After magnetic separation, the supernatant is discarded, and the BcMag beads are suspended in 300 μL of MES buffer containing 260 μg of SIRT6 protein.
3. 50 μL of a mixture of 10 mg of EDC and 15 mg of sulfo-NHS in 1 mL of water is added to the reaction mixture and vortexed for 5 min and left for 3 h at 4 °C with gentle rotation.
4. 20 μL of 1 M hydroxylamine is added for 30 min at 4 °C with gentle rotation.
5. The supernatant is discarded and the SIRT6 (CT)-MB is rinsed three times with 1 mL of SB.(*see* **Notes** 7 and **8**)
6. SIRT6 (CT)-MB is incubated with 50 μL of HDAC assay buffer containing 5 μg of H3K9Ac for 4 h at 37 °C.
7. After magnetic separation, the supernatant is collected and analyzed using a HPLC coupled to a 5500 QTRAP.
8. The chromatographic separation of H3K9 and acetylated H3K9 is achieved as described in Subheading 3.1, **steps 8** and **9**.

3.4 SIRT6 Frontal Chromatographic Analysis [15]

1. An open tubular capillary (30 cm × 100 μm i.d.) is washed with MES buffer for 20 min using a Rabbit peristaltic pump with a setting of 85.
2. A 1 mL solution of MES containing 700 μL of SIRT6 (44 μg/mL) with 100 μl of EDC (500 mg/mL) and 50 μl of Sulpho-NHS (340 mg/mL) is passed through the column.
3. Both tips of the capillary are submerged into the solution for 18 h at 4 °C. After which MES buffer is passed through for 10 min. The column is now ready for analysis.
4. The SIRT6(CT)-OT column is connected to the Agilent HPLC-MS system.
5. Frontal mobile phase containing a series of concentrations of the tested compounds is delivered at 0.05 μL/min at room temperature for determination of the binding affinity (Kd).
6. In order to study the quercetin binding site, a series of concentrations of the tested compounds is placed in the frontal mobile

phase containing 5 μM Quercetin (Ki). Quercetin is monitored in the negative ion mode using single ion monitoring at *m/z* = 301.00 [MW-H], with the capillary voltage at 3000 V, the nebulizer pressure at 35 psi, and the drying gas flow at 11 L/min at a temperature of 350 °C.

3.5 SIRT6 In Silico-Screen [7]

A [7]

1. The homology model for SIRT6 is used for molecular docking studies.
2. The centroid of the grid box defining the docking region is determined with Trp187 and Pro219. The constraint for hydrogen bonding is set with carbonyl oxygen of Leu184 in order to ensure the correct binding pose for acetylated lysine. Molecular docking is carried out with Schrödinger Glide SP (Standard Precision) version (Small-Molecule Drug Discovery Suite 2011: Glide, version 5.7, Schrödinger, LLC, New York, 2011).
3. Docking poses are visually inspected.

B [17]

1. The modified crystal structure of SIRT6 is utilized for docking.
2. The centroid of grid box is set based on ADP-ribose ligand present in the crystal structure and all water is removed.
3. Molecular docking is performed with SP (Standard Precision) protocol of Schrödinger Glide version (Small-Molecule Drug Discovery Suite 2012: Glide, version 5.8, Schrödinger, LLC, New York, 2012) for screening.
4. Based on Glide score top 1500 compounds are further docked with Glide's XP (Extra Precision) and top 500 molecules of them are visually inspected and further manually filtered.
5. Pan Assay Interference Compounds (PAINS) using FAF-Drugs2 is used to filter the final active compounds.

4 Notes

1. NAD+ solution should be prepared fresh for each experiment.
2. Avoid repeated cycles of freezing and thawing of GST-Tag SIRT6
3. Addition of NAD to the mobile phase to study the SIRT6-OT is necessary for correct ranking
4. The ranking of a series of compounds on the SIRT6-OT column can only be determined by looking at the change in

retention volume of a marker ligand, for example, quercetin, and not based on retention of the compound tested.

5. The SIRT6-magnetic beads should be tested periodically (once a week) to confirm that the SIRT6 activity is maintained at similar levels
6. The immobilization of SIRT6-GST is not recommended for ligand fishing experiments, as the GST fusion protein may result in retention of a significant amount of compounds that have no affinity for the SIRT6 protein.
7. DMSO should be kept below 2 %.
8. The use of black well plate is recommended for fluorogenic assay.

References

1. Haigis MC, Sinclair DA (2010) Mammalian sirtuins: biological insights and disease relevance. Annu Rev Pathol 5:253–295. doi: 10.1146/annurev.pathol.4.110807.092250
2. Cheng HL, Mostoslavsky R, Saito S, Manis JP, Gu Y, Patel P, Bronson R, Appella E, Alt FW, Chua KF (2003) Developmental defects and p53 hyperacetylation in Sir2 homolog (SIRT1)-deficient mice. Proc Natl Acad Sci U S A 100(19):10794–10799. doi:10.1073/pnas.1934713100
3. Liu Y, Xie QR, Wang B, Shao J, Zhang T, Liu T, Huang G, Xia W (2013) Inhibition of SIRT6 in prostate cancer reduces cell viability and increases sensitivity to chemotherapeutics. Protein Cell. doi:10.1007/s13238-013-3054-5
4. Kim HS, Xiao C, Wang RH, Lahusen T, Xu X, Vassilopoulos A, Vazquez-Ortiz G, Jeong WI, Park O, Ki SH, Gao B, Deng CX (2010) Hepatic-specific disruption of SIRT6 in mice results in fatty liver formation due to enhanced glycolysis and triglyceride synthesis. Cell Metab 12(3):224–236. doi:10.1016/j.cmet.2010.06.009
5. He B, Hu J, Zhang X, Lin H (2014) Thiomyristoyl peptides as cell-permeable Sirt6 inhibitors. Org Biomol Chem 12(38):7498–7502. doi:10.1039/c4ob00860j
6. Feldman JL, Baeza J, Denu JM (2013) Activation of the protein deacetylase SIRT6 by long-chain fatty acids and widespread deacylation by mammalian sirtuins. J Biol Chem 288(43):31350–31356. doi:10.1074/jbc.C113.511261
7. Kokkonen P, Rahnasto-Rilla M, Kiviranta PH, Huhtiniemi T, Laitinen T, Poso A, Jarho E, Lahtela-Kakkonen M (2012) Peptides and pseudopeptides as SIRT6 deacetylation inhibitors. ACS Med Chem Lett 3(12):969–974. doi:10.1021/ml300139n
8. Hu J, He B, Bhargava S, Lin H (2013) A fluorogenic assay for screening Sirt6 modulators. Org Biomol Chem 11(32):5213–5216. doi:10.1039/c3ob41138a
9. Kokkonen P, Rahnasto-Rilla M, Mellini P, Jarho E, Lahtela-Kakkonen M, Kokkola T (2014) Studying SIRT6 regulation using H3K56 based substrate and small molecules. Eur J Pharm Sci 63:71–76. doi:10.1016/j.ejps.2014.06.015
10. Li Y, You L, Huang W, Liu J, Zhu H, He B (2015) A FRET-based assay for screening SIRT6 modulators. Eur J Med Chem 96:245–249. doi:10.1016/j.ejmech.2015.04.008
11. Yasuda M, Wilson DR, Fugmann SD, Moaddel R (2011) Synthesis and characterization of SIRT6 protein coated magnetic beads: identification of a novel inhibitor of SIRT6 deacetylase from medicinal plant extracts. Anal Chem 83(19):7400–7407. doi:10.1021/ac201403y
12. Borra MT, O'Neill FJ, Jackson MD, Marshall B, Verdin E, Foltz KR, Denu JM (2002) Conserved enzymatic production and biological effect of O-acetyl-ADP-ribose by silent information regulator 2-like NAD+-dependent deacetylases. J Biol Chem 277(15):12632–12641. doi:10.1074/jbc.M111830200
13. Rahnasto-Rilla M, Kokkola T, Jarho E, Lahtela-Kakkonen M, Moaddel R (2015) Ethanolamides binds to the SIRT6., Under the review
14. Pan PW, Feldman JL, Devries MK, Dong A, Edwards AM, Denu JM (2011) Structure and biochemical functions of SIRT6. J Biol Chem 286(16):14575–14587. doi:10.1074/jbc.M111.218990

15. Singh N, Ravichandran S, Norton DD, Fugmann SD, Moaddel R (2013) Synthesis and characterization of a SIRT6 open tubular column: predicting deacetylation activity using frontal chromatography. Anal Biochem 436(2):78–83. doi:10.1016/j.ab.2013.01.018
16. Ravichandran S, Singh N, Donnelly D, Migliore M, Johnson P, Fishwick C, Luke BT, Martin B, Maudsley S, Fugmann SD, Moaddel R (2014) Pharmacophore model of the quercetin binding site of the SIRT6 protein. J Mol Graph Model 49:38–46. doi:10.1016/j.jmgm.2014.01.004
17. Parenti MD, Grozio A, Bauer I, Galeno L, Damonte P, Millo E, Sociali G, Franceschi C, Ballestrero A, Bruzzone S, Del Rio A, Nencioni A (2014) Discovery of novel and selective SIRT6 inhibitors. J Med Chem 57(11):4796–4804. doi:10.1021/jm500487d

Chapter 18

Molecular, Cellular, and Physiological Characterization of Sirtuin 7 (SIRT7)

Jiyung Shin and Danica Chen

Abstract

Sirtuin 7 (SIRT7), a histone 3 lysine 18 (H3K18) deacetylase, functions at chromatin to suppress endoplasmic reticulum (ER) stress and mitochondrial protein folding stress (PFS^{mt}), and prevent the development of fatty liver disease and hematopoietic stem cell aging. In this chapter, we provide a methodology to characterize the molecular, cellular, and physiological functions of SIRT7.

Key words SIRT7, Myc, NRF1, Mitochondrial UPR (UPR^{mt}), Mitochondrial protein folding stress, ER UPR (UPR^{er}), ER stress, Fatty liver, Hematopoietic stem cell, Aging

1 Introduction

Silent information regulator 2 (Sir2) proteins, or sirtuins, are a class of highly conserved proteins found in organisms ranging from bacteria to humans that possess nicotinamide adenine dinucleotide (NAD^+)-dependent protein deacetylase activity, including deacetylase activity [1–3]. There are seven mammalian sirtuins, SIRT1-7, that localize to various cellular compartments [4–15]. SIRT7 is a chromatin binding protein that deacetylates H3K18 at specific gene promoters to repress transcription [4]. A major class of SIRT7 target genes is ribosomal proteins and mitochondrial ribosomal proteins [4]. SIRT7 does not have a known DNA binding domain and is recruited to chromatin through its interaction with transcription factors, such as Myc, NRF1, and ELK4 [4, 10, 15]. SIRT7 expression is induced upon various stress conditions, such as ER stress and PFS^{mt}, and represses the expression of ribosomal proteins and mitochondrial ribosomal proteins to alleviate stresses [10, 15]. Mouse genetics studies reveal that SIRT7 prevents the development of fatty liver disease and hematopoietic stem cell aging [10, 15].

Sibaji Sarkar (ed.), *Histone Deacetylases: Methods and Protocols*, Methods in Molecular Biology, vol. 1436,
DOI 10.1007/978-1-4939-3667-0_18, © Springer Science+Business Media New York 2016

Here, we describe the methods we utilized to characterize the molecular, cellular, and physiological role of SIRT7. We describe a co-immunoprecipitation (co-IP) method to identify SIRT7-interacting transcription factors. We present a method to assess the effect of SIRT7 on PFS^{mt} using an aggregation prone mutant mitochondrial protein ornithine transcarbamylase (ΔOTC) as a marker. Finally, we provide a guide to study SIRT7 physiology with a focus on hepatic lipid metabolism and hematopoietic stem cell aging.

2 Materials

2.1 Molecular and Cellular Characterization of SIRT7

2.1.1 Endogenous Co-immunoprecipitation

1. High salt lysis buffer: 10% glycerol (w/v), 50 mM Tris–HCl (pH 7.5), 300 mM NaCl, 1% NP-40, 20 mM $MgCl_2$. Freshly add protease inhibitor cocktail before use.
2. Co-IP buffer: 10% glycerol (w/v), 0.6% Triton™ X-100 in PBS. Freshly add protease inhibitor cocktail before use.
3. Laemmli sample buffer (5×): 10% SDS (w/v), 10 mM dithiothreitol, 20% glycerol (w/v), 0.2 M Tris–HCl (pH 6.8), 0.05% bromophenol blue.

2.2 Physiological Characterization of SIRT7

1. Homogenization buffer: 1:2 ratio of 100% MeOH and 100% chloroform.
2. Reconstitution buffer: 1% Triton™ X-100 in 100% EtOH.

2.2.1 Physiological Function of SIRT7 in the Liver

Triglyceride (TG) Quantification in the Livers of Wild Type (WT) and $SIRT7^{-/-}$ Mice

Histological Analyses of the Livers of WT and $SIRT7^{-/-}$ Mice

1. Paraformaldehyde: 4% solution in water.
2. Oil Red O Solution: Add 0.7 g of Oil Red-O powder into 100 mL of isopropanol. Stir overnight in a glass bottle. Filter the mixture through two layers of Whatman papers. Store at 4 °C. Before staining, prepare a fresh Oil Red-O working solution (WS) by adding 6.0 mL of stock to 4.0 mL of DD water. Then filter the mixture through two layers of Whatman papers (*see* **Note 1**).
3. Mayer's hematoxylin.
4. Blocking buffer: 5% FBS, 0.3% Triton™ X-100 in PBS.
5. Antibody dilution buffer: 2.5% FBS, 0.3% Triton™ X-100 in PBS.
6. Rat anti-mouse F4/80 antibody: diluted 1:200 in antibody dilution buffer.
7. Donkey anti-rat IgG antibody: diluted 1:500 in antibody dilution buffer.
8. DAPI: 10 mg/mL in H_2O.
9. Prolong Gold antifade reagent.

2.2.2 Functional Assessment of SIRT7 in Hematopoietic Stem Cell (HSC) Aging

Infection of HSCs with Control or SIRT7 Overexpression Lentivirus

1. HEK-293T growing media: 10% fetal bovine serum in high-glucose Dulbecco's modified eagle medium
2. Stem cell stimulation media: 9 mL of StemSpan SFEM, 1 mL of stem cell fetal bovine serum, 100 μL Pen-Strep, cytokine mix: 100 μL SCF, 20 μL TPO, 25 μL Flt3, 10 μL IL3, 10 μL IL6 (all stocks at 10 μg/mL, all cytokines murine).
3. Syringe filter: 0.45 μm, nitrocellulose.
4. Polybrene: 10 mg/mL, used in 1:1000 dilution.

3 Methods

3.1 Molecular and Cellular Characterization of SIRT7

3.1.1 Endogenous Co-immunoprecipitation

1. The day before starting with the experiment, seed 1×10^7 HEK-293T cells in 15 cm dish.
2. Aspirate media off of the cells and wash cells with ice-cold PBS.
3. Lyse cells by adding 700 μL of high salt lysis buffer directly to the plate.
4. Incubate the cells in the plate on a shaker in the cold room (4 °C) for 30 min.
5. Use cell scraper to dislodge all cells and transfer to 1.5 mL tubes. Centrifuge the cell lysate for 10 min at 16,100 × *g* at 4 °C.
6. Transfer supernatant to new 1.5 mL tube.
7. Measure protein concentration by BCA assay.
8. Dilute 1 mg of protein to 1 mL total volume with co-IP buffer (*see* **Note 2**).
9. Pre-clear lysates by adding 15 μL of washed protein A beads.
10. Place samples in a rotator in the cold room for 30 min at 4 °C.
11. Spin down at 1000 × g for 5 min at 4 °C.
12. Transfer supernatant to new tubes. Reserve a 50 μL aliquot in a separate new 1.5 mL tube for input controls and set aside.
13. For the rest of the supernatant add either 1 μg of IgG control antibody or 1 μg of SIRT7 antibody.
14. Place the samples in a rotator overnight at 4 °C.
15. Wash 20 μL worth of protein A/G bead slurry with 1 mL Co-IP buffer.
16. Spin down for 30 s at 1000 × *g* at 4 °C.
17. Aspirate supernatant, and add 950 μL of protein + antibody that had been incubating the night before to the washed protein A/G bead slurry.
18. Place the samples in the rotator for 3 h at 4 °C.
19. Spin down for 30 s at 1000 × *g* at 4 °C.
20. Aspirate off supernatant. Add 1 mL Co-IP buffer.

21. Spin down for 30 s at 1000 × *g* at 4 °C.
22. Repeat **steps 20** and **21** twice more.
23. Using 25 G needle, pierce a hole on the lid of 1.5 mL tube and a second "half" hole at the bottom, only going half the way through the tube bottom for the elution of the protein off the beads. Place the tube with holes on top of a new 1.5 mL tube.
24. Spin down for 30 s at 1000 × *g* at 4 °C. Repeat this step until the beads are dry (*see* **Note 3**).
25. Add 40 μL of 100 mM glycine (pH 3.0) to the beads. Incubate for 10 min at room temperature, flicking intermittently to mix.
26. Place the tube on top of a new 1.5 mL tube. Spin down for 30 s at 1000 × *g* at 4 °C (*see* **Note 4**).
27. Add 10 μL 5× Laemmli sample buffer (for a final concentration of 1×) to the Input and IP eluents.
28. Boil samples at 95 °C in a heat block for 10 min.
29. Continue with Western blotting and probe for SIRT7 interacting transcription factors.

3.1.2 Overexpression of ΔOTC in SIRT7 Knockdown Cells

1. The day before starting with the experiment, seed 5×10^5 of control or SIRT7 knockdown (KD) HEK-293T cells in a 6 well dish.
2. Transfect OTC or ΔOTC constructs into control or SIRT7 KD cells. Mix 2 μg of DNA with 250 μL of Opti-MEM and 6 μL of Lipofectamine 2000 in a sterile 1.5 mL tube.
3. Incubate for 20 min at room temperature.
4. Add dropwise to cells.
5. Collect cells at 48 h post transfection.
6. Continue with Western blotting and probe for OTC.

3.2 Physiological Characterization of SIRT7

3.2.1 Physiological Function of SIRT7 in Liver

TG Quantification in the Livers of WT and SIRT7$^{-/-}$ Mice

1. Cut 50–100 mg of liver samples on dry ice and weigh.
2. Homogenize samples in 1 mL of homogenization buffer using a motorized pestle in a 1.5 mL tube (*see* **Note 5**).
3. Rotate samples in fume hood at RT for 2 h.
4. Spin down homogenates at 15,000 × *g* for 5 min in a tabletop centrifuge.
5. Transfer supernatant to glass GC/MS tubes.
6. Evaporate under N_2 gas for ~5–20 min and ~10–40 μL of viscous fat layer will remain.
7. Reconstitute with 200–500 μL of reconstitution buffer.
8. Perform enzymatic TG measurement assay according to manufacturers' instructions.

Histological Analyses of the Livers of WT and SIRT7$^{-/-}$ Mice

Oil Red O Staining

1. Embed liver tissues in Optimal Cutting Temperature compound (OCT) and make 5 μm sections using the Cryostat.
2. Fix for 40 min in 4% paraformaldehyde (*see* **Note 6**).
3. Wash with PBS.
4. Immerse into Oil Red-O working solution (WS) for 30 min.
5. Wash briefly in 60% isopropanol.
6. Stain with hematoxylin for 1 min.
7. Wash with H_2O.
8. Observe under fluorescent microscope with magnifications of ×20.

Immunohistochemistry for F4/80

1. Embed liver tissues in OCT and make 16 μm sections using the Cryostat.
2. Fix for 40 min in 4% paraformaldehyde.
3. Wash with 1× PBS.
4. Permeabilize in 1% Triton™ X-100 in PBS for 15 min.
5. Rinse once in PBS for 1 min.
6. Block for 1 h in blocking buffer
7. Incubate sections with rat anti-mouse F4/80 antibody diluted 1:200 in antibody dilution buffer overnight at 4 °C.
8. Rinse three times in PBS, 5 min each wash.
9. Incubate sections with secondary antibody diluted 1:500 in antibody dilution buffer for 1 h at room temperature (*see* **Note 7**).
10. Rinse three times in PBS, 5 min each wash.
11. Incubate sections with DAPI diluted 1:500 in PBS for 5 min at room temperature.
12. Rinse once in PBS for 5 min.
13. Add coverslip with anti-fade reagent, avoiding air bubbles.
14. Observe under microscope.

3.2.2 Functional Assessment of SIRT7 in Hematopoietic Stem Cell Aging

Infection of Enriched HSCs with Control and SIRT7 Overexpression Lentivirus

Day 1: Transfect ~90% confluent HEK-293T cells in 10 cm dish

1. Add 1 mL serum free Opti-MEM to 1.5 mL tube.
2. Add 6 μg each of: pFUGW/pFUGW-SIRT7; pMDLg-pRRE; pRSV-Rev; pMD2.G (Addgene).
3. Mix and add 60 μL of Lipofectamine 2000 to the tube and vortex briefly.
4. Incubate at room temperature for 20 min.
5. Add the mixture dropwise into 10 cm dish.
6. Mix by gentle swirling of the plate.
7. Change media the next day (~18 h after transfection).

Day 3: Infection

1. Plate 50,000 of sorted Lineage− c-Kit+Sca-1+ (LSK) cells from aged mice (18–24 months old) in a round bottom 96-well plate and incubate in stem cell stimulating media for 3 h.
2. Filter viral media from 10 cm dish through 0.45 μm syringe filter into centrifuge bottles.
3. Add 10 mL of HEK-293T growing media back to viral producing 293T plates for a second round of viral infection to be carried out the next day.
4. Equilibrate weights of bottles with sterile PBS and cap.
5. Spin bottles in Sorval Centrifuge, using SS34 rotor, at 20,000 × *g* for 90 min at 4 °C.
6. Gently pour off supernatant and use aspirator at the mouth of the bottle to remove any remaining supernatant (*see* **Note 8**).
7. Resuspend invisible pellet with 200 μl of Stem cell stimulating media.
8. Add polybrene to the resuspended media (final concentration of polybrene in the media is 10 μg/mL).
9. Transfer media containing sorted LSK cells from 96 well round-bottom dish to 96 well flat-bottom dish (*see* **Note 9**).
10. Add resuspended virus to each well.
11. Spin plate at 270 × *g* for 90 min at room temperature.
12. Return to 37 °C CO_2 incubator and incubate for 24 h.

Day 4: Second round of infection

1. Repeat **steps 2–12** on *Day 3*.
2. Give the cells a few hours to rest. Transduced LKS can be used for transplantation on this day.

4 Notes

1. It is important to remove any precipitates.
2. You would need one IgG control IP and one SIRT7 IP.
3. The beads will turn white once they are dry. If the beads remain transparent and wet after spinning down, you would need to make your "half" hole bigger and spin down again.
4. Make sure eluents are clear of beads, but bead pellet in the eluted tube is dry.
5. Prepare a mastermix of 1:2 MeOH–chloroform freshly before cutting samples.

6. The sections can be stored in −20 °C freezer for later use.
7. Protect from light whenever handling the secondary antibody, as well as your samples after you have added the secondary antibody.
8. Since the viral pellet is invisible, care should be taken while aspirating, such as tilting the centrifuge tubes and aspirating near the neck of the tubes.
9. Enriched HSCs are usually plated in round-bottom plates, however, we have found more consistent transduction when it is carried out in flat-bottom plates.

Acknowledgement

We thank M. Mohrin and H. Luo for comments. This work was supported by National Science Foundation GRFP (JS) and NIH AG040990, DK101885, Ellison Medical Foundation, Glenn Foundation, and PackerWentz endowment to DC.

References

1. Finkel T, Deng CX, Mostoslavsky R (2009) Recent progress in the biology and physiology of sirtuins. Nature 460(7255):587–591
2. Hirschey MD (2011) Old enzymes, new tricks: sirtuins are NAD(+)-dependent de-acylases. Cell Metab 14(6):718–719
3. Imai S, Guarente L (2010) Ten years of NAD-dependent SIR2 family deacetylases: implications for metabolic diseases. Trends Pharmacol Sci 31(5):212–220
4. Barber MF et al (2012) SIRT7 links H3K18 deacetylation to maintenance of oncogenic transformation. Nature 487(7405):114–118
5. Brown K et al (2013) SIRT3 reverses aging-associated degeneration. Cell Rep 3(2):319–327
6. Chen D et al (2005) Increase in activity during calorie restriction requires Sirt1. Science 310(5754):1641
7. Jeong SM et al (2014) SIRT4 protein suppresses tumor formation in genetic models of Myc-induced B cell lymphoma. J Biol Chem 289(7):4135–4144
8. Lo Sasso G et al (2014) SIRT2 deficiency modulates macrophage polarization and susceptibility to experimental colitis. PLoS One 9(7):e103573
9. Michishita E et al (2008) SIRT6 is a histone H3 lysine 9 deacetylase that modulates telomeric chromatin. Nature 452(7186):492–496
10. Mohrin M et al (2015) Stem cell aging. A mitochondrial UPR-mediated metabolic checkpoint regulates hematopoietic stem cell aging. Science 347(6228):1374–1377
11. Mostoslavsky R et al (2006) Genomic instability and aging-like phenotype in the absence of mammalian SIRT6. Cell 124(2):315–329
12. Park J et al (2013) SIRT5-mediated lysine desuccinylation impacts diverse metabolic pathways. Mol Cell 50(6):919–930
13. Qiu X et al (2010) Calorie restriction reduces oxidative stress by SIRT3-mediated SOD2 activation. Cell Metab 12(6):662–667
14. Rardin MJ et al (2013) SIRT5 regulates the mitochondrial lysine succinylome and metabolic networks. Cell Metab 18(6):920–933
15. Shin J et al (2013) SIRT7 represses Myc activity to suppress ER stress and prevent fatty liver disease. Cell Rep 5(3):654–665

Part III

Histone Deacetylase Inhibitors

Chapter 19

HDAC Inhibitors

Heidi Olzscha, Mina E. Bekheet, Semira Sheikh, and Nicholas B. La Thangue

Abstract

Lysine acetylation in proteins is one of the most abundant posttranslational modifications in eukaryotic cells. The dynamic homeostasis of lysine acetylation and deacetylation is dictated by the action of histone acetyltransferases (HAT) and histone deacetylases (HDAC). Important substrates for HATs and HDACs are histones, where lysine acetylation generally leads to an open and transcriptionally active chromatin conformation. Histone deacetylation forces the compaction of the chromatin with subsequent inhibition of transcription and reduced gene expression. Unbalanced HAT and HDAC activity, and therefore aberrant histone acetylation, has been shown to be involved in tumorigenesis and progression of malignancy in different types of cancer. Therefore, the development of HDAC inhibitors (HDIs) as therapeutic agents against cancer is of great interest. However, treatment with HDIs can also affect the acetylation status of many other non-histone proteins which play a role in different pathways including angiogenesis, cell cycle progression, autophagy and apoptosis. These effects have led HDIs to become anticancer agents, which can initiate apoptosis in tumor cells. Hematological malignancies in particular are responsive to HDIs, and four HDIs have already been approved as anticancer agents. There is a strong interest in finding adequate biomarkers to predict the response to HDI treatment. This chapter provides information on how to assess HDAC activity in vitro and determine the potency of HDIs on different HDACs. It also gives information on how to analyze cellular markers following HDI treatment and to analyze tissue biopsies from HDI-treated patients. Finally, a protocol is provided on how to detect HDI sensitivity determinants in human cells, based on a pRetroSuper shRNA screen upon HDI treatment.

Key words Histone acetyltransferase, Histone deacetylase, HDAC inhibitor, Biomarker, shRNA screen, HDAC activity, Western Blot, Immunoprecipitation, Immunohistochemistry, Vorinostat

1 Introduction

Lysine acetylation is one of the prominent posttranslational modifications in all three domains of life. Acetylation and deacetylation of proteins at the ε-amino group of proteinogenic lysine residues are achieved by histone acetyltransferases (HATs) and their functional antagonists, the histone deacetylases (HDACs) [1]. The names HATs and HDACs have historic reasons, and it is well established that a plethora of non-histone proteins are acetylated

Sibaji Sarkar (ed.), *Histone Deacetylases: Methods and Protocols*, Methods in Molecular Biology, vol. 1436,
DOI 10.1007/978-1-4939-3667-0_19, © Springer Science+Business Media New York 2016

and deacetylated by HATs and HDACs [2, 3]. Reversible histone acetylation and deacetylation are important epigenetic regulation mechanisms, besides other posttranslational modifications (PTMs) such as histone methylation, phosphorylation, ubiquitination and sumoylation [1, 4]. The histone PTMs and their various combinations can be very complex and can be seen as a so called "histone code", which is recognized by reader proteins such as bromodomain containing proteins (BRDs). HATs are often referred to as the "writers" of acetylation and HDACs as the "erasers" of this code [5]. Therefore, histone acetylation and deacetylation have a great impact on chromatin remodeling and epigenetics. Chromatin consists of DNA which is wrapped around a histone core complex, containing two copies of the histones H2A, H2B, H3, and H4 [6]. Lysine acetylation can neutralize the positive charge of the proteinogenic lysine residues and can cause greater chromatin relaxation as the negatively charged DNA is not bound so tight. This condition is also referred to as euchromatin and enables transcription and gene expression [1, 7]. Vice versa, deacetylation of lysine residues leads to chromatin condensation, visible as heterochromatin, which is associated with repression of transcription [8].

Eighteen different human HDACs have been identified to date. They can be classified into four classes according to their homology with yeast proteins and functional criteria [9]. The four classes can be divided into Zn^{2+}-dependent classes (class I, II, and IV, *see* Table 1) and NAD-dependent classes (class III). Class I consists of the members HDAC1, 2, 3, and 8. Class II HDACs are again subdivided into class IIa consisting of the HDACs 4, 5, 7, and 9 and into class IIb with the HDACs 6 and 10, whereas class IV has only one member, HDAC11 [10]. Class III HDACs are also referred to as sirtuins according to their homology with Sir2,

Table 1
Human Zn^{2+}-dependent HDAC superfamily organized according to their sequence similarities and homologies to the respective yeast proteins. The size of the HDACs is given in amino acid residues

Class	HDAC	Size	Chromosomal locus	Cellular localization	Substrate example
I	1	482	1p34.1	Nuclear	MyoD
	2	488	6p21	Nuclear	Histone H3
	3	428	5q31	Nuclear/cytoplasmic	RelA
	8	377	Xq13	Nuclear	HoxA5
II	4	1084	2q372	Nuclear/cytoplasmic	RunX2 TF
	5	1122	17q21	Nuclear/cytoplasmic	Slit 2
	6	1215	12q13	Cytoplasmic	alpha-tubulin
	7	855	7p21-p15	Nuclear/cytoplasmic	Mef2D TF
	9	1011	Xp11.22–33	Nuclear/cytoplasmic	FoxP3
	10	669	22q13.31–33	Cytoplasmic	Hsp90
IV	11	347	3p25.2	Nuclear	Nup98

and encompasses SIRT1–SIRT7. Although this great variety of HDACs exists in human cells, their redundancy for certain substrate proteins seems to be limited [8]. HDAC1 and HDAC2 for instance share a sequence identity of 86 %, however, they have only minor overlapping functions and cannot replace each other [11].

HDACs are involved in the genesis and progression of cancer. There are several reports that increased HDAC levels are found in certain types of cancer, for instance raised levels of HDAC1 in gastric cancer and hormone refractory prostate cancer [12, 13]. Higher expression of HDAC2 and HDAC3 has been observed in colon cancer [14, 15], whereas sequencing studies identified HDAC4 mutations in breast cancer samples [16] which caused reduced HDAC4 levels.

These and other examples, together with the modulating effect on chromatin suggested that HDACs may be appropriate anticancer targets, and currently, four different HDIs are approved by the FDA for treatment of hematological malignancies. Generally, HDIs are a group of small molecules that block HDACs and therefore increase the acetylation level of cellular proteins through inhibition of HDAC activity [17].

HDIs can be classified according to their chemical structure or their specificity for HDAC sub-types or classes. Thus, if classified by structure, HDIs include hydroxamates (e.g., SAHA/vorinostat), cyclic peptides (e.g., FK228/depsipeptide/romidepsin), benzamides (e.g., MS-275/entinostat), and aliphatic acids (e.g., valproic acid) [18, 19]. Alternatively, HDIs can be classified by which class of HDAC is preferentially inhibited. For example, SAHA and TSA are regarded as pan-HDAC inhibitors, whereas tubastatin A specifically targets HDAC6, romidepsin acts on class I HDACs and panobinostat inhibits HDAC classes I and II [18, 20]. Table 2

Table 2
Examples of HDIs, chemical class, HDAC specificity, and clinical trial status

Chemical class	Example (INN)	HDAC specificity	Clinical trial stage
Hydroxamates	SAHA (vorinostat)	Pan-inhibitor	Approved for CTCL
	PXD101 (belinostat)	Pan-inhibitor	Approved for PTCL
	LBH589 (panobinostat)	Classes I and II	Approved for multiple myeloma
	ITF2357 (givinostat)	Pan-inhibitor	Phase II alone or in combination
	4SC-201 (resminostat)	Pan-inhibitor	Phase II alone or in combination
	PCI 24781	Classes I and II	Phase II alone or in combination
	CXD101	Class I	Phase I
Cyclic peptides	Depsipeptide/FK228 (romidepsin)	Class I	Approved for CTCL and PTCL
Benzamides	MS-275 (entinostat)	Class I	Phase III alone or in combination
	MGCD0103 (mocetinostat)	Class I	Phase II alone or in combination
Aliphatic fatty acids	Valproic acid	Classes I and IIa	Phase III alone or in combination
	Butyrate	Classes I and IIa	Phase II alone or in combination

illustrates how some of these compounds are used in clinical trials, both as single agents or as part of a combination therapy.

Given that HDACs deacetylate thousands of cellular proteins which are involved in many different pathways, it is conceivable that these many pathways are affected by HDIs and could contribute to their cytotoxicity. A number of HDACs that are involved in cell cycle progression and proliferation, are targeted in cancer therapy with HDI treatment causing cell cycle arrest [11, 15, 21, 22]. HDI can disrupt DNA repair in different ways, including DNA double-strand breaks (DSB) stabilization, downregulation of expression of DNA repair factors, and ROS accumulation [23]. This might explain the synergistic effects of HDIs and DNA-damaging agents.

HDIs act on the protein quality control systems in human cells, including the ubiquitin proteasome system (UPS), autophagy, and molecular chaperones [24, 25]. Especially HDAC6 plays a major role, since it can bind ubiquitinated proteins and recruits them to dynein motors for transport to aggresomes [26, 27] and has been shown to interact with the proteasomal shuttling factor HR23B [28].

HDIs also impact on cytokine signaling and the JAK-STAT pathway, which is frequently aberrant in hematological malignancies [29]. HDACs dynamically regulate STAT acetylation, and HDI treatment causes STAT hyperacetylation which has effects on tumor cell survival, suggesting that cytokine signaling pathways are important targets of HDI treatment [30]. The HDAC6-specific HDI tubastatin showed significant inhibition of the inhibitory cytokines IL-6 and TNF in human macrophages [31].

In order to streamline treatment with HDIs it would be valuable to have predictive biomarkers. To determine protein levels upon HDI treatment and the amount of acetylated marker proteins, cell extraction and isolation with subsequent immunodetection methods can be applied, as described in this chapter. Also, immunohistochemistry of patient samples can be an approach to monitor the outcome of HDI treatment (e.g., by staining tissue biopsies before and after treatment with an HDI). The search for HDI sensitivity factors is a further approach which might lead to the discovery of new mechanisms through which HDIs act and rational combination treatments with other chemotherapeutic agents (*see* Fig. 1).

To analyze HDAC activity and the potency of HDIs, single reagent luminescent assays can be applied that measure the relative activity of HDAC enzymes in intact cells, extracts or purified enzyme sources. These types of assays use an acetylated, live-cell-permeant, luminogenic peptide which serves as a substrate and can be deacetylated by HDACs. The deacetylation reaction is measured using a coupled enzymatic system in which a protease cleaves the peptide from luciferin, which is processed itself and quantified with luciferase.

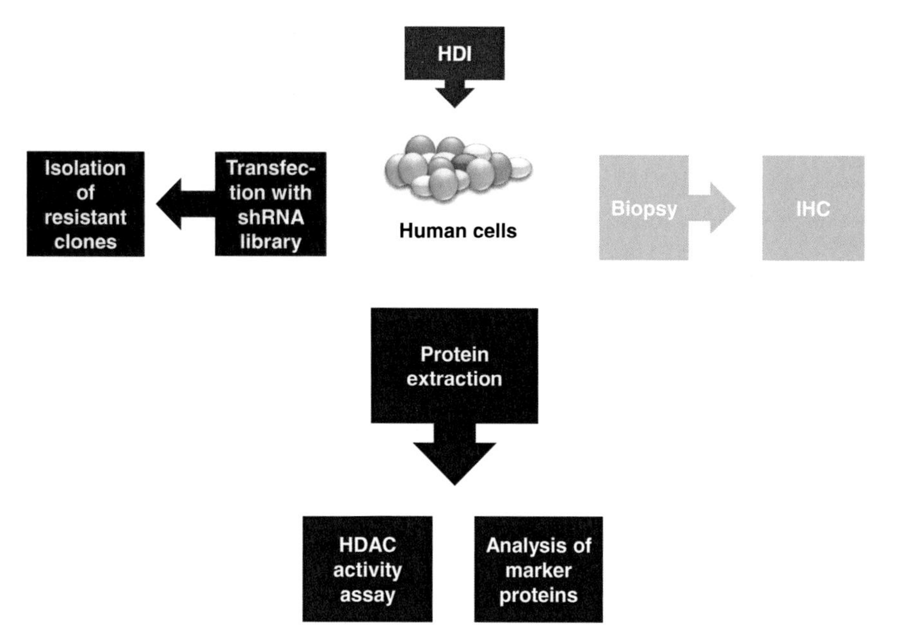

Fig. 1 Different approaches to analyze HDAC inhibition and its consequences on human cells. After treatment of human cells with HDIs, a screen can be applied in order to detect HDI-sensitive/resistant genes and proteins. For this purpose, the cells are transfected with a shRNA library and treated with the respective HDAC inhibitor. Surviving cell clones are selected and the underlying DNA is isolated to identify the respective gene and therefore protein which is an HDAC sensitivity determinant. Upon cell extraction, HDAC activity can be determined in vivo, and immunological detection methods, for instance immunochemistry and Western Blot, can be applied to determine the level of histone acetylation, acetylation of other marker proteins or the level of apoptosis caused by HDIs. Biopsies from patients treated with HDIs can be paraffin-embedded and analyzed for biomarkers and other proteins which show prognostic or predictive value upon HDI treatment

The HDAC-dependent (and also HDI-dependent, if present) signal is proportional to the HDAC activity [32, 33] (*see* Fig. 2).

The amount of acetylation of different marker proteins upon HDI application can be measured after protein extraction from cells with subsequent immunoblotting, such as acetylated histone 3. Immunoprecipitation (IP) is a technique that is used to determine the molecular weight of protein antigens, in this context especially of their acetylated species, study protein–protein interactions, monitor the acetylation status and most importantly, enrich low abundant proteins [34] (*see* Fig. 3).

Immunohistochemistry (IHC) uses antibodies to detect specific cellular antigens in tissues, for example clinical patient samples. This method allows the evaluation of antigen expression in the context of tissue architecture and cell morphology at the same time. It is therefore mostly used in the diagnosis and classification of malignancies, but the identification of cancer-specific biomarkers has also made it possible to use IHC as a tool for identifying therapeutic targets and predicting response to a given drug

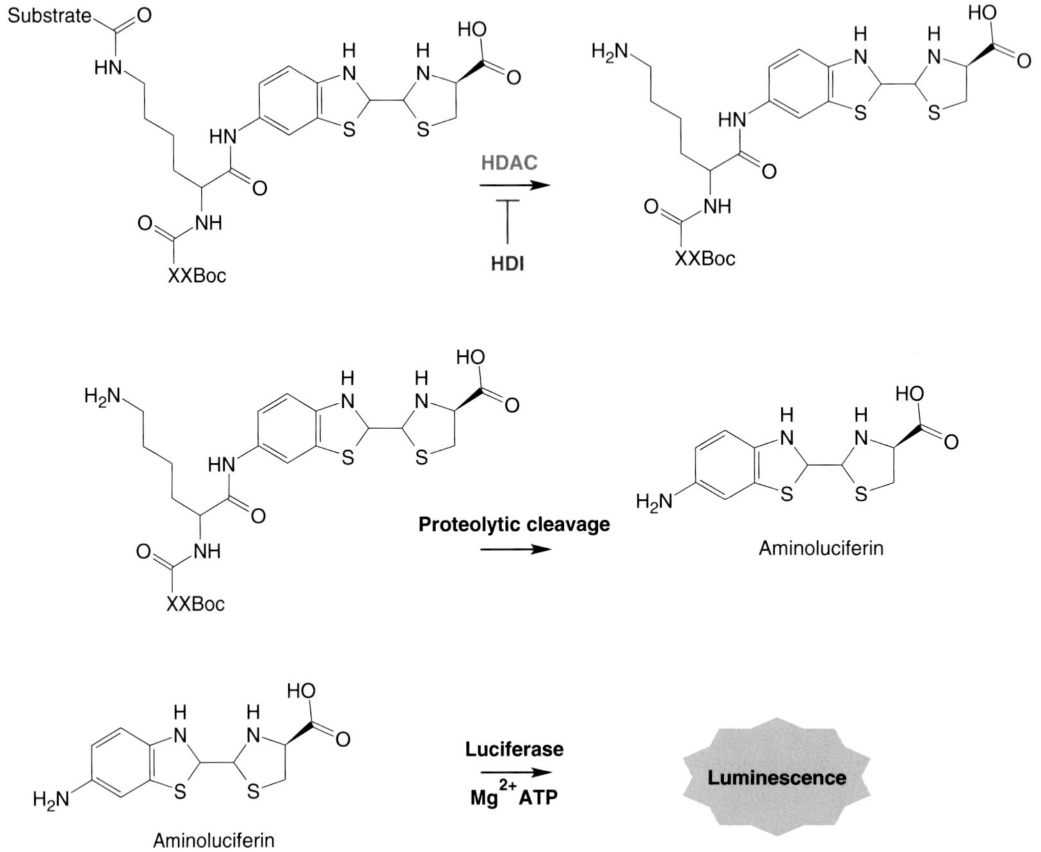

Fig. 2 HDAC activity assay to determine the potency of HDAC inhibitors on cell extracts or intact cells. HDACs deacetylate the luminogenic HDAC substrate, enabling the peptide to be cleaved by specific proteases, mediated by added reagent, and liberate aminoluciferin. Aminoluciferin is then enzymatically processed by luciferase and produces stable and persistent light which can be measured. *Boc* represents an amino-terminal blocking group, protecting the substrate from nonspecific cleavage and *XX* represents an in length optimized linker molecule between the protective Boc group and the actual cleavable unit

therapy or prognosis. One example is the identification of HR23B as a marker of response to HDIs in patients with cutaneous T-cell lymphoma and hepatocellular carcinoma: high expression of HR23B in clinical patient samples pre-therapy was correlated with a response to HDIs and subsequent disease stabilization, whereas low HR23B expression predicted a poor response to therapy with an HDI [35, 36]. Most antibodies that are used in IHC are monoclonal with a unique specificity for a given antigen. An enzyme label on the secondary antibody allows conversion of a chromogenic substrate which indicates that antibody-antigen binding has taken place. Horseradish peroxidase (HRP) is a

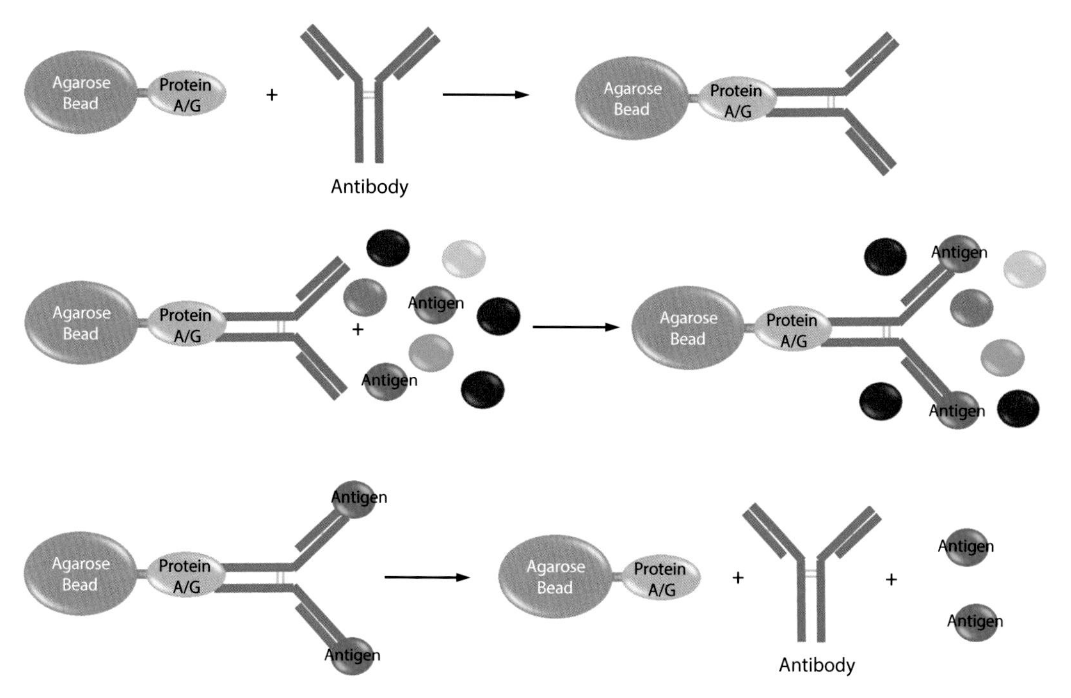

Fig. 3 Schematic representation of an immunoprecipitation (IP). Antibodies against the antigen or protein of interest are bound to protein A or G, which is itself linked to agarose beads. The antigen is bound to the antibody, whereas most other proteins are washed away under stringent washing conditions. Then, only the antigen-antibody-protein A/G-bead complex is present. The antigen can be eluted and analyzed, e.g. for acetylation by Western Blot

commonly used enzyme, and 3,3′-diaminobenzidine (DAB) is usually the chromogenic substrate which gives a brown color reaction. For optimal results, a clinical sample of adequate size and fixation is required, and a histopathologist trained in the interpretation of IHC samples should be involved in validating histoscores on stained tissue samples.

The method for stable expression of short interfering RNAs in mammalian cells was developed in 2002 [37]. It involves the use of an shRNA library targeting ~8000 human genes containing three different shRNA constructs for each gene. Strong suppression of particular gene expression can be achieved via siRNAs produced from the shRNAs which can be induced by retroviral integration into the host cell genome. Stable expression of siRNAs using the pRetroSuper vector mediates selective suppression of gene expression over prolonged periods of time. This allows analysis of loss-of-function phenotypes in long term assays. This screen was used to identify some of the genes responsible for mediating susceptibility to suberoylanilide hydroxamic acid (SAHA)-mediated apoptosis (*see* Fig. 4) [38].

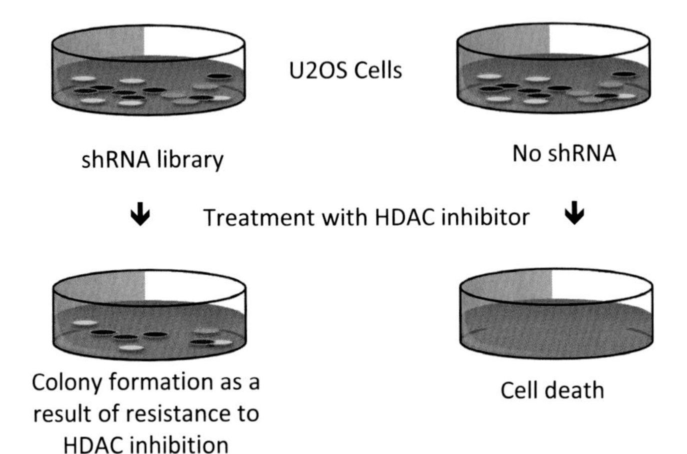

Fig. 4 Schematic presentation of shRNA screen to analyze HDI sensitivity determinants. U2OS cells were transfected with a library targeting a large number of genes involved in human cancer alongside an untransfected control. The cells were then treated with the pan-HDI SAHA at 2 μM and allowed to grow for 25 days. The colonies which survived where those in which a potential SAHA sensitivity determinant gene was silenced

2 Materials

2.1 HDAC Activity Assay

1. HDAC assay buffer: 25 mM Tris–HCl, pH 8.0, 150 mM NaCl, 3 mM KCl, 1 mM $MgCl_2$ thereafter called solution I (*see* **Note 1**).
2. HDAC substrate: 4 mM acetylated fluorometric substrate in dimethylsulfoxide thereafter called solution II (*see* **Note 1**).
3. HDAC developer thereafter called solution III (*see* **Note 1**).
4. Trichostatin A at a concentration of 10 mM in DMSO.
5. Nuclear extracts from e.g., HeLa cells at a concentration of 5 mg/ml.
6. 96-well microplates, white-walled, tissue culture plates, compatible with luminometer (*see* **Note 2**).
7. Luminescence meter.
8. Multichannel pipette.
9. Reagent reservoirs.
10. Orbital shaker.
11. 20% Triton X-100 prepared using ultrapure water.
12. HDAC enzyme source, for instance intact cells, cell extracts or purified HDAC enzyme.
13. Multiparameter, iterative, nonlinear regression compatible program, for instance SigmaPlot.

2.2 Analysis of Acetylated Marker Proteins Upon HDAC Inhibition in Human Cells

2.2.1 Cell Culture

1. Phosphate-buffered saline (PBS) without calcium and magnesium, pH 7.2 (*see* **Note 8**).
2. Dulbecco's modified Eagle's medium (DMEM, containing 10 % fetal bovine serum, and 1 % penicillin–streptomycin used for culture of different human cell lines).
3. Trypsin–EDTA solution: 0.05 % trypsin and 0.53 mM EDTA.
4. Human cells and cell lines can be purchased from the American Type Culture Collection http://www.lgcstandards-atcc.org/.

2.2.2 Drug Treatment

1. HDAC inhibitors: SAHA (suberoylanilide hydroxamic acid, vorinostat, TSA (7-[4-(dimethylamino) phenyl]-N-hydroxy-4,6-dimethyl-7-oxo-2,4-heptadienamide belinostat (PXD101) (*see* **Note 10**).

2.2.3 Cell Lysis

1. Phosphate-buffered saline (PBS) without calcium and magnesium, pH 7.2.
2. Lysis buffer: 50 mM Tris at pH 8, 120 mM NaCl, 0.5 % NP-40, 0.5 % Triton X-100, 1 mM DTT, and protease inhibitors and SAHA (*see* **Note 11**).

2.2.4 Protein Concentration Determination by Bradford Assay

1. Bradford reagent.
2. BSA protein standard solution.
3. 96-well plate.
4. Plate reader for absorbance in VIS.

2.2.5 Immunoprecipitation

1. Lysis buffer: 50 mM Tris at pH 8, 120 mM NaCl, 0.5 % NP-40, 0.5 % Triton X-100, 1 mM DTT, and protease inhibitors and SAHA.
2. Protein sepharose A or G beads.
3. Rotator.
4. Primary antibodies, for instance anti-HDAC 6 anti-histone 3 (Cell Signalling), anti-acetyl-histone 3.
5. 2× SDS-loading buffer: 0.5 M Tris–HCl pH 6.8, 10% SDS, 50 % glycerol, 0.01 % bromophenol blue.
6. DTT.

2.2.6 SDS-PAGE and Western Blot

1. 2× SDS-loading buffer: 0.5 M Tris–HCl pH 6.8, 10% SDS, 50 % glycerol, 0.01 % bromophenol blue, 5 % DTT.
2. SDS-PAGEs according to appropriate protocols or pre-casted gels.
3. Ponceau S.
4. Nonfat dry milk powder.
5. Tris-buffered saline.

6. Tris-buffered saline with 0.5 % Tween (TBST).
7. Primary antibodies, e.g., anti-HDAC 6, anti-histone 3, anti-acetyl-histone 3 anti-cleaved caspase 3 anti-PARP.
8. Secondary antibodies conjugated to HRP.
9. ECL reagents.
10. Developer.

2.3 IHC to Determine Protein Levels in Individual Cells Upon HDI Treatment

1. Histo-Clear.
2. Ethanol: 100 %, 70 %; 400 ml each.
3. Antibody retrieval solution: EDTA buffer (1 mM EDTA, 0.05 % Tween 20, pH 8.0), or sodium citrate buffer (10 mM Sodium Citrate, 0.05 % Tween 20, pH 6.0).
4. ImmEdge hydrophobic barrier pen.
5. 6 % H_2O_2 in methanol.
6. Phosphate buffered saline (PBS) without calcium and magnesium, pH 7.2 (*see* **Note 8**).
7. PBST (PBS with 0.5 % Tween 20).
8. Vectastain ABC kit (for mouse and/or rabbit, depending on antibodies used).
9. Primary antibodies e.g., anti-acetylated H3 (mouse).
10. Secondary antibodies: mouse 1 mg, rabbit 1 mg, goat 10 mg.
11. DAB substrate kit.
12. Mayer's hematoxylin.
13. Aquatex.
14. Glass coverslips.
15. Light microscope with 20× and 40× magnification.

2.4 shRNA Screen for the Identification of HDI Sensitivity Determinants

2.4.1 Cell Culture

1. Phosphate-buffered saline (PBS) without calcium and magnesium, pH 7.2 (*see* **Note 8**).
2. Dulbecco's modified Eagle's medium (DMEM) containing 10 % fetal bovine serum, and 1 % penicillin–streptomycin used for culture of different human cell lines.
3. Trypsin–EDTA solution: 0.05 % trypsin and 0.53 mM EDTA.
4. U2OS cells and U2OS cells expressing the murine ectopic receptor (U2OSEcR).
5. Human cells and cell lines can be purchased from the American Type Culture Collection: http://www.lgcstandards-atcc.org/.

2.4.2 Drug Treatment

1. HDAC inhibitors: SAHA (suberoylanilide hydroxamic acid, vorinostat).

2.4.3 shRNA (in pRetroSuper) Library and PCR of Genomic DNA

1. shRNA (in pRetroSuper) library available from [37] (*see* **Note 31**).
2. Lysis buffer (100 mM Tris pH 8.5, 0.2 % SDS, 200 mM NaCl, and 100 μg/ml proteinase K).
3. Expand Long Template PCR system.
4. Standard Thermocycler.
5. PCR reaction tubes.
6. Agarose.
7. Ethidium bromide.

3 Methods

3.1 HDAC Activity Assay

3.1.1 Determination of the Linear Range of HDAC Activity

1. Prepare a dilution of HDAC using buffer I. If using nuclear extracts, dilute the cell extract 1:3000 in buffer I. If the source is purified HDAC protein, dilute the enzyme to a final concentration of 1–5 μg/ml in buffer I (*see* also **Note 3**). If cells are used, they should be diluted in either serum-free medium or buffer I to a concentration of 1×10^5 cells/ml (*see* **Note 4**).
2. Prepare a series of twofold dilutions of the HDAC enzyme source in buffer I in rows A-D of the specified 96-well plate along the rows, the last dilution would then be in wells A11, B11, C11, and D11 1:1014, wells in column 12 are enzyme free controls. The final volume of the diluted enzyme in each well should be 100 μl for 96-well plates.
3. Equilibrate solutions I, II, and III to room temperature (*see* **Note 5**).
4. Prepare the assay reagent with solution III in a 1:1000 dilution in solution II.
5. Add equal volumes of the assay reagent to each assay well (100 μl for a 96-well plate).
6. Mix the plate at room temperature for 1 min using an orbital shaker at 500–700 rpm. Incubate the plate then for further 30 min at room temperature.
7. Measurement of the luminescence has to be carried out at a steady state signal (30 min after incubation).
8. Plot the data using a program compatible with different regression modes, such as SigmaPlot. On the *x*-axis, the concentration of HDAC enzyme or protein concentration should be plotted, e.g., in ng/ml in logarithmic scale, whereas on the *y*-axis the signal-to-noise ratio should be plotted in arbitrary units, but also in logarithmic scale. This should result in a line with positive slope. Data should represent the mean plus/minus standard deviation of the four replicates.

3.1.2 Determination of HDAC Inhibitor Potency

1. Dilute the unknown compound with potential HDI ability and the known HDAC inhibitor trichostatin A in buffer I. The dilution should be twofold in four technical replicates in a specified 96-well plate. The final volume in each well should be 50 μl. In rows A-D the unknown test compound is diluted to column 10 to a 1:512 dilution. In rows E–H trichostatin A with known concentration between 50 nM (column 1) and 0.098 nM (column 10) is diluted. Buffer I without inhibitor but with HDAC is added to column 11 to serve as a positive control, whereas neither inhibitor nor enzyme is added to column 12 to serve as a negative control.
2. Dilute the HDAC enzyme source in buffer I to a concentration, which was determined in Subheading 3.1.1, being within the linear range determined (*see* **Note 6**).
3. Add 50 μl of HDAC enzyme dilution to each well of the diluted inhibitors from **step 1** and non-inhibitor controls (column 11). Add buffer I without HDAC to column 12.
4. Agitate the 96-well plate at room temperature for 1 min by shaking in an orbital shaker at 500–700 rpm.
5. Incubate the plate at room temperature for at least 30 min, but not longer than 2 h.
6. Dilute solution III 1:1000 in solution II to produce the HDAC assay reagent. Mix and equilibrate to room temperature (*see* **Note 7**).
7. Add an equal volume of HDAC assay reagent to each well (100 μl).
8. Agitate the plate at room temperature for 1 min using an orbital shaker at 500–700 rpm. Incubate then at room temperature for 30 min.
9. Place the plate in a luminometer and measure the luminescence at signal steady-state 30 min after adding the HDAC assay reagent.
10. Plot the data using a program compatible with different regression modes, such as SigmaPlot. On the *x*-axis, the concentration of the inhibitor should be plotted, e.g., in μM or nM in logarithmic scale, whereas on the *y*-axis the relative luminescence units (RLU) should be plotted in linear scale. Use nonlinear regression to fit the data to the log (inhibitor) vs. response (variable slope) curve (*see* Fig. 5). Data should display the mean plus/minus standard deviation of the four replicates. The IC50 represents the concentration of inhibitor, where the half-maximal response is obtained.

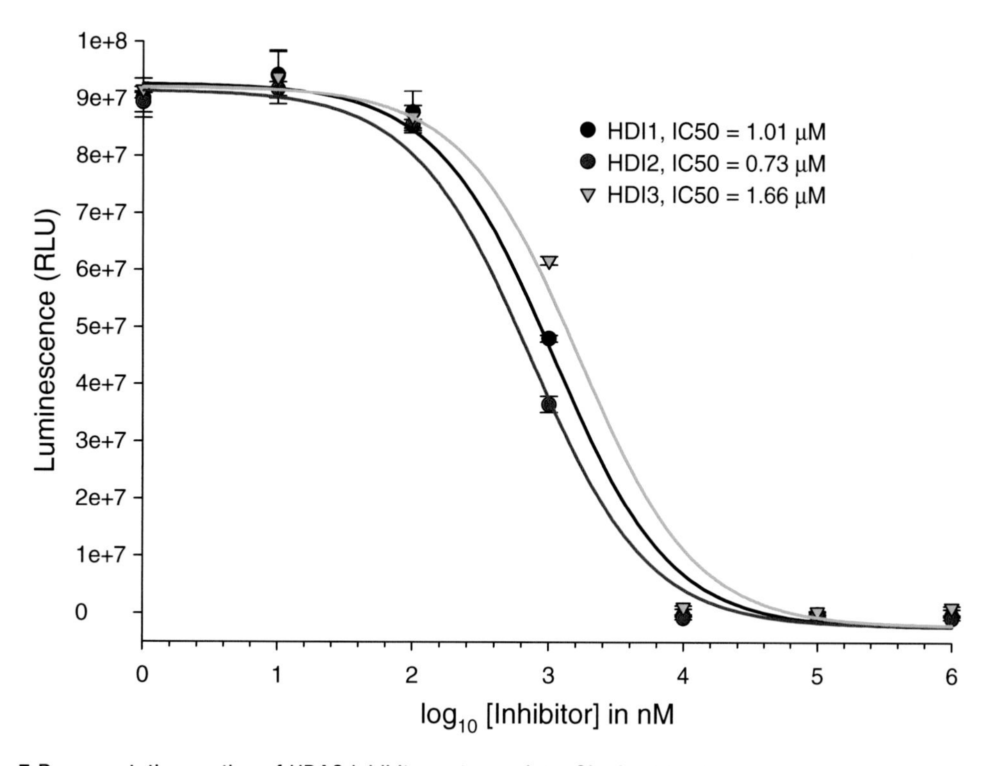

Fig. 5 Representative section of HDAC inhibitor potency data. Obtained luminescence values were averaged and plotted against the inhibitor concentration (log scale). After a nonlinear data fit in SigmaPlot, IC50s of the three different compounds were determined, displayed in the graph

3.2 Analysis of Acetylated Marker Proteins Upon HDAC Inhibition in Human Cells

3.2.1 Cell Culture

1. Use human cells of different cell lines (e.g., HeLa, HEK293T, U2OS) cultured at 37 °C in a humidified 5% CO_2 incubator in DMEM containing 10% FCS and 1% penicillin–streptomycin. In order to split the cells, wash them once with 1× PBS and treat them then with trypsin (*see* **Note 9**).

3.2.2 Drug Treatments

1. Treat cells were with the respective HDIs, for instance trichostatin A (TSA), suberoylanilide hydroxamic acid (SAHA) and tubastatin at the indicated concentrations and for the indicated times before harvesting. All compounds are solubilised in dimethylsulfoxide (DMSO) and added to media at a 1:1000 dilution to minimize vehicle effects.

3.2.3 Cell Lysis

1. Wash cells with ice-cold 1× PBS, remove the washing solution.
2. Add again ice-cold PBS and scrape the cells from the bottom of the dish.
3. Transfer the cell/PBS suspension in an Eppendorf tube and centrifuge for 2.5 min at 500 ×*g*, 4 °C.
4. Remove and discard the supernatant.

5. Add lysis buffer to the cells (*see* **Note 12**) and put them on ice for 30–60 min (*see* **Note 13**).
6. Centrifuge the lysate in order to remove cell debris at 15,000 × g for 10 min, 4 °C.
7. Remove the supernatant and transfer it in a new Eppendorf tube. Discard the cell pellet.
8. Determine the protein concentration (*see* separate Subheading 3.2.4).
9. Freeze the lysate for later analysis in −70 °C. If used for immunoprecipitation, the cell lysate should not be frozen and used immediately (*see* **Note 14**).

3.2.4 Protein Concentration Determination by Bradford Assay

1. 1 ml of the Bradford reagent is filled per Eppendorf tube, ten in total for the determination of the protein concentration range. In addition, fill 1 ml Bradford reagent in Eppendorf tubes, as many as samples have to be measured plus one additional for the lysis buffer which serves as a negative control (*see* **Note 15**).
2. Add the following amount of BSA stock solution (1 mg/ml) to the Eppendorf tubes: 0, 1, 2.5, 5, 10, 20, and 40 μl to get final concentrations of 1, 2.5, 5, 10, 20, and 40 μg/ml for the calibration curve.
3. Add 3–5 μl of the protein samples to the Eppendorf tubes with 1 ml Bradford solution, depending on the concentration (*see* **Note 16**).
4. Add 3–5 ml (the same volume as for the protein samples under **step 3** of Subheading 3.2.4) of lysis buffer to one Eppendorf tube with 1 ml Bradford solution content (serves as a negative control).
5. Mix the content of all Eppendorf tubes well, pipet 200 μl into a 96-well plate and measure with a wavelength of 595 nm in an appropriate plate reader.
6. Calculate the protein concentration of the samples from the calibration curve (*see* **Note 17**).

3.2.5 Immunoprecipitation

1. 20 μl of a 50% Protein A or G sepharose (*see* **Note 18**) per sample is washed twice in 1× PBS. Add 1 ml PBS to the sepharose beads, centrifuge it for 1 min at 14,000 × g at 4 °C. Then remove the supernatant from the pelleted sepharose and repeat the washing (*see* **Note 19**).
2. Dilute 10 μg of the respective primary antibody in 200 μl lysis buffer and add it to the sepharose. Rotate for 3–4 h at 4 °C to bind the antibody to the protein A or G (*see* **Note 20**).
3. Wash the solution three times as described in Subheading 3.2.5, **step 1**.

4. After the last washing step, discard the supernatant and add similar protein amounts (may be different volumes) of the respective protein lysates from Subheading 3.2.3, protein concentration determined according to Subheading 3.2.4.
5. Rotate overnight in 4 °C (*see* **Note 21**).
6. Wash the sepharose beads with bound protein A or G, bound primary antibody and the protein of interest (antigen) four times in lysis buffer.
7. Elute the proteins with SDS-containing 2× SDS-loading buffer: Add 2× of the 2× buffer without reducing agent to the beads (e.g., 20 μl if 20 μl bead suspension was used) (*see* **Note 22**).
8. Heat the SDS-loading buffer solution with the beads for 5 min at 95 °C, vortex then rigorously after the heating and spin down quickly.
9. Transfer the supernatant with the eluted protein to a fresh Eppendorf tube and use similar volumes for the subsequent SDS-PAGE and Western Blot analysis. DTT can be added now in a 1:20 dilution.

3.2.6 SDS-PAGE and Western Blot

1. Dilute equal amounts of proteins in 2× SDS loading buffer and boil at 95 °C for 5 min or use the samples obtained from the immunoprecipitation as carried out in Subheading 3.2.5.
2. Separate nuclear extract or histones on SDS-PAGE gels and transfer them onto nitrocellulose membrane according to appropriate protocols [39].
3. Visualize the blotted protein amount via Ponceau S staining and scan the membrane in to ensure equal loading of the gel and sufficient transfer of the proteins.
4. Block the membrane in TBS with 5% nonfat dry milk for 1 h at room temperature.
5. Add the primary antibodies in TBS with 5% nonfat dry milk (*see* **Notes 23** and **24**) and incubate overnight at 4 °C.
6. Wash the membrane with TBS on a shaker with at 40–50 rpm for 3 × 10 min.
7. Incubate blot with secondary antibody conjugated to HRP at a concentration of 1:3000 in TBS containing 5% nonfat dry milk at room temperature for 2 h.
8. Wash the blot two times for 10 min in TBST and once for 10 min in 1× TBS.
9. Visualize the bands of the respective proteins with the ECL kit (*see* Fig. 6).

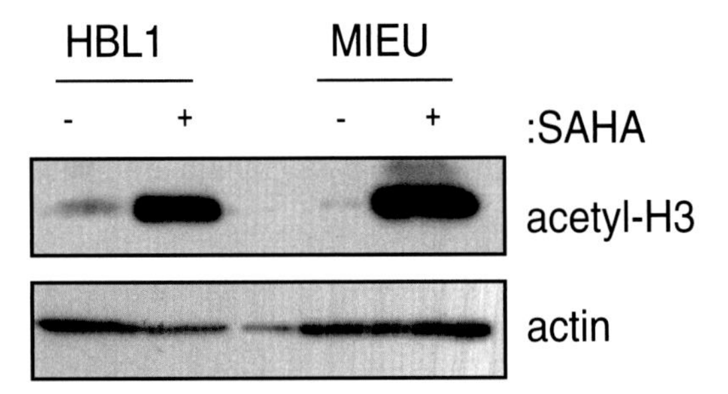

Fig. 6 Example of human cell protein extract treated with a pan-HDI. Lymphoma cell lines HBL-1 and MIEU were treated with SAHA at a concentration of 5 μM. The Western Blot shows an increase in acetylation following treatment with SAHA. Actin served as the loading control

3.3 IHC to Determine Protein Levels in Individual Cells Upon HDI Treatment

1. Switch on water bath and set the temperature to 99 °C (*see* **Note 25**).
2. Fill water bath container rack with antigen retrieval solution (EDTA or sodium citrate) and place into the water bath.
3. Label sample slides with pencil (*see* **Note 26**).
4. Put slides into slide rack and immerse completely in Histo-Clear for 5 min.
5. Immediately transfer slide rack with slides into 100% ethanol for 3 min, followed by 3 min in 70% ethanol and another 3 min in fresh solution of 70% ethanol (*see* **Note 27**).
6. Rinse slides in tap water for 5 min.
7. Remove water bath container slide rack and place slides into this. Then replace this in the water bath container which is already filled with the antigen retrieval solution.
8. Proceed with antigen retrieval and leave slides at 99 °C for 20 min. Then remove water bath container and leave on bench to cool for approximately 1 h.
9. Remove slides from container and draw around tissue with ImmEdge pen (*see* **Note 28**).
10. Wash slides in distilled water three times, for 2 min each time.
11. Add a few drops of freshly prepared 6% H_2O_2/methanol solution to each slide and leave for 15 min at room temperature.
12. Wash slides in tap water for 1 min.
13. Wash slides in PBST for 5 min.
14. Add 2–3 drops of diluted normal blocking serum (make up fresh in PBS) to each slide (*see* **Note 29**). Leave for 20 min.
15. Replace slides into slide rack and wash in PBST for 5 min.
16. Dilute primary antibody to correct concentration in PBS (*see* **Note 30**).
17. Tap excess liquid off slides and place onto slide stain tray.

18. Add one or two drops of primary antibody solution (usually, depending on size of tissue section, 400 μL are needed) so that the entire tissue section encircled by ImmEdge pen is completely covered.
19. Add a small amount of water to the base of the slide stain tray (to prevent drying out of slides), cover tray with lid and incubate the slides at 4 °C over night.
20. The next day, tap off primary antibody solution. Wash slides in PBST for 5 min.
21. Prepare biotinylated secondary antibody using the appropriate reagent kit (three drops normal blocking serum, one drop secondary antibody serum in 5–10 ml of PBST) and at the same time ABC reagent (two drops reagent "A" and two drops reagent "B" in 5 ml PBST)—leave at room temperature for 30 min before adding (*see* **step 24** below).
22. Add secondary antibody to washed slides as before and leave at room temperature for approximately 30 min.
23. Wash slides in PBST for 5 min.
24. Add a few drops of ABC reagent to slides after tapping off excess liquid and leave for 30 min.
25. Wash slides in PBST for 5 min.
26. Make up DAB substrate: four drops DAB, two drops H_2O_2 (both supplied in DAB substrate kit) in 5 ml distilled water. Mix well.
27. Place slides on a white background (to be able to observe color change) and add DAB substrate to each slide. Allow 10 min for incubation on every slide (this should be uniform across all slides so as to allow a comparison of the depth of staining). A change in color in the tissue sections will become apparent.
28. After 10 min, rinse slides in distilled water and place in rack for counterstaining.
29. Counterstain with hematoxylin for 60 s.
30. Rinse slides thoroughly with tap water and tap off excess liquid.
31. Mount coverslips with Aquatex and allow drying for a minimum of 20 min.
32. Place specimens under the microscope, record pictures and score slides. Each slide is scored at the same magnification, for instance 40×.

3.4 shRNA Screen for the Identification of HDI Sensitivity Determinants

1. Infect U2OS cells expressing the murine ectopic receptor (U2OSEcR) with the 83 viral pools where one pool of the library contains 96 siRNAs per well, overnight, to allow viral infection (*see* **Note 32**).
2. Infect U2OS cells with virus containing pLibGFP and GFP levels (*see* **Note 33**).

3. Allow cells to recover for 72 h, keep them at 37 °C in the incubator (*see* **Note 34**).
4. Plate 40000 of the recovered cells overnight, and then add 2 μM of the pan-HDI SAHA to each plate (*see* **Note 35**).
5. Replace SAHA containing media every 3 days for 18-30 days until the appearance of colonies on plates treated with SAHA and viral library.
6. Pick and expand colonies to allow isolation of total genomic DNA and total protein (*see* below).
7. Store cells at −80 °C to allow maintenance and future use of these stable cell lines.
8. Lyse cells using lysis buffer in order to obtain genomic DNA and incubate at 37 °C for 30 min with shaking to allow DNA to precipitate (*see* **Note 36**).
9. Add one volume of isopropanol to the lysate and dissolve DNA precipitate in 10 mM Tris pH 7.5 to allow it to be used in PCR to determine the identity of the gene in question.
10. Recover the genomic insert by PCR using the primers: pRS forward: 5′-CCCTTGAACCTCCTCGTTCGACC-3′ and pRS reverse: 5′-GAGACGTGCTACTTCCATTTGTC-3′, using the Expand Long Template PCR system (Roche) to amplify 200 ng/μl genomic DNA.
11. Prepare the reaction mixture to contain PCR mix A (5 μl buffer 3, 2 μl MgCl2 (50 μM), 16.5 μl H_2O, 1.5 μl enzyme mix), reaction mix B (10 μM pRS forward, 10 μM pRS reverse 10 μM dNTPs, to 20 μl with water), and 2 μl genomic DNA (400 ng); combine in PCR reaction tubes.
12. Carry on the reaction as following: cycle 1 (10 min at 94 °C), cycle 2 (20 s at 94 °C, 30 s at 62 °C, 90 s at 68 °C; 30 times), and cycle 3 (10 min at 68 °C).
13. Analyze each PCR on a 1.2% 1× TBE/agarose gel containing ethidium bromide.
14. Sequence the PCR product then using the primer pRS SEQ (5′- GGAAGCCTTGGCTTTTG-3′).
15. Identify the gene of interest via a BLAST search (http://blast.ncbi.nlm.nih.gov/Blast.cgi) of the shRNA insert identified from the PCR.

4 Notes

1. Store HDAC assay components and substrates at −20 °C. Store the cell extracts and nuclear extracts at −70 °C. For best performance, the assay buffer, substrate and developer should be

used on the day prepared. If the reagents cannot be used immediately to perform the assay, it should be stored on ice until use. Then equilibrate all reagents to room temperature before use.

2. Different compositions of the plastic 96-well plates can affect the luminescence values.
3. Different HDACs may vary in their specific activities; therefore, the determination of the linear range may also vary and has to be determined before the potency of the enzymes is analyzed.
4. Different sera used to supplement tissue culture medium can contain detectable levels of HDAC activity, which could influence the measurements. Therefore, it is not recommended to use serum-supplemented medium for the assay. Also, phenol red-containing medium may affect the luminescence by quenching the signal and it is recommended to use phenol red-free media.
5. Centrifuge the solutions briefly after they are completely thawed.
6. Cellular or nuclear extracts containing HDACs should be diluted 1:3000 and/or adjusted to the linear range determined in Subheading 3.1.1.
7. Buffer II cannot be used immediately to assay HDAC activity, it should be stored on ice until use and then shortly before use equilibrated to room temperature.
8. Calcium and magnesium ions can interfere with the EDTA in trypsin–EDTA solution.
9. The time for treatment with trypsin–EDTA varies in the different cell lines. We found it useful to monitor the detachment process under the binocular.
10. Compounds solubilized in DMSO should be stored at −20 °C, and repeated freeze–thaw cycles should be avoided.
11. To determine acetylation on proteins, it is crucial to have a pan-HDAC inhibitor present in the lysis buffer, to maintain this PTM.
12. The amount of lysis buffer is also dependent how many cells to lyse, for one million cells use for example 100 μl of lysis buffer.
13. Also time for lysis depends on the purpose, if for example nuclei are to be isolated, the lysis time should not exceed 30 min.
14. If the cell lysate is frozen, there is the risk that weak protein–protein interactions fall apart.

15. It is crucial to have the lysis buffer with the inhibitors used as a negative control, since some protein or peptide based inhibitors can give a strong background signal.
16. The emerging blue color should not be more intensive than the one from the highest concentrated samples of the calibration curve, in order to avoid out-of-range values.
17. Calculate the linear function from the calibration curve in Excel and set in the measured values, after deduction of the negative control (Bradford reagent with lysis buffer).
18. Whether to use protein A or G sepharose is dependent on the primary antibody, which should bind to protein A or G. The affinity of the primary antibody depends on the Fc part of the respective antibody, the class of the antibody (for instance IgG or IgM) and the respective species (for instance differences in mouse and human). The table from NEB gives a good overview:

https://www.neb.com/tools-and-resources/selection-charts/affinity-of-protein-ag-for-igg-types-from-different-species.

19. Washing is important to remove ethanol or other preservatives which might affect the native binding of the proteins in the lysate.
20. We found this amount sufficient for most IPs; however, the optimal amount must be determined by the experimenter.
21. The incubation period of the protein lysate with the primary antibodies bound to the beads can be varied, dependent how much background is obtained from 1 h to overnight.
22. In order not to break the disulfide bonds of the antibodies, reducing agents are only then added, when the beads and bound antibodies are removed.
23. Start with a dilution which is recommended by the vendor of the respective antibody. If the background is too high, then reduce the concentration or the incubation time of the primary and secondary antibody.
24. To analyze protein acetylation in general and upon treatment with HDIs, a pan-lysine acetylation antibody can give a good overview of the overall acetylation events in cells. Antibodies against specific lysine residues on specific proteins can be useful to determine the influence of different HDACs and different HDIs, such as acetylated histone 3. Cleaved caspase 3 or cleaved PARP could give an indication, whether the respective HDI works and has cytotoxic effects on cells.
25. Heating the water bath can take up to an hour. While this reaches preset temperature, prepare slides.
26. The information needed is usually the date of procedure, which antibody is used, at what concentration/dilution and which patient or cohort.

27. Make sure the slides are immersed completely each time.
28. Make sure to encircle tissue completely, thus creating a hydrophobic barrier which is essential for uniform tissue staining.
29. Take care to use the correct blocking serum depending on the type of primary/secondary antibody used and make sure that tissue section is completely covered.
30. Make sure to include appropriate positive and negative controls where available.
31. Viral pools are produced by transfection of Phoenix packaging cells with the pRetroSuper NKI shRNA library using calcium phosphate precipitation.
32. Test several different concentrations of cells (from 5000 to 100,000 cells per 10 cm plate) to determine the optimal density which would decrease the chances of multiple viral infections in a single cell.
33. These constructs can be visualized by microscopy in order to ascertain the efficiency of viral infection.
34. This is to allow for siRNA expression and reduction of expression of target genes.
35. Treat cells at this optimal density with a various low concentrations of SAHA to determine a concentration of SAHA which would allow the cells to survive for a short time (about 5 days) before causing apoptosis in treated cells.
36. Volume of lysis buffer has to be adjusted to the number of cells, e.g., a count of one million cells corresponds to 200 μl of lysis buffer.

Acknowledgements

We thank the MRC, CRUK (Programme Grant C300/A13058), Rosetrees Trust, and the 7th European Framework Programme with Eurocan. Heidi Olzscha was supported by an EMBO long-term fellowship and by CRUK, Semira Sheikh by an OCRC fellowship, and Mina Elisha Bekheet by a Yossef Jameel Fellowship.

References

1. Inche AG, La Thangue NB (2006) Chromatin control and cancer-drug discovery: realizing the promise. Drug Discov Today 11(3-4):97–109
2. Choudhary C, Kumar C, Gnad F, Nielsen ML, Rehman M, Walther TC, Olsen JV, Mann M (2009) Lysine acetylation targets protein complexes and co-regulates major cellular functions. Science 325(5942):834–840
3. Johnstone RW, Licht JD (2003) Histone deacetylase inhibitors in cancer therapy: is transcription the primary target? Cancer Cell 4(1):13–18

4. Bannister AJ, Kouzarides T (2011) Regulation of chromatin by histone modifications. Cell Res 21(3):381–395
5. Strahl BD, Allis CD (2000) The language of covalent histone modifications. Nature 403(6765):41–45
6. Khorasanizadeh S (2004) The nucleosome: from genomic organization to genomic regulation. Cell 116(2):259–272
7. Kouzarides T (2007) Chromatin modifications and their function. Cell 128(4):693–705
8. Haberland M, Montgomery RL, Olson EN (2009) The many roles of histone deacetylases in development and physiology: implications for disease and therapy. Nat Rev Genet 10(1):32–42
9. Trapp J, Jung M (2006) The role of NAD+ dependent histone deacetylases (sirtuins) in ageing. Curr Drug Targets 7(11):1553–1560
10. Olzscha H, Sheikh S, La Thangue NB (2015) Deacetylation of chromatin and gene expression regulation: a new target for epigenetic therapy. Crit Rev Oncog 20(1-2):1–17
11. Yamaguchi T, Cubizolles F, Zhang Y, Reichert N, Kohler H, Seiser C, Matthias P (2010) Histone deacetylases 1 and 2 act in concert to promote the G1-to-S progression. Genes Dev 24(5):455–469
12. Choi JH, Kwon HJ, Yoon BI, Kim JH, Han SU, Joo HJ, Kim DY (2001) Expression profile of histone deacetylase 1 in gastric cancer tissues. Jpn J Cancer Res 92(12):1300–1304
13. Halkidou K, Gaughan L, Cook S, Leung HY, Neal DE, Robson CN (2004) Upregulation and nuclear recruitment of HDAC1 in hormone refractory prostate cancer. Prostate 59(2):177–189
14. Zhu P, Martin E, Mengwasser J, Schlag P, Janssen KP, Gottlicher M (2004) Induction of HDAC2 expression upon loss of APC in colorectal tumorigenesis. Cancer Cell 5(5):455–463
15. Wilson AJ, Byun DS, Popova N, Murray LB, L'Italien K, Sowa Y, Arango D, Velcich A, Augenlicht LH, Mariadason JM (2006) Histone deacetylase 3 (HDAC3) and other class I HDACs regulate colon cell maturation and p21 expression and are deregulated in human colon cancer. J Biol Chem 281(19):13548–13558
16. Sjoblom T, Jones S, Wood LD, Parsons DW, Lin J, Barber TD, Mandelker D, Leary RJ, Ptak J, Silliman N, Szabo S, Buckhaults P, Farrell C, Meeh P, Markowitz SD, Willis J, Dawson D, Willson JK, Gazdar AF, Hartigan J, Wu L, Liu C, Parmigiani G, Park BH, Bachman KE, Papadopoulos N, Vogelstein B, Kinzler KW, Velculescu VE (2006) The consensus coding sequences of human breast and colorectal cancers. Science 314(5797):268–274
17. New M, Olzscha H, La Thangue NB (2012) HDAC inhibitor-based therapies: can we interpret the code? Mol Oncol 6(6):637–656
18. Marks PA (2010) The clinical development of histone deacetylase inhibitors as targeted anticancer drugs. Expert Opin Investig Drugs 19(9):1049–1066
19. Finnin MS, Donigian JR, Cohen A, Richon VM, Rifkind RA, Marks PA, Breslow R, Pavletich NP (1999) Structures of a histone deacetylase homologue bound to the TSA and SAHA inhibitors. Nature 401(6749):188–193
20. Butler KV, Kalin J, Brochier C, Vistoli G, Langley B, Kozikowski AP (2010) Rational design and simple chemistry yield a superior, neuroprotective HDAC6 inhibitor, tubastatin A. J Am Chem Soc 132(31):10842–10846
21. Zupkovitz G, Tischler J, Posch M, Sadzak I, Ramsauer K, Egger G, Grausenburger R, Schweifer N, Chiocca S, Decker T, Seiser C (2006) Negative and positive regulation of gene expression by mouse histone deacetylase 1. Mol Cell Biol 26(21):7913–7928
22. Wilson AJ, Byun DS, Nasser S, Murray LB, Ayyanar K, Arango D, Figueroa M, Melnick A, Kao GD, Augenlicht LH, Mariadason JM (2008) HDAC4 promotes growth of colon cancer cells via repression of p21. Mol Biol Cell 19(10):4062–4075
23. Khan O, La Thangue NB (2012) HDAC inhibitors in cancer biology: emerging mechanisms and clinical applications. Immunol Cell Biol 90(1):85–94
24. Bali P, Pranpat M, Bradner J, Balasis M, Fiskus W, Guo F, Rocha K, Kumaraswamy S, Boyapalle S, Atadja P, Seto E, Bhalla K (2005) Inhibition of histone deacetylase 6 acetylates and disrupts the chaperone function of heat shock protein 90: a novel basis for antileukemia activity of histone deacetylase inhibitors. J Biol Chem 280(29):26729–26734
25. Nimmanapalli R, Fuino L, Bali P, Gasparetto M, Glozak M, Tao J, Moscinski L, Smith C, Wu J, Jove R, Atadja P, Bhalla K (2003) Histone deacetylase inhibitor LAQ824 both lowers expression and promotes proteasomal degradation of Bcr-Abl and induces apoptosis of imatinib mesylate-sensitive or -refractory chronic myelogenous leukemia-blast crisis cells. Cancer Res 63(16):5126–5135
26. Lee JY, Koga H, Kawaguchi Y, Tang W, Wong E, Gao YS, Pandey UB, Kaushik S, Tresse E, Lu J, Taylor JP, Cuervo AM, Yao TP (2010) HDAC6 controls autophagosome maturation

essential for ubiquitin-selective quality-control autophagy. EMBO J 29(5):969–980

27. Kawaguchi Y, Kovacs JJ, McLaurin A, Vance JM, Ito A, Yao TP (2003) The deacetylase HDAC6 regulates aggresome formation and cell viability in response to misfolded protein stress. Cell 115(6):727–738
28. New M, Olzscha H, Liu G, Khan O, Stimson L, McGouran J, Kerr D, Coutts A, Kessler B, Middleton M, La Thangue NB (2013) A regulatory circuit that involves HR23B and HDAC6 governs the biological response to HDAC inhibitors. Cell Death Differ 20(10):1306–1316
29. Ward AC, Touw I, Yoshimura A (2000) The Jak-Stat pathway in normal and perturbed hematopoiesis. Blood 95(1):19–29
30. Kramer OH, Knauer SK, Greiner G, Jandt E, Reichardt S, Guhrs KH, Stauber RH, Bohmer FD, Heinzel T (2009) A phosphorylation-acetylation switch regulates STAT1 signaling. Genes Dev 23(2):223–235
31. Vishwakarma S, Iyer LR, Muley M, Singh PK, Shastry A, Saxena A, Kulathingal J, Vijaykanth G, Raghul J, Rajesh N, Rathinasamy S, Kachhadia V, Kilambi N, Rajgopal S, Balasubramanian G, Narayanan S (2013) Tubastatin, a selective histone deacetylase 6 inhibitor shows anti-inflammatory and anti-rheumatic effects. Int Immunopharmacol 16(1):72–78
32. Auld DS, Southall NT, Jadhav A, Johnson RL, Diller DJ, Simeonov A, Austin CP, Inglese J (2008) Characterization of chemical libraries for luciferase inhibitory activity. J Med Chem 51(8):2372–2386
33. Shinde R, Perkins J, Contag CH (2006) Luciferin derivatives for enhanced in vitro and in vivo bioluminescence assays. Biochemistry 45(37):11103–11112
34. Harlow E, Lane D (2006) Lysing tissue-culture cells for immunoprecipitation. CSH Protoc 2006(4)
35. Khan O, Fotheringham S, Wood V, Stimson L, Zhang C, Pezzella F, Duvic M, Kerr DJ, La Thangue NB (2010) HR23B is a biomarker for tumor sensitivity to HDAC inhibitor-based therapy. Proc Natl Acad Sci U S A 107(14): 6532–6537
36. Yeo W, Chung HC, Chan SL, Wang LZ, Lim R, Picus J, Boyer M, Mo FK, Koh J, Rha SY, Hui EP, Jeung HC, Roh JK, Yu SC, To KF, Tao Q, Ma BB, Chan AW, Tong JH, Erlichman C, Chan AT, Goh BC (2012) Epigenetic therapy using belinostat for patients with unresectable hepatocellular carcinoma: a multicenter phase I/II study with biomarker and pharmacokinetic analysis of tumors from patients in the Mayo Phase II Consortium and the Cancer Therapeutics Research Group. J Clin Oncol 30(27):3361–3367
37. Brummelkamp TR, Bernards R, Agami R (2002) A system for stable expression of short interfering RNAs in mammalian cells. Science 296(5567):550–553
38. Fotheringham S, Epping MT, Stimson L, Khan O, Wood V, Pezzella F, Bernards R, La Thangue NB (2009) Genome-wide loss-of-function screen reveals an important role for the proteasome in HDAC inhibitor-induced apoptosis. Cancer Cell 15(1):57–66
39. Kurien BT, Scofield RH (2009) Nonelectrophoretic bidirectional transfer of a single SDS-PAGE gel with multiple antigens to obtain 12 immunoblots. Methods Mol Biol 536:55–65

Chapter 20

Assessment of the Antiproliferative Activity of a BET Bromodomain Inhibitor as Single Agent and in Combination in Non-Hodgkin Lymphoma Cell Lines

Elena Bernasconi, Eugenio Gaudio, Ivo Kwee, and Francesco Bertoni

Abstract

To evaluate the antiproliferative activity of a novel BET Bromodomain inhibitor as single agent and in combination with the BTK inhibitor ibrutinib in non-Hodgkin lymphoma cell lines, we performed the MTT proliferation assay. This assay is based on the direct correlation between absorbance (measured colorimetrically at a wavelength of 570 nm) and cell proliferation. Thiazolyl Blue Tetrazolium Blue (MTT) is a yellowish solution that distinguishes between proliferating and dead cells since it is converted to water-insoluble MTT-formazan of dark blue color by mitochondrial dehydrogenases of living cells only.

Key words BET inhibitor, BRD inhibitor, Lymphoma, Ibrutinib, MTT

1 Introduction

Non-Hodgkin lymphomas are a group of cancers originating from lymphoid cells [1] and are among the commonest cancers, accounting for 4–5 % of all new cases of cancers in adults and up to 25 % in children and adolescents [2, 3]. In general, lymphomas are among the tumors with the highest cure rates, however the prognosis is very different among the individual histologies [1], and there is still a big need to improve our therapeutic approach. An example is given by diffuse large B-cell lymphoma (DLBCL), the most frequent lymphoma comprising 30–40 % of all lymphomas: still 30–40 % of adults with DLBCL present a refractory tumor or relapses after a first complete response [1].

Here, we present a protocol we have used to assess the preclinical anti-lymphoma activity of a new molecule, OTX015 (MK-8628) [4], belonging to a novel class of compounds, the BET Bromodomain inhibitors, which are believed to act by blocking

Sibaji Sarkar (ed.), *Histone Deacetylases: Methods and Protocols*, Methods in Molecular Biology, vol. 1436,
DOI 10.1007/978-1-4939-3667-0_20, © Springer Science+Business Media New York 2016

transcription [5]. The MTT proliferation assay is a robust technique, largely used to evaluate the anti-proliferative activity of compounds, although it has the disadvantage of not discriminating between a cytotoxic or cytostatic effect [6].

2 Materials and Methods

2.1 Instruments

- Cell culture hood (i.e., laminar-flow hood or biosafety cabinet).
- Incubator for cell culture (humid CO_2 incubator).
- Water bath.
- Inverted microscope.
- Centrifuge.
- Cell imaging multi-mode microplate reader.
- Fridges and freezers.

2.2 Material

- Flasks.
- Pipettes.
- Pipette controllers.
- Fast-Read 102 10-chamber counting grid.
- Tips.
- Multichannel pipette.
- Eppendorf tubes.
- Trypan blue.
- Falcon sterile tubes.
- Non-tissue culture 96-well treated plate.
- Tissue culture test plates 96U.
- Reservoir.
- Masterblock 96 well.
- RPMI1640, RPMI1640 Red Phenol Free mediums.
- Fetal bovine serum.
- Penicillin–streptomycin–neomycin.
- DMSO.
- OTX015 (MK-8628).
- Ibrutinib.
- MTT (3-(4,5-dimethylthiazol-2-yl)-2,5-diphenyl tetrazolium bromide) 5 mg/ml.
- Lysis Buffer (25% SDS, 0.025 N HCl in water).

3 Proliferation Assay for a Single Agent

3.1 Protocol

3.1.1 Day 1

Established human cell line TMD8, derived from DLBCL of the activated B-cell like (ABC-DLBCL) subtype, is cultured according to the recommended conditions. RPMI 1640 medium is supplemented with fetal bovine serum (20 %), 1 % penicillin–streptomycin–neomycin (~5000 U penicillin, 5 mg streptomycin, and 10 mg neomycin/ml) and 1 % L-glutamine.[1] TMD8 cells are adjusted at the concentration of 5×10^5/ml.

3.1.2 Day 2: Experimental Plate

Seed 10^4 cells[2] suspended in 100 μl of medium in each well of a 96-well plate (five replicates)[3] using a multichannel pipette. Put medium only in the peripheral wells (Fig. 1). Absorbance got from these wells will serve as blank. OTX015 is dissolved in dimethyl sulfoxide (DMSO) and serially diluted in the appropriate tissue culture medium, at a range of 15.6–1000 nM (*see* the dilution plate preparation).

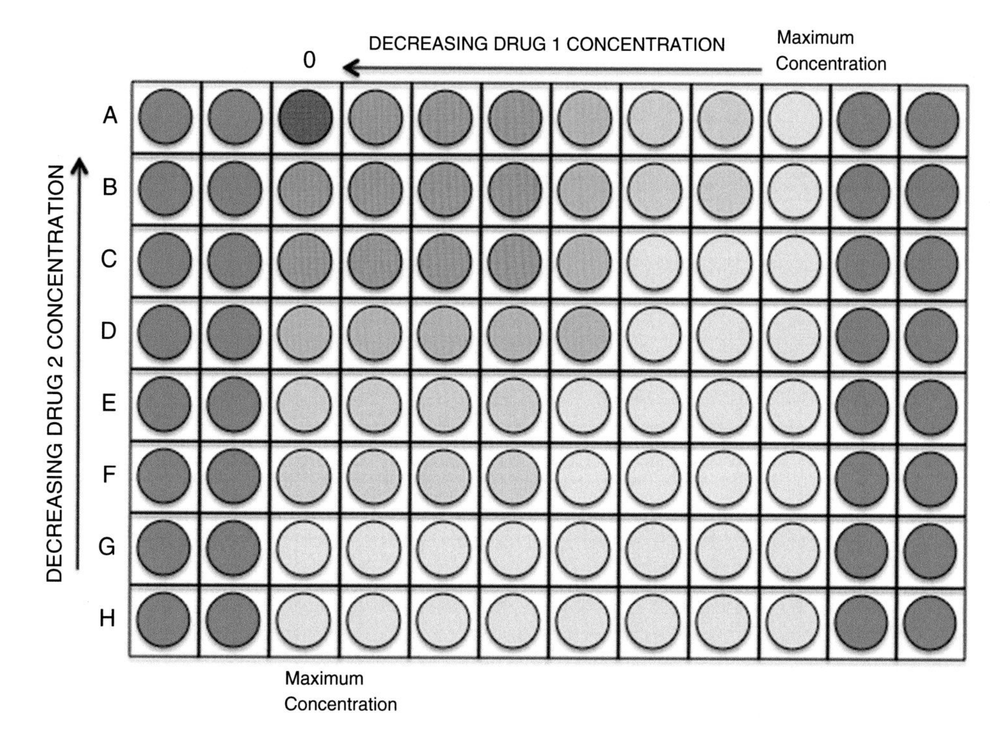

Fig. 1 Experimental plate. 10^4 cells suspended in 100 μl of medium will be seeded in each well of a 96-well plate (five replicates). Add 100 μl of DMSO dilution in the first line and add 100 μl of each drug concentration from the dilution plate to the respective wells in the experimental plate

[1] RPMI1640, IMDM, and DMEM with 10–20 % of FBS are used as requested by specific cell lines. For cell lines requiring RPMI 1640, use phenol red-free RPMI 1640 to perform MTT tests.

[2] The number of cells to be seeded is adjusted in relation to the plate (e.g., 96- or 384-well plate) and to the specific cell line.

[3] At least three replicates are suggested.

3.1.3 Day 2: Time Zero (t0) Plate

Seed 10^4 cells[2] suspended in 200 μl of medium (RPMI 1640 Red Phenol Free, with 20% FBS) in each well of a 96-well plate (five replicates). After approximately 2 h add 20 μl of MTT to each well, incubate at 37 °C, 5% CO_2 for 4 h until purple precipitate is visible. Add 50 μl of SDS lysis buffer and put the plate back in the incubator. Read absorbance after 2 days.

3.1.4 Day 2: Dilution Plate

In a tissue culture test-plate 96U put 125 μl of media in lines from B to G (Fig. 2). In a tube (number 1) prepare 2 ml of media with maximum concentration of OTX015 (2× concentrated). In a tube (number 2) prepare 2 ml of media with DMSO at the same drug's concentration. Prepare both compound and DMSO controls at 2× concentration. In the line H add 250 μl from tube 1. Perform a serial dilution: take 125 μl from the first well (line H) and add it to the next well (line G), mix by pipetting. The process is repeated step by step until line B (Fig. 2). In the line A add the control (tube 2). In the plate with the cells add 100 μl of DMSO in the first line and add 100 μl of each OTX015 concentration to the respective wells (Fig. 1). Return plates to incubator (keep for 72 h).

3.1.5 Day 4

Read absorbance of t0 plate with cell imaging multi-mode microplate reader: shake 5 s (low), read absorbance 560 nm.

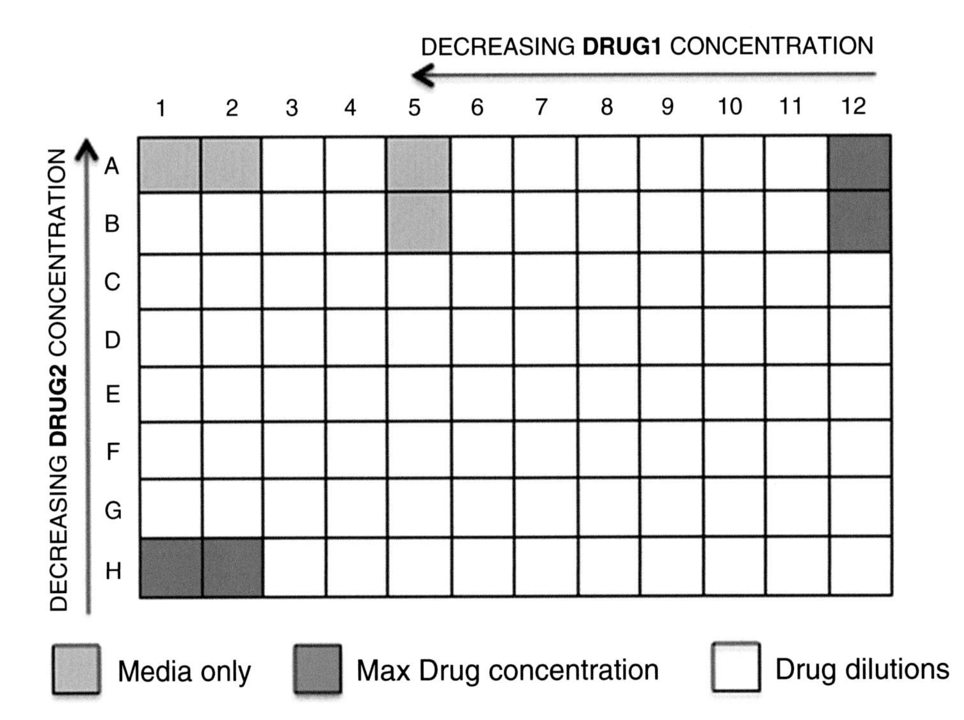

Fig. 2 Dilution plate. In a tissue culture plate put 125 μl of media in lines/wells from B to G. Prepare both compound and DMSO controls at 2× concentration. In the line H add 250 μl from tube 1 (maximum concentration of the drug). Perform a serial dilution: take 125 μl from the first well (line H) and add it to the next well (line G), pipette to mix. The process is repeated step by step until line B. In the line A add the DMSO control (tube 2)

3.1.6 Day 5

Add 20 μl of MTT to wells of the experimental plate. Incubate in 37 °C, 5 % CO_2 incubator for 4 h until purple precipitate is visible. Add 50 μl of SDS lysis buffer. Return to the incubator and leave for 2 days.

3.1.7 Day 7

Read absorbance of the experimental plate with cell imaging multimode microplate reader: shake 5 s (low), read absorbance 560 nm.

4 Proliferation Assay for the Combination of Two Compounds

4.1 Protocol

4.1.1 Day 1

Adjust TMD8 ABC-DLBCL cell line at the concentration of 500,000/ml.

4.1.2 Day 2: Experimental Plate

Seed 10^4 cells[2] suspended in 100 μl of RPMI 1640 Phenol Red Free medium (supplemented with 20 % FBS) in each well of a 96-well plate using multichannel pipette. Seed cells in three plates[3] for each cell line to test with compounds in combination. Put medium only in the peripheral wells (Fig. 3). Absorbance read

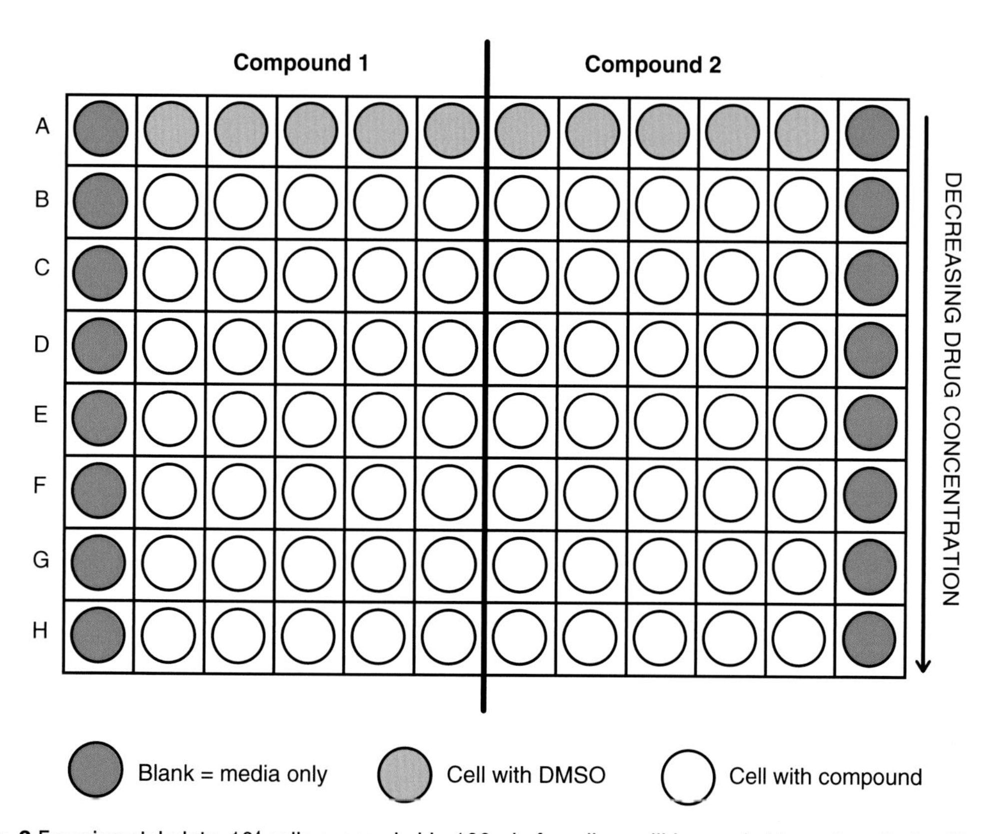

Fig. 3 Experimental plate. 10^4 cells suspended in 100 μl of medium will be seeded in each well of a 96-well plate using multichannel pipette. Seed cells in three plates for each cell line to test with compounds in combination: each well will contain 200 μl of volume, comprising seeded cells (100 μl) and two drugs or regular medium (50 μl each)

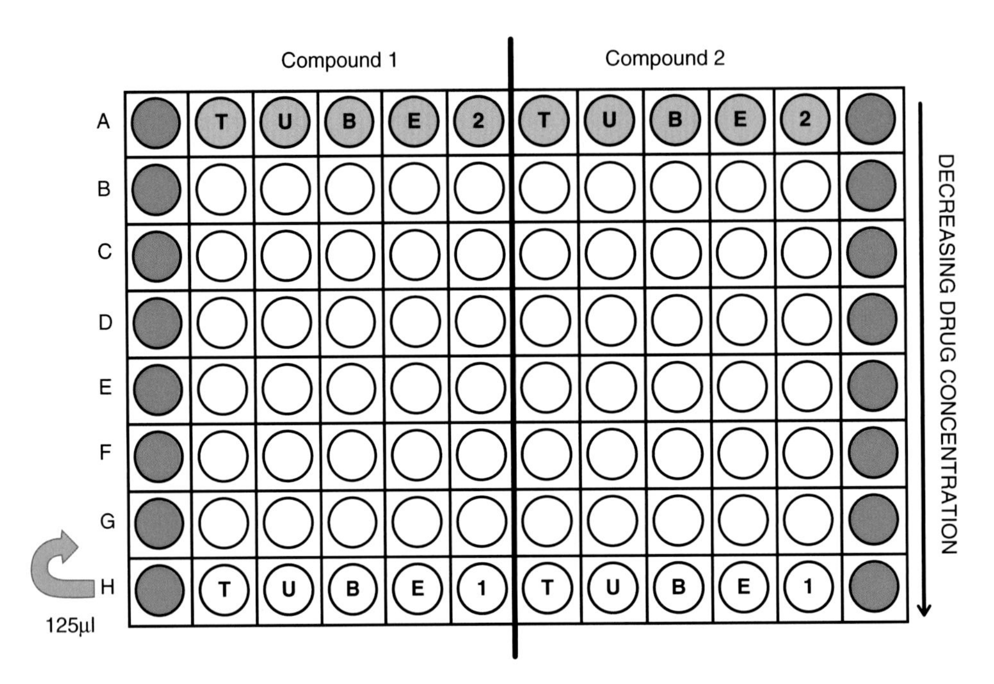

Fig. 4 Dilution plate. Drugs are diluted (4×) in a masterblock plate as explained in the text

from these wells will serve as blank. Basically you will have 200 μl of volume in each wells as sum of seeded cells (100 μl) and two drugs (OTX015 and ibrutinib) or regular medium (50 μl each) (Fig. 4). Return plates to incubator and keep there for 72 h.

4.1.3 Day 2: Dilution Plate

In a masterblock plate add 1 ml of media with the multichannel pipette in each well of the columns 1 and 2 from rows A to G (to perform the dilutions of compound 2) and do the same with rows A and B from column 5 to 11 (to perform the dilutions of compound 1) (Fig. 4).

In H1 and H2 prepare 2 ml of media each with maximum concentration of the drug1 (here, OTX015) 4× the final concentration since 50 μl of compound will be added to 100 μl of cells. (For a final concentration of 1000 nM just prepare 4× concentration, 4000 nM, in the Masterblock plate.) Using multichannel pipette perform a serial dilution: take 1 ml from the first well (line H) and add it to the next well (line G), pipetting to mix. The process is repeated until well in line B (Fig. 4).

In the other two empty wells (well A12 and B12) prepare 2 ml each of media with maximum concentration of the drug2 (here, ibrutinib) 4× the final concentration since 50 μl of compound will be added to 100 μl of cells. (For a final concentration of 10 nM just prepare 4× concentration, 40 nM, in the Masterblock plate.) Perform a serial dilution: take 1 ml well A12 and add it to the next well A11, pipette to mix. The process is repeated up to well A6 (Fig. 4).

4.1.4 Day 5

Add 20 μl of MTT to wells of the experimental plate. Incubate in 37 °C, 5 % CO_2 incubator for 4 h until purple precipitate is visible. Add 50 μl of SDS lysis buffer. Return to the incubator and leave for 2 days.

4.1.5 Day 7

Read absorbance of the experimental plate with cell imaging multi-mode microplate reader: shake 5 s (low), read absorbance 560 nm.

5 Data Analysis

5.1 Single Agent Proliferation Assay

For each cell line and biological replicate, a "time zero" (t0) plate is seeded with the appropriate number of cells per well and processed immediately to allow a time zero (baseline) density to be determined. The IC50 (50 % inhibitory concentration), GI50 (50 % growth inhibition), LC50 (50 % lethal concentration: the drug concentration that kills 50 % of the cells that were present at the time of drug addition), and TGI (total growth inhibition: the drug concentration that yields no net growth over the course of the assay) [7] were calculated from absorbance values obtained from the treatment plates alone (IC50) or from the treatment plates and time zero plates (GI50, LC50, and TGI). As the GI50, LC50, and TGI take into consideration the concentration of cells that were seeded at the beginning of the assay before the addition of drug or vehicle alone, they also give an indication of the cytostatic and cytotoxic activities of the drug. The doses corresponding to the IC50, GI50, LC50, and TGI were estimated by fitting a sigmoidal model through the dose response curve using the R statistical package (www.r-project.org).

5.2 Combination of Two Compounds Proliferation Assay

The combinations are evaluated using the Chou–Talalay Combination Index (CI) [8], calculated with the Synergy R package [9]. The effect of the combinations is defined synergistic (CI < 0.9), additive (CI, 0-9–1.1), or antagonist (CI > 1.1).

References

1. Swerdlow SH, Campo E, Harris NL et al (2008) WHO classification of tumours of haematopoietic and lymphoid tissues. IARC Press 2:439
2. Stewart B, Wild CP (2014) World cancer report 2014. IARC Nonserial Publication 2014: 630
3. Siegel RL, Miller KD, Jemal A (2015) Cancer statistics, 2015. CA Cancer J Clin 65:5
4. Boi M, Gaudio E, Bonetti P et al (2015) The BET bromodomain inhibitor OTX015 affects pathogenetic pathways in preclinical B-cell tumor models and synergizes with targeted drugs. Clin Cancer Res 21:1628
5. Filippakopoulos P, Knapp S (2014) Targeting bromodomains: epigenetic readers of lysine acetylation. Nat Rev Drug Discov 13:337
6. Plumb JA (2004) Cell sensitivity assay: the MTT assay. In: Langdon SP (ed) Cancer cell culture: methods and protocols, vol 88, Methods in

molecular medicine. Humana Press Inc., Totowa, NJ, p 165

7. Boyd MR, Paull KD (1995) Some practical considerations and applications of the national cancer institute in vitro anticancer drug discovery screen. Drug Dev Res 34:91
8. Chou TC (2008) Preclinical versus clinical drug combination studies. Leuk Lymphoma 49:2059
9. Lee JJ, Kong M, Ayers GD, Lotan R (2007) Interaction index and different methods for determining drug interaction in combination therapy. J Biopharm Stat 17:461

Chapter 21

Screening the Impact of Sirtuin Inhibitors on Inflammatory and Innate Immune Responses of Macrophages and in a Mouse Model of Endotoxic Shock

Eleonora Ciarlo and Thierry Roger

Abstract

The development and screening of pharmacological modulators of histone deacetylases (HDACs), and particularly sirtuins, is a promising field for the identification of new drugs susceptible to be used for treatment strategies in a large array of welfare-associated, autoimmune and oncologic diseases. Here we describe a comprehensive protocol to evaluate the impact of sirtuin-targeting drugs on inflammatory and innate immune responses in vitro and in a preclinical mouse model of endotoxemia. We first provide an overview on strategies to design in vitro experiments, then focus on the analysis of cytokine production by primary macrophages and RAW 267.7 macrophages at the mRNA and protein levels, and finally describe the setup and follow-up of a mouse model of inflammation-driven endotoxic shock.

Key words Histone deacetylases, Sirtuins, Pharmacological inhibitors, Epigenetics, Innate immunity, Cytokine, Macrophage, Endotoxemia, Sepsis

1 Introduction

The superfamily of histone deacetylases (HDACs) comprises 18 members in mammals, 11 Zn-dependent HDACs (HDAC1-11) and 7 NAD-dependent sirtuins (SIRT1-7). The first gene member of the sirtuin family, silent information regulator-2 (*SIR2*) from *S. cerevisiae*, was described more than 35 years ago as mating-type regulator 1 (*MAR1*) [1]. Yet, the role of sirtuins in physiological and pathological processes is far from being fully elucidated.

Mammalian sirtuins differ according to their sub-cellular localization, enzymatic activity and substrate specificity. SIRT1 and SIRT6 are essentially nuclear, SIRT2 cytoplasmic, SIRT3-5 mitochondrial, and SIRT7 nucleolar proteins, although the distribution of some isoforms (e.g., SIRT1, SIRT2, SIRT3, and SIRT5) is dynamic and not fully restricted to a specific subcellular compartment [2]. Sirtuins are characterized by their NAD+-dependent deacetylase activity. Similarly to other HDACs, most sirtuins remove acetyl groups from lysine

Sibaji Sarkar (ed.), *Histone Deacetylases: Methods and Protocols*, Methods in Molecular Biology, vol. 1436,
DOI 10.1007/978-1-4939-3667-0_21,

residues on histones and non-histone proteins thereby modifying their expression or activity. Yet, SIRT4 has no appreciable deacetylase activity, and SIRT6 is probably a stronger deacylase than deacetylase [3]. Both SIRT4 and SIRT6 work as ADP-ribosyltransferases [4, 5]. SIRT5 is a weak deacetylase but has efficient demalonylase, desuccinylase, and deglutarylase activities [6, 7]. Sirtuins have numerous targets such as transcription regulators, enzymes, and structural proteins, in line with the fact that more than 6800 acetylation sites have been identified in mammalian proteins [8].

Sirtuins are involved in multiple biological processes. Their dependence on NAD+ naturally links sirtuin activity to cellular metabolic status, and most sirtuins are connected to metabolic processes such as gluconeogenesis, fatty acid metabolism, oxidative phosphorylation, amino acid metabolism, urea cycle, and mitochondrial biogenesis. Sirtuins also control cell cycle, rDNA transcription, DNA repair, telomere homeostasis, microtubule stability, neuronal development, and circadian functions [2, 9, 10]. It is therefore not surprising that changes in expression or activity of sirtuins are associated with the development of metabolic (type 2 diabetes and obesity), cardiovascular, neurodegenerative (Alzheimer, Parkinson and Huntington's diseases), oncologic and other age-associated diseases [11–13]. Sirtuins have also been associated with the development of inflammatory and autoimmune diseases, but here the picture is less clear. Indeed, sirtuins have either a protective or a deletary role in experimental airway diseases, autoimmune encephalomyelitis and arthritis [14–20], possibly due to different settings and strategies used to address the role of sirtuins (inhibitory drugs, si/shRNA, germline and tissue specific knockouts).

In recent years, quite some efforts have been devoted to the discovery and design of sirtuin inhibitors. Several compounds inhibiting primarily SIRT1 and SIRT2 are promising drugs for cancer and neurodegenerative diseases such as Parkinson and Huntington's diseases [21, 22]. Although the panel of sirtuin inhibitors obtained thus far is rather restricted and mainly directed towards SIRT1 and SIRT2, there is little doubt that numerous more selective and more powerful compounds will be available in a near future. Considering that inflammation is central to the development of metabolic, oncologic, neurologic, autoimmune, and infectious diseases, sirtuin inhibitors should be systematically tested for their inflammatory activity, as we recently reported for the SIRT1/SIRT2 inhibitor cambinol [23]. Reinforcing this idea, classical HDAC share with sirtuins several targets among which some crucial regulators of inflammatory and immune responses, and inhibitors of classical HDACs have demonstrated strong anti-inflammatory and immunomodulatory properties [24–27].

Here we describe simple and comprehensive protocols to test the impact of sirtuin inhibitors on immune responses in vitro and in vivo. In vitro testing is performed using macrophages that, as a main source of cytokines, are central initiators of inflammatory-driven pathologies. The readout is the production of proinflammatory cytokines by both primary macrophages and RAW 264.7 macrophages exposed to microbial ligands such as lipopolysaccharide (LPS, also known as endotoxin), the main proinflammatory component of the outer membrane of gram-negative bacteria [28, 29]. The relevance of in vitro results is then addressed in a mouse model of endotoxemia in which animals die from overwhelming inflammation.

2 Materials

Prepare the solutions sterile, handle material with gloves. Use only ultrapure water and analytical grade reagents. In Subheadings 2.1, 2.2, 2.3, and 2.4, all products are cell culture certified or appropriately sterilized before usage. Always use polypropylene tubes for macrophage preparations in order to minimize cell adherence to the tube wall. In Subheadings 2.5 and 2.6, use only RNase free plastic disposals and solutions. Cell culture incubators are set at 37 °C with 5% CO_2, and cold means 4 °C.

2.1 Stock Solutions of Stimuli and Sirtuin Inhibitors

Stimuli and inhibitors are generally obtained as powders that are stored at +4 °C unless specified otherwise. The screening is usually performed with LPS. We propose two additional stimuli, Pam_3CSK_4 and CpG ODN, which can be used secondly to validate the results obtained with LPS. Prepare stock solutions, aliquot and store at −20 °C (*see* **Note 1**).

1. LPS (also called endotoxin, *see* **Note 2**): 1 mg/mL in phosphate-buffered saline (PBS).
2. Pam_3CSK_4 (*see* **Note 3**): 1 mg/mL in PBS.
3. CpG ODN (*see* **Note 4**): 500 μM in PBS.
4. Sirtuin inhibitors: in an appropriate vehicle—water (H_2O), ethanol (EtOH), methanol (MetOH), dimethyl sulfoxide (DMSO), etc.—taking into account compound solubility and vehicle effect/toxicity (*see* **Note 5**).

2.2 Production and Culture of Bone-Marrow Derived Macrophages (BMDM)

1. Mice (*see* **Note 6**), and a carbon dioxide (CO_2) system to euthanize mice.
2. Laminar flow, water bath, refrigerated centrifuge, full ice tray, cell culture incubator, inverted microscope, material for cell counting.
3. Scissors and forceps, pipetboy, micropipettes.
4. Sterile bacterial petri dishes (100 and 150 mm, *see* **Note 7**), 50 mL conical tubes, 20 mL syringes and 25 G needles, pipettes (5 and 10 mL), cell strainers (nylon, 100 μm), cell lifters (*see* **Note 8**), cryotubes.

5. PBS, 70% EtOH, 1× Versene solution (0.48 mM ethylenediaminetetraacetic acid tetrasodium salt (EDTA, Na_4) in PBS).
6. Red blood cells (RBC) lysis solution: 0.15 M ammonium chloride (NH_4Cl), 10 mM potassium bicarbonate ($KHCO_3$), 0.1 mM EDTA in H_2O. Sterilize by filtration (0.22 μm). Store at 4 °C.
7. Basic IMDM: IMDM (Iscove's Modified Dulbecco's Medium) containing GlutaMAX™ and 4.5 g/L glucose.
8. Complete IMDM: basic IMDM supplemented with 10% heat inactivated (*see* **Note 9**) low endotoxin (*see* **Note 10**) fetal calf serum (FCS), 100 UI/mL penicillin, 100 μg/mL streptomycin, and 50 μM 2-mercaptoethanol.
9. Differentiation IMDM: complete IMDM supplemented with 30% of L929 cell-conditioned medium (*see* **Note 11**).

2.3 Culture of RAW 264.7 Mouse Macrophages

1. Stock ampoule of RAW 264.7 macrophages (ATCC® TIB-71™).
2. Laminar flow, water bath, centrifuge, cell culture incubator, inverted microscope, material for cell counting.
3. Pipettes (5 and 10 mL), pipetboy, 30 mL sterile conical tubes, 100 mm cell culture petri dishes, cell lifter (*see* **Note 8**).
4. Complete RPMI medium: RPMI (Roswell Park Memorial Institute) medium containing GlutaMAX™ and 4.5 g/L glucose supplemented with 10% heat inactivated (*see* **Note 9**) low endotoxin FCS (*see* **Note 10**), 100 UI/mL penicillin, 100 μg/mL streptomycin.

2.4 Testing Inhibitors on Cultures of BMDM and RAW 264.7 Macrophages

1. Laminar flow, water bath, refrigerated centrifuge, full ice tray, vortex mixer, cell culture incubator.
2. Pipettes, single channel and multichannel micropipettes, filter tips, tubes, polystyrene flat bottom 96-well cell culture plates.
3. Complete IMDM or RPMI medium, PBS, stimuli, and drugs described in Subheading 2.1.

2.5 RNA Purification and cDNA Synthesis

1. RNA isolation kit (preferably using a column-based procedure).
2. Pipettes, single channel and multichannel micropipettes, filter tips, vortex mixer, refrigerated microcentrifuge, nucleic acid quantification apparatus, full ice tray, flat bottom 96-well cell culture plates, polypropylene tubes.
3. EtOH (70 and 100%), 3 M sodium acetate (NaOAC) pH 5.2, 5 mg/mL glycogen in H_2O.
4. Reverse transcription kit.

2.6 Cytokine Measurement by RT-PCR

1. Pipettes and filter tips, tubes, micropipettes, vortex mixer, refrigerated microcentrifuge, benchtop cooler, thermoblock or water bath at 42 and 95 °C, 96-well plate spinner, full ice tray.

2. Forward and reverse primers (stock solution at 100 μM).
3. DNase free H_2O, Fast SYBR® Green Master Mix.
4. MicroAmp® Fast Optical 96-Well Reaction Plate and Optical Adhesive Film.
5. Real-time PCR apparatus.

2.7 Cytokine Measurement by ELISA

1. ELISA kits for the detection of TNF, IL-6 or any other cytokine of interest. Here we will use DuoSet® kits consisting of capture antibody, biotinylated detection antibody, and streptavidin–horseradish peroxidase (HRP).
2. Multichannel micropipettes and reservoirs, flat-bottom 96-well plate, plate sealers, PBS.
3. Wash buffer: 0.05 % Tween® 20 in PBS, pH 7.2–7.4.
4. Reagent diluent: 1 % BSA (bovine serum albumin) in PBS, pH 7.2–7.4, 0.22 μm filtered.
5. TMB substrate (*see* **Note 12**).
6. Stop solution: 2 N sulfuric acid (H_2SO_4).

2.8 Cell Viability Assay Using MTT

1. MTT (3-(4,5-dimethylthiazol-2-yl)-2,5-diphenyltetrazolium bromide) powder.
2. Multichannel micropipettes and reservoirs, balance.
3. Lysis buffer (for one plate): 10 mL 2-propanol, 5 mL 20 % SDS (sodium dodecyl sulfate), 80 μL 5 M chlorhydric acid (HCl).
4. Ninety-six well plate and luminescence plate reader.

2.9 Mouse Model of Endotoxemia

1. Eight 12-weeks old female BALB/c mice. The number of animals will depend on experimental design and experiments should be planned based on ARRIVE guidelines (Animal Research: Reporting of In Vivo Experiments) and respecting the 3R principles (Reduce, Refine, and Replace) (*see* Subheading 3.9 and **Note 13**.) [30].
2. 1 mL sterile syringes, 25 G needles.
3. Drug and vehicle at an equivalent concentration (for example 0.8 mg/mL cambinol and 8 % DMSO in 0.9 % sodium chloride (NaCl)).
4. LPS: 1.4 mg/mL in 0.9 % NaCl (*see* **Note 14**).
5. Infrared lamp, scalpels, hemostatic gauze, lithium heparin (for ELISA measurements) or potassium EDTA (for multiplex assay measurement by Luminex) coated Microvette® for capillary blood collection, microcentrifuge, mice digital weighing scale, non-wetting sterile water gel for animal hydration such as HydroGel®.

3 Methods

3.1 Preparation of BMDMs

If possible, use separate laminar flows dedicated to organ collection and cell culture.

1. Day 0: put under the laminar flow an aluminum foil, scissors, and forceps and prepare three petri dishes: A and C containing 15 mL of PBS and B containing 15 mL of 70% EtOH.
2. Sacrifice the mouse with CO_2 (*see* **Note 13**).
3. Collect the hind femurs and tibias. Place the mouse on the aluminum foil and spray it with 70% EtOH. Cut with scissors the skin just above the foot and pull it over the hip. Cut the legs above the hip articulation (*see* **Note 15**). Remove the remaining skin from the feet. Cut ankle tendons and destroy the ankle articulation by gently twisting and pulling the feet. Remove the feet with scissors. Cut knee tendons and loosen the articulation by gently twisting femur and tibias. Remove as much as possible flesh from the bones, with the help of scissors and forceps.
4. Soak briefly the cleaned bones in petri dish A, petri dish B, and petri dish C.
5. Cut bone epiphyses and immediately flush the bone marrow (BM): firmly holding the bone with forceps above a 50 mL tube, introduce into the opening of the bone a 25 G needle adapted to a 20 mL syringe filled with 20 mL of basic IMDM, a volume sufficient to flush both the tibia and the femur from one leg. Flush 10 mL per bone, moving the needle up and down into the cavity of the bone. The bone becomes progressively white as the BM is expelled.
6. Homogenize the cell suspension by pipetting up and down with a 10 mL pipette.
7. Centrifuge for 7 min at $400 \times g$ and 4 °C. Discard supernatant (SN). Resuspend the pellet in 3 mL of cold RBC lysis buffer. Incubate for 5 min on ice. Add 40 mL of cold PBS.
8. Filter the suspension through a 100 μm nylon cell strainer into a new tube.
9. Centrifuge for 7 min at $400 \times g$ and 4 °C. Discard SN. Resuspend the pellet in 10 mL of complete IMDM and enumerate BM cells (*see* **Note 16**).
10. Plate 3×10^6 BM cells in a 100 mm bacterial petri dish in a volume of 15 mL of differentiation IMDM (to obtain $6–9 \times 10^6$ BMDM after 7 days of culture, *see* **Notes 17** and **18**).
11. Put the petri dish in a cell culture incubator.

12. Freeze BM cells in excess in complete IMDM containing 10% DMSO (*see* **Note 19**).
13. Day 2 and day 4: examine the culture using an inverted microscope. Add 5 mL of preheated differentiation IMDM in the petri dish (*see* **Note 20**).
14. Day 7: examine the culture using an inverted microscope to check for cell growth (the confluency of adhering macrophages should be around 100%) and contaminants. Under the laminar flow, aspirate the SN, wash the cell layer with 10 mL of cold PBS and add 3 mL of cold Versene (*see* **Note 21**).
15. Using a cell lifter strongly pressed onto the surface of the petri dish, detach the cells by performing a round movement to detach cells on the edges, and then linear one-way movements to lift cells on the remaining surface. If this step is performed properly, maximal cell recovery with minimal cell death is obtained.
16. Homogenize the cell suspension with a 5 mL pipette and transfer to a 50 mL tube.
17. Add cold PBS to 40 mL. Centrifuge for 7 min at $400 \times g$ and 4 °C. Discard the SN. Resuspend the pellet in 10 mL of complete IMDM and enumerate BMDM.
18. Dilute cell suspension to a concentration of 2×10^6 cells/mL and proceed according to Subheading 3.3.

3.2 Culture of RAW 264.7 Macrophages

1. Thaw quickly (in a water bath at 37 °C) a cryotube of frozen RAW 264.7 macrophages.
2. Pipet the cell suspension in a conical tube containing 9 mL of complete RPMI.
3. Centrifuge for 7 min at $300 \times g$ at room temperature. Discard the SN. Resuspend the pellet in 10 mL of complete RPMI. Enumerate cells. Transfer 1×10^6 living cells in a 100 mm cell culture petri dish. Add complete RPMI to reach 10 mL. Incubate in a cell culture incubator.
4. Detach cells with a cell lifter when reaching 70–80% confluence (*see* **Note 22**).
5. Homogenize the cell suspension by pipetting up and down with a 10 mL pipette. Transfer to a tube. Centrifuge for 7 min at $300 \times g$. Discard the SN. Resuspend the pellet in 10 mL of complete RPMI. Enumerate cells. Transfer 1×10^6 or 2.5×10^6 cells to a 100 mm or 150 mm cell culture petri dish, respectively. Add pre-warmed complete RPMI up to 10 or 20 mL.
6. Split cells as indicated in **steps 4** and **5** every 2–4 days. Increase the number of plates to reach the number of cells required in Subheading 3.3.

3.3 Testing Inhibitors on BMDM and RAW 264.7 Macrophages

Prepare a scheme of your experimental plate as exemplified in Fig. 1a (*see* **Note 23**). Don't forget negative controls (*see* **Note 24**). First screening is performed with cells stimulated with 10 ng/mL LPS. Drug effects on cytokine production can be confirmed using cells stimulated with 10 ng/mL Pam_3CSK_4 or 1 μM CpG ODN. Test each condition at least in triplicate. We prefer to seed and treat cells all at once, and then collect samples sequentially according to the incubation time. A first screening can be performed with RAW 264.7 macrophages, which are easier to obtain than BMDMs, but with the possibility that drugs are more toxic for established cell

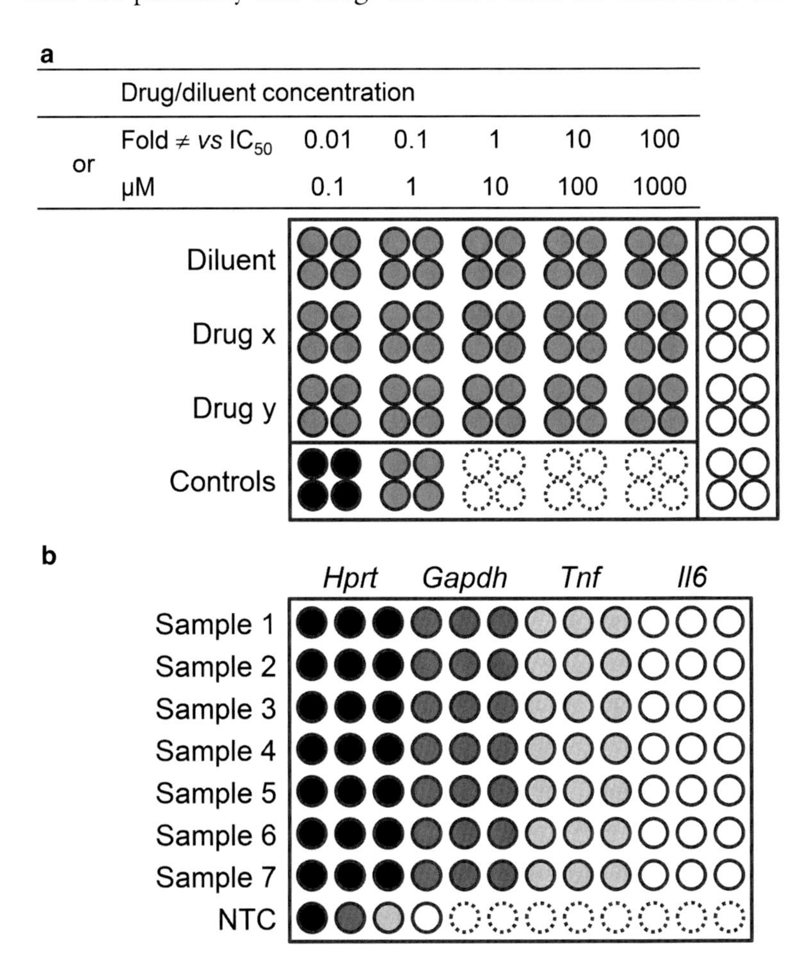

Fig. 1 Example of cell culture and RT-PCR 96-well plate schemes for testing sirtuin inhibitors on cytokine production by macrophages (**a**) Cell culture plate scheme. Cells are seeded in the *grey* or *black* wells. Cells are preincubated with diluent and drug at final concentrations corresponding to 100×, 10×, 1×, 0.1×, and 0.01× of the lowest IC50 or in range from 1 mM to 100 nM as indicated, and then stimulated with (*grey*) or without (*black*) LPS. Each condition is analyzed in quadruplicates. The two right columns are left empty to leave room for standard and blank in ELISA or Luminex assays. (**b**) RT-PCR plate scheme. Samples are distributed horizontally, gene specific mix vertically according to fill color. Each sample is measured in triplicates. NTC: No template control. The *dotted wells* are not used

lines than for primary cells. It is therefore highly recommended to screen the drugs on primary cells, either first-line or second-line, before testing the best-of-class candidate(s) in preclinical models.

1. Prepare suspensions of BMDMs or RAW 264.7 macrophages at 2×10^6 cells/mL in complete IMDM or RPMI medium. Seed 100 μL (2×10^5 cells) per well in 96-well cell culture plates (*see* **Notes 25** and **26**). Incubate in a cell culture incubator while performing **step 2**.
2. Prepare 2× concentrated working solutions of drugs since drugs will be diluted 2× when added to cell cultures. A selection of commercially available sirtuin inhibitors is presented in Table 1. When a drug is tested for the first time, a broad range of final concentrations corresponding to 100×, 10×, 1×, 0.1×, and 0.01× of the lowest IC_{50} for any sirtuin is recommended. If the IC_{50} are unknown, final concentrations should range from 1 mM to 100 nM. Less broad ranging with more frequent concentrations are tested in subsequent experiments to refine results.
3. Add 100 μL of 2× concentrated working solutions of drug to the cells according to your experimental scheme.
4. Incubate for 1 h (*see* **Note 27**).
5. During incubation, prepare 11× concentrated (110 ng/mL) working solution of LPS (*see* **Note 28**).
6. Add 20 μL of LPS working solution to the cells according to your experimental scheme.
7. Incubate cells for 1, 4, and 8 h to measure cytokine mRNA, for 8 and 18 h to measure cytokine secretion (*see* **Notes 29**and **30**), and for 18 h to assess cell viability (*see* **Note 31**).

3.4 RNA Extraction

The RNA extraction protocol is performed using a column-based method. It is adapted to the experimental scheme of Fig. 1a where four replicates are combined. The procedure is run at room temperature. Tubes are kept on ice as soon as RNA is collected.

1. With a 12-channel multipipette, transfer in a new plate 150–200 μL of SN from the plates stimulated for 8 h if you want to quantify cytokines (*see* **Note 30**). Sore at −20 °C for cytokine measurement by ELISA or Luminex. Discard the remaining liquid, and the SN from the plates stimulated for 1 h and 4 h.
2. Add 100 μL of lysis buffer to each well containing cells. Pipet up and down to homogenize. Pool the four replicates in a single pre-labeled 1.5 mL eppendorf tube.
3. Purify RNA following manufacturer's recommendations (*see* **Note 32**). Eluate RNA in 100 μL RNase-free water.
4. Place tubes containing the RNA eluate on ice. The sample can be used for RNA precipitation or stored at −80 °C.

Table 1
Selection of commercially available sirtuin inhibitors with IC_{50}

Inhibitor	IC_{50} (μM)						
	SIRT1	SIRT2	SIRT3	SIRT4	SIRT5	SIRT6	SIRT7
AGK2	>30	3.5	>50	NT	NT	NT	NT
AK-7	>50	15.5–24	>50	NT	NT	NT	NT
Cambinol	56	59	NI	NT	42 % inhibition at 300 μM	NI up to 250 μM	NT
CHIC-35	0.06–0.12	2.8	NI	?	?	?	?
EX-527	0.098–1	20–33	49	?	NI	56 % inhibition at 200 μM	?
Ginkgolic acids	119–126	80–141	NT	NT	NT	NT	NT
Inauhzin	0.7–2	NI	NI	?	?	?	?
Nicotinamide	50–100	1.2–100	30–43	?	1600	2200	?
Sirtinol	37.6–131	38–58	>50	?	?	No inhibition up to 200 μM	?
Salermide	76	25–45	>50	?	?	?	?
Splitomicin HR73	5	NI	NI	?	?	?	?
Suramin	0.297	1.15	NI	?	22	No inhibition up to 200 μM	?
Tenovin-1	70–90 % inhibition in Jurkat and MyLa cells at 25 μM	10	?	?	?	?	?
Tenovin-6	21	10	67	?	?	?	?
Urushiols	52–118	55–143	NT	NT	NT	NT	NT

NI No inhibition observed
NT Not tested
? No report (likely not tested)

3.5 RNA Precipitation and cDNA Synthesis

1. Prepare a mix containing, per sample to precipitate, 10 μL of NaOAC, 2 μL of glycogen, and 275 μL of cold 100% EtOH (*see* **Note 33**). Distribute 287 μL of the mix to each sample. Mix well by inverting the tubes. Incubate overnight at −20 °C (*see* **Note 34**).
2. Centrifuge for 15 min at 1300 × *g* and 4 °C.
3. Remove the SN without touching the pellet. Add 0.5 mL of cold 70% EtOH. Centrifuge for 5 min at 1300 × *g* and 4 °C.
4. Repeat **step 3**.
5. Remove the SN and let dry the pellet at room temperature (*see* **Note 35**).
6. Add 30 μL of RNase-free water without touching the pellet. Leave on ice for 15 min. Vortex for 2 s to dissolve the pellet and quick spin to collect the liquid at the bottom of the tube.
7. Measure yield and check quality of the RNA preparation (*see* **Note 36**).
8. Pipet 500 ng of RNA from each sample and perform reverse transcription as recommended by the manufacturer (*see* **Notes 37–39**).
9. Dilute the cDNA samples 1:5 in RNase-free water (*see* **Note 40**). Store at −20 °C until use.

3.6 Real-Time PCR (RT-PCR)

The RT-PCR is performed in 96-well plates using the Fast SYBR® Green Master Mix and FAST 7500 apparatus and analyzed using the comparative method (ΔΔCT) (*see* **Note 41**). A triplicate measure of each sample is recommended.

1. Prepare a scheme of your RT-PCR plate (a simple example is given in Fig. 1b). Indicate sample position and gene names. Do not omit "no template controls" (NTC) (*see* **Note 42**).
2. Prepare 10 μM primer working solutions by mixing in a tube 10 μL of each of the forward and reverse primer stock solutions with 80 μL of DNAse free water. A selection of validated primers for housekeeping and cytokine genes is listed in Table 2.
3. Amplification mix (per sample to analyze): 4.5 μL of H_2O, 0.5 μL of primer working solution, and 7.5 μL of Fast SYBR® Green Master Mix (*see* **Notes 33** and **43**). Vortex for 2 s. Short spin. Keep on ice protected from direct light. Repeat for each gene of interest.
4. Pipet 2.5 μL of cDNA into the bottom of the wells (*see* **Note 44**).
5. Add 12.5 μL of the amplification mix on the sidewall. Seal the plate carefully. Spin down.
6. Run the PCR in the FAST 7500 apparatus: initial step of 20 s at 95 °C and 40 amplification cycles with 3 s at 95 °C and 30 s at 60 °C. Analyze data using the FAST 7500 software v2.0.6.

Table 2
Primers used to amplify cytokine and housekeeping genes by RT-PCR

Gene	Forward primer (5′→3′)	Reverse primer (5′→3′)
Gapdh	CTC ATG ACC ACA GTC CAT GC	CAC ATT GGG GGT AGG AAC AC
Hprt	GTT GGA TAC AGG CCA GAC TTT GTT G	GAT TCA ACT TGC GCT CAT CTT AGG C
Il6	CCG GAG AGG AGA CTT CAC AG	CAG AAT TGC CAT TGC ACA AC
Il12b	GGA AGC ACG GCA GCA GAA TA	AAC TTG AGG GAG AAG TAG GAA TGG
Tnf	CCA GGC GGT GCC TAT GTC	GGC CAT TTG GGA ACT TCT CAT

3.7 ELISA

Cytokines can be quantified by ELISA or with the Luminex technology using pre-custom or personalized assays like ProcartaPlex™ Multiplex Immunoassays, Bio-Plex Pro™ Mouse Cytokine Assays or Luminex Screening Assays and Luminex Performance Assays. We do not describe Luminex-based methodology because it necessitates specific apparatus. The detection of mouse TNF and IL-6 by ELISA reported here uses DuoSet® ELISA kits but can be achieved with any ELISA kits of choice. Each brand has technical specificities thoroughly detailed in protocol handbooks. Dilutions of reagents are given by the manufacturer. All volumes are given per well.

1. Prepare a scheme of your ELISA plate with the positions of blank, standard, and samples.
2. Coat 96-well MaxiSorp™ plate(s) with 100 μL of capture antibody. Seal the plate and incubate overnight at room temperature.
3. Empty wells. Wash three times with 350 μL of wash buffer (*see* **Note 45**).
4. Add 300 μL of reagent diluent. Incubate at room temperature for 1 h.
5. During the 1 h incubation, dilute samples (*see* **Note 46**) and prepare a standard curve (starting from 1 or 2 ng/mL) made of seven twofold serial dilutions in reagent diluent.
6. Wash the plate three times with 350 μL of wash buffer. Add 100 μL of reagent diluent alone (blank), standard, or sample to the appropriate wells. Cover with an adhesive strip. Incubate for 2 h at room temperature. Wash three times with 350 μL of wash buffer.
7. Add 100 μL of biotinylated detection antibody. Cover with an adhesive strip. Incubate for 2 h at room temperature. Wash three times with 350 μL of wash buffer.
8. Add 100 μL of streptavidin–HRP. Cover with an adhesive strip. Incubate for 20 min at room temperature protected from light. Wash three times with 350 μL of wash buffer.

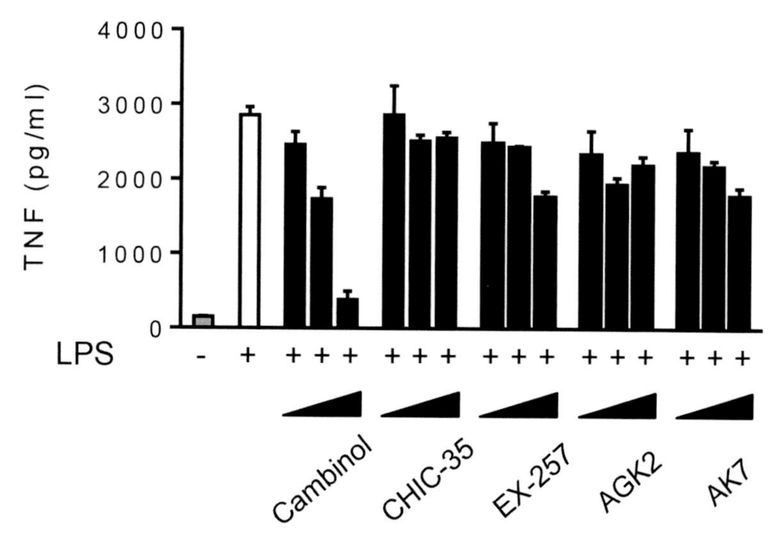

Fig. 2 Impact of sirtuin inhibitors on TNF production by RAW 264.7 macrophages. RAW 264.7 macrophages were preincubated for 1 h with or without cambinol (15, 60, and 240 μM), CHIC-35 (30, 120, and 480 nM), EX-527 (0.25, 1, and 4 μM), AGK2 (0.8, 3.5, and 14 μM), and AK-7 (6, 24, and 96 μM) before exposure for 8 h to LPS (10 ng/mL). Concentrations of TNF in cell culture SN were quantified by ELISA. Data are means ± SD of triplicate samples from one experiment representative of two experiments. The results show that the SIRT1/SIRT2 inhibitor cambinol powerfully inhibits TNF production in a dose-dependent manner whereas SIRT1 and SIRT2 selective inhibitors are much less potent. Note that drug dilutions used here are (0.25×, 1×, and 4× the IC_{50} for SIRT1 or SIRT2, *see* Table 1) were optimized based on pretests performed with a broader panel of dilutions as described in Fig. 1

9. Add 100 μL of TMB substrate solution. Incubate for around 20 min at room temperature avoiding exposure to direct light (*see* **Note 47**).
10. When the color has developed sufficiently, add 50 μL of stop solution. Shake the plate. Measure the ODs using a micro-plate reader set at 450 nm with wavelength correction set at 570 nm.
11. Analyze data using the SoftMax® Pro or equivalent software to generate a four parameter logistic (4-PL) curve-fit, calculate cytokine concentrations and plot data in a graph (an example is given in Fig. 2).

3.8 Cell Viability Assay Using MTT

The MTT assay is used to assess cell viability (*see* **Note 48**). It can be performed with dedicated cell culture plates or with cell culture plates from which the SN has been collected to measure cytokine release (*see* Subheading 3.3 and **Note 31**). To have a more accurate quantification of drug effect on cell viability, a standard curve can be established using values obtained from wells seeded with 0.5, 1,

2, 4, and 8×10^5 cells in complete medium. These cells are used only for the MTT assay (SN is not collected) and are seeded in wells "preserved" for establishing the ELISA standard (*see* **Note 23**). Quantities mentioned below are for one full 96-well plate. Scale up as necessary.

1. Weight 10 mg MTT in a 15 mL tube. Add 5 mL of PBS. Vortex and sonicate until full dissolution. Distribute 50 μL to each well. Incubate the plate for 2 h in a cell culture incubator. During the last 30 min, prepare 15 mL of cell lysis buffer.
2. Empty the plate by inversion in one single movement over a waste container. Add 150 μL of cell lysis buffer.
3. Shake the plate on an orbital shaker until precipitates are dissolved.
4. Measure OD_{570} (quantification) and OD_{690} (correction) of the samples. Report to non-treated controls representing 100% viability, or to a standard curve established with serial quantities of cells. Lower the intensity of the signal, higher is the toxicity.

3.9 Mouse Model of Endotoxemia

The preclinical model of endotoxemia is a well accepted and easy to manage model to test the impact of drugs on inflammation-driven mortality (*see* **Notes 13** and **49**). It has to be setup in a preliminary experiment to define the lethal doses 20 and 90 (LD_{20}, LD_{90}) of LPS as they vary according to the batch of LPS and mouse strain used. If the results obtained in vitro indicate that the drug of interest decreases inflammation, then animals should be challenged with a LD_{90} of LPS to demonstrate drug-mediated protection from endotoxemia. On the contrary, if the drug increases inflammation in vitro, then animals should be challenged with a LD_{20} of LPS to demonstrate drug-mediated increased morbidity/mortality. At least two groups of mice are required to test a drug: a control group injected with vehicle and an experimental group injected with the drug of interest. Power calculation should be performed to define the minimal number of animals to be used to obtain statistically significant results, according to ARRIVE guidelines [30]. Different drug concentrations or treatment schedules may also be considered, respecting the 3R principles. Results from a typical experiment showing the protective effects of cambinol are shown in Fig. 3.

1. Label mice from each cage (*see* **Note 50**). Add food and HydroGel® on the bed layer.
2. Fill 25 G needle-mounted syringes with 250 μL of vehicle or drug (*see* **Note 51**). Inject intraperitoneally into mice. Treatment may be repeated at the time of, or post-LPS challenge.
3. Fill 25 G needle-mounted syringes with 250 μL of LPS. Inject intraperitoneally 1 h after vehicle or drug challenge (*see* **Note 13**).

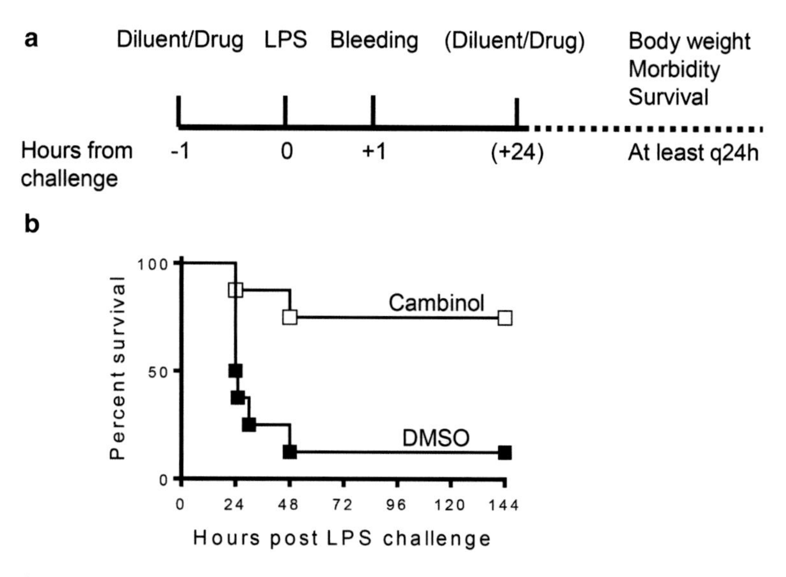

Fig. 3 Cambinol protects mice from endotoxic shock. (**a**) Flowchart of the endotoxemia model. Animals are checked for morbidity and mortality at least twice a day and for body weight once a day up to 6 days. (**b**) BALB/c mice ($n=8$ per group) were injected i.p. with 17.5 mg/Kg LPS. Cambinol (10 mg/Kg) and an equivalent amount of vehicle (DMSO) were administrated i.p. 1 h before and 24 h after LPS challenge. Survival was recorded regularly to build Kaplan–Meier plots. Differences were analyzed by the log-rank sum test ($P=0.0085$). The results show that cambinol protects from endotoxemia

4. Bleed the mice exactly 1 h after LPS challenge (*see* **Note 52**). Warm mice a few minutes with an infrared lamp. Make with a scalpel a small incision on the tail vein (*see* **Note 53**). Collect five drops (~100 μL) of blood in a 300 μL microvette. Press the tail with the haemostatic gauze to stop bleeding. Mix the tube by inversions. Keep on ice. Back to the lab, centrifuge for 10 min at 2000 ×*g*. Transfer the plasma in a new tube. Store at −20 °C until use.
5. Record weight, severity scores and survival at least once daily, preferably every 8–12 h (*see* **Note 54**).

4 Notes

1. Aliquots of inhibitors and stimuli should not be too small in order to avoid evaporation. This is particularly important if long term storage is considered. Use tubes that close tightly.
2. LPS (lipopolysaccharide, also known as endotoxin) is a major component of the outer membrane of gram-negative bacteria and one of the most powerful inflammatory stimuli [28]. LPS from *Escherichia coli* or *Salmonella minnesota* can be used indifferently. Better use ultra-pure LPS than standard LPS

well-known to contain contaminant bacterial components. Sonication of the LPS solution allows a better dissolution and exposure of the immunostimulatory acyl groups of the molecules.

3. Pam_3CSK_4 (*N*-Palmitoyl-S-[2,3-*bis*(palmitoyloxy)-(2RS)-propyl]-[R]-cysteinyl-[S]-seryl-[S]-lysyl-[S]-lysyl-[S]-lysyl-[S]-lysine) is a synthetic tripalmitoylated lipopeptide that mimics the acylated amino terminus of bacterial lipoproteins.
4. CpG ODN (oligodeoxynucleotides) are short single-stranded synthetic phosphorothioated oligonucleotides that contain unmethylated CpG dinucleotides. CpG ODN mimic microbial DNA. Different classes (A, B, and C) of CpG ODN are commercially available. Class B mouse specific CpG ODN 1826 is a better stimulus of mouse macrophages.
5. Drug stock solutions should be in the mM range (1–100 mM) as far as possible. When drugs are tested in cell culture, vehicles are diluted enough to avoid side effects, especially toxicity. In general, we try to reach a final concentration of diluents that is ≤0.1 %.
6. Any immunocompetent mouse strain (such as BALB/c and C57BL/6) mice can be used. Older (12–20 weeks), male, mice give generally a better yield.
7. The usage of bacterial instead of cell culture petri dishes is mandatory in order to reduce cell adherence and facilitate cell recovery.
8. Prefer a soft polyethylene cell lifter to prevent cell disruption.
9. The activity of the complement system, that affects cell growth and immune responses, is destroyed by incubation for 30 min at 56 °C. Immerse a bottle of room temperature-equilibrated FCS into a water bath pre-warmed at 56 °C. Swirl the bottle every 10 min. The incubation period has to be adapted (around 1 h) to effectively reach 56 °C for 30 min inside the bottle. Immerse the bottle in ice. Aliquot and store the FCS at −20 °C.
10. Using low endotoxin or endotoxin free reagents is mandatory since minute amounts of endotoxin stimulate macrophages and introduce a bias in the results.
11. L929 cells produce macrophage-colony stimulating factor (M-CSF) required for stimulating the differentiation and growth of BMDM. To produce L929 cell-conditioned medium, seed 3×10^6 L929 cells in 150 cm^2 flasks with 100 mL of Dulbecco's Modified Eagle Medium (DMEM) containing GlutaMAX and 4.5 g/L glucose and supplemented with 10 % heat inactivated low endotoxin FCS, 100 UI/mL of penicillin and 100 μg/mL of streptomycin. Incubate cells in a cell culture incubator for 1–2 weeks, in any case until cells start to die and the cell culture medium turns yellow. Collect the supernatant,

centrifuge for 10 min at 1800×*g*, filter (0.22 μM), aliquot in 50 mL polypropylene tubes and stock at −20 °C. Once an aliquot is thawed, keep it at +4 °C for a maximum of 7 days. It is recommended to prepare a large stock of conditioned medium by multiplying the number of flasks seeded in parallel with L929 cells as each stock of cell-conditioned medium has to be checked for growth factor properties. If low numbers of BMDM are recovered, the stock has to be discarded and a new batch performed (probably with longer cell incubation period).

12. TMB substrate solution is a 1:1 mixture of hydrogen peroxide (H_2O_2) and 3,3′,5,5′-tetramethylbenzidine. The two reagents are stored separately and are mixed just before usage. Many ready-to-use solutions are commercially available.
13. Only accredited experimenters can perform experimentation on living animals. Animal experiments have to be approved by the relevant office and performed according to institution guidelines. Other methods of euthanasia are possible.
14. The concentration of the working solution of LPS should be adjusted according to the activity of your batch of LPS and the mouse strain used. In our hands, LD_{20} and LD_{90} are around 5 and 17.5 mg/Kg in BALB/c mice, and 10 and 25 mg/Kg in C57BL/6 mice, respectively. *See* also Subheading 3.9.
15. It is important to cut the leg above the hip articulation to avoid rupture of femur.
16. The number of BM cells varies according to the strain, age, and sex of mice. Around 25–100 millions BM cells are obtained from the two tibias and two femurs of one mouse.
17. If a larger number of BMDMs is needed, seed several 100 mm bacterial petri dishes or seed 1–1.4 × 10^7 BM cells in 150 mm petri dishes with 30 mL of differentiation IMDM in order to obtain 3–9 × 10^7 BMDM at day 7.
18. Daily gentle shaking of petri dishes favors homogeneous distribution of growing cells.
19. Prepare cryotubes containing 1–1.5 × 10^7 BM cells. Store in liquid nitrogen. Cells can be thawed to start a new differentiation process.
20. If more practical, differentiation IMDM can be added at day 3 and 5. A better yield is obtained when medium is added three times at day 1, 3, and 5. If cells are seeded in a 150 mm petri dish, 10 mL of differentiation medium are added each feeding.
21. If working with 150 mm petri dishes, use 5 mL of Versene.
22. We suggest splitting cells every 2–4 days according to the time needed to reach 70–80 % confluence, which depends on the batch of cells and the quality of FCS. To secure the culture, two plates with different cell numbers can be seeded.

When you have a good idea of cell proliferation rate, splitting can be performed based on dilution factors such 1/4-or 1/10 without counting the cells. Cells are maintained in culture for 5–8 weeks, after what restart from an aliquot of frozen RAW 264.7 macrophages. Restart culture also if cells have overgrown, detached from the plastic surface or changed of shape. Record the number of passages.

23. To facilitate cytokine quantification by ELISA, leave 16 empty wells in your plate. Indeed, the ELISA method described in this chapter requires a standard curve made of eight points (including the blank) performed in duplicate.
24. Perform negative controls: (a) cells without treatment, and (b) cells exposed to vehicle, as vehicle may be cytotoxic. Treatment of cells with inhibitor only may also be tested.
25. The usage of a repetitive pipette is suggested to quicken and standardize the seeding process.
26. To save time the day of cell stimulation, it is possible to recover and plate BMDM in differentiation medium at the end of day 6. Day 7, substitute the medium with 100 μL of fresh, 37 °C pre-warmed complete IMDM.
27. In our experience, 1 h of preincubation with inhibitors prior to cell stimulation is sufficient to reveal an effect on cytokine production [23]. Yet when testing a drug for the first time, different (pre)incubation schedules are recommended (for example –4, –1, 0 h).
28. Stimuli should be 11× concentrated since 20 μL of stimuli are added on the top of 200 μL (100 μL of cells plus 100 μL of drug or vehicle). In the case of Pam_3CSK_4 and CpG ODN, 11× concentrated working solution are 110 ng/mL or 11 μM, respectively.
29. These time points are classically used to analyze cytokine gene and protein expression [26, 31]. Yet it might be worthwhile to perform a more complete pilot kinetic experiment at first test.
30. BMDMs and RAW 264.7 macrophages adhere firmly to cell culture treated plastic. After treatment, some cells may become loosely adherent or detach. Centrifugation of the plate for 7 min at 400 × *g* prior to collecting SN is recommended.
31. The same plate incubated for 18 h can be used to collect SN and quantify cytokine release and assess cell viability using MTT. In that case, collect 100 μL SN and proceed with the cell viability assay (*see* Subheading 3.8).
32. If the volume of lysate/mixture is larger than the column capacity, deposit one part on the column, centrifuge the column, discard the flow-through, and refill the column with the remaining mixture.

33. For all solutions, always prepare at least 10% more volume than what needed.
34. Precipitation of RNA can be also performed at −20 °C for 1 h or at −80 °C for 30 min.
35. RNA pellets can be quick dried in a vacuum concentrator, but in this case complete dissolution in water will take more time.
36. The ratio of absorbance 260/280 nm and 260/230 nm should be 1.8–2.0 and 2.0–2.2. If lower values are measured, contaminants (protein, phenol, EDTA, carbohydrates, ...) are present in your preparation and the precipitation protocol should be repeated.
37. If the volume of RNA to reach 500 ng is incompatible with kit guidelines, reduce the quantity of RNA up to 100 ng. In this case use undiluted sample for RT-PCR.
38. cDNA synthesis kits usually include a step to destroy DNA contaminants before cDNA synthesis. Some RNA isolation kits also provide material to destroy DNA contaminants.
39. The reverse transcriptase is usually active at 42 °C and inactivated at 94 °C.
40. The 1:5 dilution is optimized for cytokine gene measurement in macrophages, but should be adapted according to the expression level of the gene of interest.
41. SYBR Green is a cyanine dye that binds double-stranded DNA molecules. During the annealing step of each cycle, the SYBR Green emission signal is registered and allows quantification of DNA.
42. The NTC should be prepared for each gene measured in the assay and consist of the mix prepared in **step 3** without DNA. The NTC should not emit any signal. Otherwise, contamination should be considered and the measure repeated with new reagents.
43. The RT-PCR protocol described here uses 96-well plates. It can be convenient to use the 384-well format if large series are to be tested. In this case, the volumes of reagents need to be adjusted. For example, we use 1.25 μL of cDNA to which we add a mix composed of 1.25 μL of H_2O, 0.625 μL of primers, and 3.125 μL of Fast SYBR® Green Master Mix and run reactions using a QuantStudio™ 12 K Flex system. The usage of electronic pipettes is highly recommended for filling 384-well plates.
44. The amplification mix can be distributed in the plate before step 4. The proposed flow allows checking where samples have been filed by watching at the plate from below, which is reassuring particularly for beginners.

45. Wash by filling each well with 350 μL of wash buffer. If available, use an autowasher. After the last wash, tap the inverted plate on absorbent paper to removal remaining liquid.
46. Samples are diluted to get measures in the range of the standard curve. The dilution factor depends on several parameters (time of stimulation, volumes used, cell type, cytokine of interest, ...) and can be determined empirically in a small pilot experiment (testing a few presumed positive samples and a negative sample).
47. TMB (3, 3′, 5, 5′-tetramethyl benzidine) is a chromogenic substrate. TMB oxidized by HRP in the presence of H_2O_2 forms a water-soluble blue product. Addition of stop solution (sulfuric acid) stops the enzymatic reaction and turns TMB to yellow.
48. MTT is a pale yellow compound that is cleaved by living cells through NADPH-dependent cellular oxidoreductase enzymes to yield a dark blue formazan product. The process requires metabolically active mitochondria of live cells.
49. Experimental entotoxemia is a severe condition during which mice die of overwhelming inflammation.
50. Labeling with a permanent marker should be repeated every 2–3 days. Chip-based identification systems can also be used if allowed by the veterinary office for short-term experiments. Weak septic animals have to have access to food and water; it is therefore mandatory to place food and gel water inside the cage.
51. Drug concentration has to be adapted to its efficacy. In the example given in Fig. 3, 250 μL of cambinol (0.8 mg/mL) corresponds to 200 μg, i.e., 10 mg/Kg (8–12-week-old mice weight around 20 g).
52. To draw blood from mice exactly 1 h after LPS challenge, prepare all the materials needed in advance and collect blood in the same order as mice were challenged. One hour post-challenge is normally the optimal timing to detect circulating TNF by ELISA or Luminex (*see* Subheading 3.7) [31]. A 1/4 dilution of the serum is usually then used but should be adapted.
53. If the operator is not comfortable with mouse holding, a restraint system can be used. Our experience is that mice left free on a grid during the bleeding procedure are less stressed and aggressive.
54. We have established a severity score graded from 1 to 5 as follows: grade 1, ruffled fur; grade 2, ruffled fur plus either mobility disturbance, conjunctivitis or diarrhea; grade 3, ruffled fur, plus mobility disturbance, plus either conjunctivitis or

diarrhea; grade 4, moribund; and grade 5, death [32]. The score, and more generally the whole follow-up of in vivo study, should be performed blinded. Humane endpoints based on severity score may be applied when you have standardized your model. Measurement of animal weight and temperature are giving additional criteria of follow-up.

Acknowledgments

T.R. is supported by grants from the Swiss National Science Foundation (SNF 138488, 146838, 145014, and 149511) and an interdisciplinary grant from the Faculty of Biology and Medicine of the University of Lausanne (Switzerland).

References

1. Klar AJ, Fogel S, Macleod K (1979) MAR1-a regulator of the HMa and HMalpha Loci in SACCHAROMYCES CEREVISIAE. Genetics 93:37–50
2. Houtkooper RH, Pirinen E, Auwerx J (2012) Sirtuins as regulators of metabolism and healthspan. Nat Rev Mol Cell Biol 13: 225–238
3. Jiang H, Khan S, Wang Y et al (2013) SIRT6 regulates TNF-alpha secretion through hydrolysis of long-chain fatty acyl lysine. Nature 496:110–113
4. Huang JY, Hirschey MD, Shimazu T et al (1804) Mitochondrial sirtuins. Biochim Biophys Acta 2010:1645–1651
5. Carafa V, Nebbioso A, Altucci L (2012) Sirtuins and disease: the road ahead. Front Pharmacol 3:4
6. Du J, Zhou Y, Su X et al (2011) Sirt5 is a NAD-dependent protein lysine demalonylase and desuccinylase. Science 334:806–809
7. Tan M, Peng C, Anderson KA et al (2014) Lysine glutarylation is a protein posttranslational modification regulated by SIRT5. Cell Metab 19:605–617
8. Rauh D, Fischer F, Gertz M et al (2013) An acetylome peptide microarray reveals specificities and deacetylation substrates for all human sirtuin isoforms. Nat Commun 4:2327
9. Choi JE, Mostoslavsky R (2014) Sirtuins, metabolism, and DNA repair. Curr Opin Genet Dev 26:24–32
10. Choudhary C, Weinert BT, Nishida Y et al (2014) The growing landscape of lysine acetylation links metabolism and cell signalling. Nat Rev Mol Cell Biol 15:536–550
11. Cencioni C, Spallotta F, Mai A et al (2015) Sirtuin function in aging heart and vessels. J Mol Cell Cardiol 83:55–61
12. Herskovits AZ, Guarente L (2014) SIRT1 in neurodevelopment and brain senescence. Neuron 81:471–483
13. Kumar S, Lombard DB (2015) Mitochondrial sirtuins and their relationships with metabolic disease and cancer. Antioxid Redox Signal 22:1060–1077
14. Zhang J, Lee SM, Shannon S et al (2009) The type III histone deacetylase Sirt1 is essential for maintenance of T cell tolerance in mice. J Clin Invest 119:3048–3058
15. Grabiec AM, Krausz S, de Jager W et al (2010) Histone deacetylase inhibitors suppress inflammatory activation of rheumatoid arthritis patient synovial macrophages and tissue. J Immunol 184:2718–2728
16. Kim SR, Lee KS, Park SJ et al (2010) Involvement of sirtuin 1 in airway inflammation and hyperresponsiveness of allergic airway disease. J Allergy Clin Immunol 125:449–460, e414
17. Legutko A, Marichal T, Fievez L et al (2011) Sirtuin 1 promotes Th2 responses and airway allergy by repressing peroxisome proliferator-activated receptor-gamma activity in dendritic cells. J Immunol 187:4517–4529
18. Niederer F, Ospelt C, Brentano F et al (2011) SIRT1 overexpression in the rheumatoid arthritis synovium contributes to proinflammatory cytokine production and apoptosis resistance. Ann Rheum Dis 70:1866–1873
19. Hah YS, Cheon YH, Lim HS et al (2014) Myeloid deletion of SIRT1 aggravates serum

transfer arthritis in mice via nuclear factor-kappaB activation. PLoS One 9:e87733
20. Lim HW, Kang SG, Ryu JK et al (2015) SIRT1 deacetylates RORγt and enhances Th17 cell generation. J Exp Med 212(6):973
21. Villalba JM, de Cabo R, Alcain FJ (2012) A patent review of sirtuin activators: an update. Expert Opin Ther Pat 22:355–367
22. Mellini P, Valente S, Mai A (2015) Sirtuin modulators: an updated patent review (2012–2014). Expert Opin Ther Pat 25:5–15
23. Lugrin J, Ciarlo E, Santos A et al (1833) The sirtuin inhibitor cambinol impairs MAPK signaling, inhibits inflammatory and innate immune responses and protects from septic shock. Biochim Biophys Acta 2013:1498–1510
24. Lugrin J, Ding XC, Le Roy D et al (2009) Histone deacetylase inhibitors repress macrophage migration inhibitory factor (MIF) expression by targeting MIF gene transcription through a local chromatin deacetylation. Biochim Biophys Acta 1793:1749–1758
25. Mombelli M, Lugrin J, Rubino I et al (2011) Histone deacetylase inhibitors impair antibacterial defenses of macrophages. J Infect Dis 204:1367–1374
26. Roger T, Lugrin J, Le Roy D et al (2011) Histone deacetylase inhibitors impair innate immune responses to Toll-like receptor agonists and to infection. Blood 117:1205–1217
27. Ciarlo E, Savva A, Roger T (2013) Epigenetics in sepsis: targeting histone deacetylases. Int J Antimicrob Agents 42(Suppl):8–12
28. Heumann D, Roger T (2002) Initial responses to endotoxins and Gram-negative bacteria. Clin Chim Acta 323:59–72
29. Savva A, Roger T (2013) Targeting toll-like receptors: promising therapeutic strategies for the management of sepsis-associated pathology and infectious diseases. Front Immunol 4:387
30. Cressey D (2015) UK funders demand strong statistics for animal studies. Nature 520:271–272
31. Roger T, Froidevaux C, Le Roy D et al (2009) Protection from lethal gram-negative bacterial sepsis by targeting Toll-like receptor 4. Proc Natl Acad Sci U S A 106:2348–2352
32. Roger T, Delaloye J, Chanson AL et al (2013) Macrophage migration inhibitory factor deficiency is associated with impaired killing of gram-negative bacteria by macrophages and increased susceptibility to Klebsiella pneumoniae sepsis. J Infect Dis 207:331–339

ERRATUM TO

Chapter 11
Functional Analysis of Histone Deacetylase 11 (HDAC11)

Jie Chen*, Eva Sahakian*, John Powers, Maritza Lienlaf, Patricio Perez-Villarroel, Tessa Knox, and Alejandro Villagra

Sibaji Sarkar (ed.), *Histone Deacetylases: Methods and Protocols*, Methods in Molecular Biology, vol. 1436, DOI 10.1007/978-1-4939-3667-0_11, © Springer Science+Business Media New York 2016

DOI 10.1007/978-1-4939-3667-0_22

A footnote indicator (*) and its description meant for the authors Prof. Jie Chen and Prof. Eva Sahakian, were omitted on the title page of chapter 11 in the initial version published online and in Print. The list of author names on the title page, as appears now in both the print and online version of the book, is:

Jie Chen*, Eva Sahakian*, John Powers, Maritza Lienlaf, Patricio Perez-Villarroel, Tessa Knox, and Alejandro Villagra

*Authors contributed equally to this manuscript

The updated online version of the original chapter can be found at http://dx.doi.org/10.1007/978-1-4939-3667-0_11

Sibaji Sarkar (ed.), *Histone Deacetylases: Methods and Protocols*, Methods in Molecular Biology, vol. 1436, DOI 10.1007/978-1-4939-3667-0_22, © Springer Science+Business Media New York 2016

Index

Sibaji Sarkar (ed.), *Histone Deacetylases: Methods and Protocols*, Methods in Molecular Biology, vol. 1436,
DOI 10.1007/978-1-4939-3667-0,

T

U

V

W